Wiley Series in Remote Sensing
Jin Au Kong, Editor

Tsang, Kong, Shin
Theory of Microwave Remote Sensing

Hord
Remote Sensing: Methods and Applications

Elachi
Physics and Techniques of Remote Sensing
(in preparation)

Wait
Complex Resistivity of Geophysical Media
(in preparation)

Remote Sensing

Methods and Applications

Remote Sensing

Methods and Applications

R. MICHAEL HORD

A Wiley-Interscience Publication
JOHN WILEY & SONS
New York · Chichester · Brisbane · Toronto · Singapore

Library of Congress Cataloging in Publication Data:

Hord, R. Michael, 1940-
  Remote sensing.

  (Wiley series in remote sensing)
  "A Wiley-Interscience publication."
  1. Remote sensing.   I. Title.   II. Series.
G70.4.H67   1986   621.36'78   86-9064
ISBN 0-471-82824-6

Printed in the United States of America

10 9 8 7 6 5 4 3 2 1

For Susan

Preface

Digital image processing of remotely sensed data has grown into a mature discipline. Thousands of workers daily apply the methods and equipment of this field for agricultural, geological, medical, military, maritime, environmental, astronomical, industrial, and other purposes.

Moreover, the field is growing. A vigorous international research and development effort to enhance and extend our capabilities with these tools for numerically manipulating pictures continues to attract many of the technical community's best talents. These technologists are represented in industry, government, and universities. The continuing development points strongly to more widespread use of digital image processing in the future.

As this field has developed, the technical literature has reflected this growth. Many hundreds of papers, books, and reports are published each year in the United States alone. However, a review of this literature reveals a deficiency that this volume attempts to redress, or at least begin to redress. That deficiency is the absence of surveys of practical topics.

There seems to be an abundance of books and surveys on specific theoretical topics. Introductory texts are numerous. Usually a particular successful implementation of digital image processing methods is documented in an individual report. But collections of pragmatic information are lacking. This state of affairs is unfortunate for two reasons. The first is that workers in the field, when faced with the implementation of a system in response to an application need, may well reinvent solutions already known elsewhere or even settle on a design that is inferior to previous designs. The other is that cross-fertilization is inhibited. Opportunities for advances in any particular application area can easily be missed unless convenient access is afforded to the tricks of the trade that have been learned in other application areas.

This volume is addressed to the experienced worker in the field of digital image processing of remotely sensed data. It discusses the material at a level that presumes familiarity with the topics normally considered in an introductory text: the Landsat Multispectral Scanner, Fourier transforms, multispectral classification by maximum likelihood estimators, image contrast enhancement, edge detection, and the like. Consequently, this volume can also serve as a text for a second course in a university program on remote sensing and for industry-oriented intensive short courses on the topic.

The content is drawn from a wide range of papers, reports, and government documents. In most cases the material has been edited down to retain only the substance of general interest for the intended audience. Meanwhile, an effort has been made at preserving sufficient detail for that audience to use whatever may be pertinent to their concerns.

The volume has been structured into three broad subjects: sensors, processing and analysis techniques, and applications. The chapter on sensors establishes the clear emphasis on civilian spaceborne image data sources, and this emphasis is continued throughout the book. The chapter on techniques is a potpourri of algorithms and methods for converting bits into pictures, extracting features of the patterns in those pictures, recognizing those patterns, and evaluating the results. The chapter on applications extends from mineral exploration and weather observation to farming and seagoing operations. The Appendix provides a list of pixel values for a window of a Landsat Multispectral Scanner (MSS) image so the reader may try some processing on real data.

As a collection the book suffers as all collections must. The treatment of any topic of particular interest to the reader is not as exhaustive as would be a book devoted to that topic alone. Yet if that reader is stimulated to seek out such a specific book as a result of encountering the topic here, that is all to the good.

R. MICHAEL HORD

Arlington, Virginia
July 1986

Contents

Chapter 2   Processing and Analysis Techniques

Chapter 3   Applications

Appendix   Landsat Pixel Printout:
Woodside, California   **327**

Index   **359**

List of Tables

Remote Sensing

Methods and Applications

Introduction

For the past 20 years digital image processing of remotely sensed data has evolved through the stages of novelty, infancy, research, and most recently development. Today it is routinely employed on specific application projects that are motivated as much by economics as by scientific interest.

An overview of remote sensing is presented in Tables 1–8. Table 1 compares wavelength regions by detectable parameters and constraints. Tables 2–6 systematically describe application areas and observable parameters for passive ultraviolet, passive visible, passive infrared, passive microwave, and active microwave sensors. Observational needs are cataloged in Table 7. Planned U.S. and foreign operational and research satellites for observing the earth are listed in Table 8. A discussion of the Electronically Scanned Microwave Radiometer can be found in Section 3.1.1.

Tables 1–6 were obtained from NASA Headquarters, Code RSI (OAST). Tables 7 and 8 are from NASA Goddard Space Flight Center, *Earth Observing System,* Technical Memo 86129, Volume 1, Part 1, Greenbelt, Md., August 1984.

TABLE 1. Relationship between Wavelength Region and Target Detection Mechanism

Wavelength Region	Detectable Parameter	Specific Detection Constraints
Ultraviolet (0.1–0.4 μm)	Reflectance differential	Strong atmosphere absorption band below 0.29 μm
	Fluorescence	Atmospheric scattering
Visible (0.4–0.7 μm)	Reflectance differential	Illumination required
		Atmospheric haze
Infrared (.07–1000 μm)		
Near IR (0.6–3 μm)	Reflectance differential	Under illumination condi-
Far IR (>3 μm)	Thermal emission differential	tions, reflectance differential masks out thermal effects below 3 μm
		Atmospheric haze and clouds
		Nonhomopolar molecular absorption
Microwave region (1 mm to 1 m)	Emissive differential	Rain clouds
	Reflection/scattering cross section	Nonhomopolar molecular absorption
		Polarization bias

TABLE 2. Passive Ultraviolet Applications Areas and Observable Parameters

User Areas	Measurement Accuracy	User Priorities	Desired Spatial Resolution	Wavelength (μm)	Sensitivities	Revisit Time	User Information Needs
Stratosphere, Mesosphere,	1%	O_3 Profiles	12°	0.25–0.34	$\dfrac{2 \times 10^{-4} \text{ erg}}{\text{sec Å cm}^2 \text{ sr}}$		Total ozone column amounts and the vertical profile above the ozone maximum
Lower thermosphere		Backscattered UV	$4 \times \frac{1}{4}°$	0.16–0.40			
Typical user agency:		NO		0.215	$\dfrac{10^7 \text{ photons}}{\text{sec Å cm}^2 \text{ sr}}$		Solar spectral irradiance
NOAA				0.219			Thermospheric nitric oxide
			17	0.236			

TABLE 3. Passive Visible Applications Areas and Observable Parameters

User Areas	Measurement Accuracy (rms)	User Priorities	Desired Spatial Resolution (km)	Wavelength (μm)	Sensitivities	Revisit Time	User Information Need
Natural vegetation	4%						
Typical user agencies:		Range improvement	0.03	0.45–0.52 0.52–0.60		Yearly	Maps of brush density groupings
USDA		Range biomass productivity	0.1	0.63–0.69		Summer biweekly	Biomasss/density
USDI						Winter monthly	Vegetation complex/quality
HUD		Range inventory, Forest assessment,	0.1 0.02–0.1	0.58–0.63		5–10 yr	Maps and area tables of vegetation, types and site class boundaries
		Tropical forest Inventory	0.03–0.1	0.63–0.60		10–15 yr	Potential agricultural production areas
Cultivated vegetation							
Typical user agencies:		Crop productivity estimates	0.03	0.52–0.60		Biweekly	Crop identification and inventorying
USDA		Identification of stress	0.02	0.74–0.80		Weekly	Crop moisture and health monitoring
USDI		Crop identification	0.03	0.63–0.69		Biweekly	
		Vegetation moisture	0.03–0.2	0.80–0.91		2–10 days	
Water Resources							
Typical user agencies:		Watershed runoff estimation	0.03	0.45–0.52		Weekly	Watershed characteristics
EPA		Surface water and flood mapping	0.01–0.03	0.90–0.91		Yearly	Stream flow forecast/erosion impact
NOAA							
USACE		Snow field mapping	0.02–0.1	0.6–0.7			Snow field mapping
USCG		Water color	0.01–0.1	0.6–1.1 0.6–0.8			
USDI							
USDOE							
Mineral energy resources							
Typical user agencies:		Rock type	0.02	0.4–0.5		Annually	Mineral exploration
EPA		Surface alteration	0.03	0.6–0.7 0.5–1.1		One time	Energy resource exploration
NOAA							
USGS							
USDI							
USDOE							

4

Oceanographic applications				Sea state/wind velocity
Typical user agencies:				Location and tracking of iceberg and flows
EPA	Ship navigation and routing	0.06–0.1	0.7–1.1	12–40 hr	
NOAA	Pollution monitoring	0.02–0.1	0.6–0.68		Oil spills and waste dumping
USCG	Ocean engineering hazard	0.01–0.03	0.7–1.1	Daily	
USDA	Fishing (commercial)	0.01		3 days	Wave forces—current boundaries
USDI	Sea ice	0.1	0.7–1.1	Weekly	Numbered location of fishing vessels within legal zone
USDOE					Map ice type and boundaries
	Freshwater ice	0.5	0.7–1.1	Weekly	
	Sediment transport	0.01	0.6–0.8	5 days	Extent and direction sediment growth
Hazardous survey	Flood mapping	0.01–0.03		ORC*	Maps of flood extent
Typical user agencies:	Hurricane damage assessment	0.01–0.02		ORC	Flood extent vegetation wind damage,
EPA					
HUD	Tornado damage assessment	0.02–0.03	0.45–0.52	ORC	vegetation flood damage, urban damage
NOAA			0.6–0.7		
USCA			0.7–0.8		
USDA	Forest and range fire damage assessment	0.03–0.05	0.8–1.1	Hourly	Burn determination
USDI	Landslide/earth slippage	0.02–0.03		ORC	Coastal motion detection
USDOE					
Land use	Existing land cover map	0.011	0.41–0.45	36 mo	Transportation
Typical user agencies:			0.5–0.6		Urban change
EPA			0.6–0.6		Residential–rural boundaries
HUD			0.5–1.1		
USDA					Land–water boundaries
USDI					
USDC					
USDOT					
USDOE					

*On Request Coverage

TABLE 4. Passive Infrared Applications Areas and Observable Parameters

User Areas	Measurement Accuracy	User Priorities	Desired Spatial Resolution (km)	Wavelength (μm)	Sensitivities	Revisit Time	User Information Need
Natural vegetation							
Typical user agencies:							
USDA		Range improvement	0.03			Yearly	Maps of brush density groups
USDI		Range biomass productivity	0.1			Summer biweekly	Biomass/density
		Range inventory	0.1			Winter monthly	Vegetation complex/quality
		Forest assessment	0.02–0.1			5–10 yr	Maps and area tables of vegetation types and site class boundaries
		Tropical forest inventory	0.03–0.1			10–15 yr	Potential agricultural production areas
Cultivated vegetation							
Typical user agencies:		Crop productivity	0.03				Crop identification and inventorying
USDA		Identification of stress	0.02	1.56–1.75			Crop moisture and health monitoring
USDI		Crop identification	0.03	2.0–2.6			
		Soil moisture		4.5–5.5			
Water Resources							
Typical user agencies:		Watershed runoff estimation	0.02–0.1	0.8–1.1		Weekly	Watershed characteristics
EPA		Surface water and flood mapping	0.01–0.03			Yearly	Stream flow forecast/erosion impact
NOAA		Snow field mapping	0.02–0.1	1.55–1.75		Weekly	Snow mapping
USACE		Alaskan lakes mapping	0.03	8–13		Monthly	Frozen lake mapping/ice thickness
USCG				0.7–1.1			
USDI							
USDOE							
Mineral energy resource							
Typical user agencies:		Rock types	0.02	0.8–1.1		Annually	Mineral exploration
EPA		Surface alteration	0.03	1.1–1.35		One-time	Ferric and ferrous iron
				1.55–1.75			Copper sulfate

NOAA		2.05–2.35		0.8–1.1 μm
USCG		8.3–9.3		Soil types
USDA				1.1–1.35 μm
USDI				Gibbsite
USDOE				1.55–1.75 μm
				Clay mineral for soil identification
				2.05–2.36 μm
				Changes in ratio of 8.3–9.3 μm and 10.5–12.5 μm indicate migration of restrahlen
				Energy exploration
Oceanographic applications				
Typical user agencies:				
EPA				
NOAA				
USCG				
USDA				
USDI				
USDOE				
Ship navigation and routing	0.05–0.1		12–48 hr	Sea state/wind velocity Location and tracking of icebergs and flows
Pollution monitoring	0.02–0.1		Daily	Oil spill waste and dumping
Ocean engineering hazards	0.01–0.03			Offshore structure design parameters
Fishing (commercial)	0.01	10–12.5	3 days	Number and location of fishing vessels wihtin legal zone
	0.1			
Sea ice			Weekly	Map ice type and boundaries
Hazard survey				
Typical user agencies:				
EPA				
HUD				
NOAA				
USCG				
Flood mapping	0.01–0.03	0.8–1.1	ORC	Maps of flood extent
Hurricane damage assessment	0.01–0.02	0.8–1.1	ORC	Flood extent, vegetation wind damage, vegetation flood damage, urban damage
Tornado damage assessment	0.02–0.03	2.0–2.6 4.5–5.5	ORC	

TABLE 4. Passive Infrared Applications Areas and Observable Parameters—*Continued*

User Areas	Measurement Accuracy	User Priorities	Desired Spatial Resolution (km)	Wavelength (μm)	Sensitivities	Revisit Time	User Information Need
USDA USDI USDOE		Forest and range fire damage assessment	0.03–0.05	10.5–12.5		Hourly	Burn determination
		Landslide/earth slippage	0.02–0.03	8.3–9.3		ORC	Crustal motion detection
		Earthquake prediction	0.01–0.03	10.5–12.5		Yearly	Fault detection and terrain analysis
Land use Typical user agencies: EPA HUD USDA USDOT USDOE		Existing land coverage	0.03–0.05	0.8–1.1 2.0–2.6 4.5–5.5 8.3–9.3 10.5–12.5		3–4 mo	Transportation Urban change Residential–urban boundary Land–water boundary Variations in geological structure
Severe storms Typical user agencies: EPA HUD NOAA USACE USDA USDI		Temperature profile	3–30	6.5–7.0		1–6 hr	Improved precipitation and position forecasts
		Water vapor profile	3–15	10.5–12.5			Spatial and temporal distribution of latent heat of weather system
		Cloud tops	3–30	10.5–12.5 3.5–4.0 6.5–7.0 10.2–11.2		1–6 hr	Sea surface temperature mapping
		Sea surface temperature	3–30	11.9–12.9 4.28		6–36 hr	Improved cloud movement and wind forecast
		Cloud movements	3–30	6.71 7.25 11.1 12.7–14.7		1–6 hr	

Stratosphere, mesosphere, and lower thermosphere
Typical user agency:
 NOAA

HF		2.4	Vertical distribution HF, HCl, CH₄, and H₂O above 12 km
H₂Cl		3.4	
CH₄		3.5; 7.6–7.3	
H₂O		5.9	
H₂O profile	15	3.7	H₂O vapor profile up to 40 km in presence of clouds
		6.7	
		8.6	
		11.1	
NO₂		6.22; 7.6–7.8	Limb measurements Map vertical profiles of temperature between 46 and 70 km and concentrations of NO₂, H₂O, CH₄, H₂O₃, CO₂, and N₂O in upper troposphere to middle stratosphere with extention to stratosphere for H₂O vapor and lower mesosphere for temperature and O₃
H₂O	15–90	2.7; 25–100	
O₃ profile	15–90	6.75; 22–30	
HN₃		9.54; 8.5–10.3	
		11.3	
CO₂ profile		14.93; 4.3	
		15.21	
NO		5.3	
CO	15–120	4.5	Measure zonal wind shift in 50- to 120-km region using Doppler shift emission lines

TABLE 5. Passive Microwave Applications Areas and Observable Parameters

User Areas	Measurement Accuracy	User Priorities	Desired Spatial Resolution (km)	Frequencies (GHz)	Sensitivities (°K rms)	Revisit Time (hr)	User Information Needs
Sea and land ice Typical user agencies: NOAA USACE USDA USDI	~2.3% Several types	Sea ice concentration	0.1–5	6–100	<1	3–12	Marine navigation and shore facility development
		Ice type	1–5	10–100 (3 frequencies)	<1	6–13	Heat budget and climatology
		Land ice properties	10–50		<1	12–36	Ocean and atmospheric dynamics
Land parameters Typical user agencies: NOAA USACE USDA		Soil moisture (low-frequency)	3–25	0.6	<1	6–36	Flood potential Watershed yield
		Soil moisture (high-frequency)	3–25	1.37–1.43	<1	6–36	Agricultural stress
		Snow cover and type, frozen ground	3–25	3–80 (~4 frequencies)	<1	12–36	Climate
Ocean surface Typical user agencies: EPA NOAA USCG	2–5 m/sec 2–5 m/sec 0.5–2°K 0.5–1 ppt — —	Surface wind velocity	2–50	1.4–10	1	3–12	Physical mechanism of coastal currents
		Surface wind velocity (no precipitation)	10–50	(3 frequencies)	1	3–12	Weather forecasting
		Sea surface temperature	1–50		0.3	6–36	Sea surface temperature
		Salinity		1.4–10 (3 frequencies)	0.3	3–12	Salinity measurements
		Surface and volume pollution	0.5	+10.37	0.3–2	1–12	Coastal baseline studies
		Freshwater ice	0.5	1.4	2	Weekly	Coastal ecosystem studies Ship routing
Severe storms Typical user agencies: EPA HUD NOAA USACE USDA USDI	2°C	Temperature profile	3–30	50–70; 118	0.3	1–6	Improve precipitation and position forecasts
		Water vapor profile	3–15	22.2; 183.3	0.5	1–6	Spatial and temporal distributions of latent heat in weather system
		Water vapor profile (non-tropical)	6–15 3–15		0.5	1–6	
		Liquid water abundance rain rate	2–10	18.6–18.8	2	1–6	Temperature anomaly detection
		Sea surface temperature	3–30		0.3	6–36	Thunderstorm buoyancy
		Sea surface wind (magnitude)	3–30		1	3–12	Rain and hail detection
		Sea surface wind (no precipitation)			1	3–12	

Synoptic meteorology and climatology
Typical user agencies:
NOAA
USCG
USDA
USDI
USDOT

Parameter	Precision		0.2	Frequencies (GHz)	3–12	Requirements
Temperature profile	1–6°K	50		~05 or 110 (~5–12 frequencies)		Atmospheric temperature (thickness) and geostrophic wind determinations
Water vapor profile	0.1–0.4 cm	15		~22 (1 or 2 frequencies)		
Water vapor profile (nontropical)	—	15		~183 (3–5 frequencies)		Atmospheric water vapor and liquid water determination over oceans
Liquid water abundance	0.008–0.012 cm	2–10		+2 frequencies		
rain				15–90 GHz (e.g., 19 and 37)		
Sea surface temperature	~0.5°K	50		1.4 and 10		
Sea surface wind	2 m/sec	50		1.4 and 6		

Stratosphere, mesosphere, and lower thermosphere
Typical user agency:
NOAA

Parameter			Frequencies (GHz)	Requirements
Temperature		15–100		Stratosphere
O$_2$ density		90–120	118	OH and atomic O with 10% precision
Wind		70–100		O$_3$ profiles with 1% precision
Magnetic field		70–100		Temperature with 2°K precision
H$_2$O		15–80	184	10, H$_2$, CO, H$_2$O, N$_2$O, NO with 10% or better precision
O$_3$		15–50	205	Mesosphere and lower thermosphere
O$_3$ profiles		15–50	230	OH, O, O$_2$, NO, CO, O$_3$, H$_2$O with 10% or better precision
CO		15–120	100–3000	Temperature with 5°K or better precision
Atmosphere gas abundance				Winds with 10 m/sec or better precision
Temperature, winds, and magnetic fields				Magnetic field with 0.1% or better precision
				Limb measurements

TABLE 6. Active Microwave Applications Areas and Observable Parameters

User Areas	Measurement Accuracy	User Priorities	Desired Spatial Resolution (km)	Frequencies (GHz)	Sensitivities (°K rms)	Revisit Time	User Information Needs
Natural Vegetation							
Typical user agencies:		Range improvement	0.03	>8		Yearly	Maps of brush density groupings
USDA		Range biomass productivity	0.1			Summer biweekly Winter monthly	Biomass/density
USDI		Range inventory	0.1			Yearly	Vegetation complex quality
HUD		Forest assessment	0.02–0.1	To be determined		8–10 yr	Maps and area tables of vegetation types and site class boundaries
		Tropical forest inventory	0.03–0.1			10–15 yr	Potential agricultural production areas
Cultivated vegetation		Crop productivity estimates	0.03			Biweekly	Crop identification and inventorying
Typical user agencies:		Identification of stress	0.02				Crop moisture and health monitoring
USDA		Crop identification	0.03	4 and 5			
		Soil moisture	0.03–0.2				
Water resources		Watershed runoff estimation	0.02–0.1	To be determined		Weekly	Watershed characteristics
Typical user agencies:		Surface water and flood mapping	0.01–0.03	1–10 and 36		Yearly	Streamflow forecast/erosion impact
EPA							
NOAA							
USACE		Snow field mapping	0.02–0.1	L and K bands		Weekly	Snow depth and water content
USCG							
USDI		Alaskan lakes mapping	0.03	Polarization: HH and HV		Monthly	Frozen lake mapping/ice thickness
USDOE							
Mineral–energy resources		Rock types	0.02	To be determined		Annually	Mineral exploration
Typical user agencies:		Surface alteration	0.03	Polarization: HH, HV, VV		One-time	Energy resource exploration
EPA							

12

Category / Typical user agencies	Application				Products
NOAA USCG USDI USDOE					
Oceanographic applications					
Typical user agencies: EPA NOAA USCG USDA USDA USDOE	Ship navigation and routing	0.00–0.1	K and KU bands	12–48 hr	Sea state/wind velocity
	Pollution monitoring	0.02–0.1	>5	Daily	Location and tracking of icebergs and flows
	Ocean engineering hazards	0.01–0.03	To be determined	3 days	Oil spills and waste dumpings
	Fishing (commercial)	0.01	To be determined	Weekly	Wave forces—current boundaries
	Sea ice	0.1	To be determined		Numbered location of fishing vessels within legal zone
					Mapping of ice type and boundaries
Hazards survey					
Typical user agencies: EPA HUD NOAA USCG USDA USDI USDOE	Flood mapping	0.01–0.03	K and L bands	ORC	Maps of flood extent
	Hurricane damage assessment	0.01–0.02	<3	ORC	Flood extent
	Tornado damage assessment	0.02–0.03	To be determined	ORC	Flood extent, vegetation wind damage, vegetation flood damage, urban damage
	Forest and range fire damage assessment	0.03–0.06	To be determined	Hourly	Burn determination
	Landslide/earth slippage	0.02–0.03	VHF, L and K bands	ORC	Crustal motion detection
	Earthquake prediction	0.01–0.03	Multipolarization	Yearly	Fault detection and terrain analysis
Land use					
Typical user agencies: EPA HUD USDA USDI USDOT USDOE	Existing land cover map	0.03–0.06	>8	3–4 mo	Transportation Urban change Residential–rural boundaries

Table 7. Observational Needs

| Parameter | Application | Accuracy | | Approach | Spatial Resolution | Observation Frequency | Spectral Resolution |
		Desired	Required				
Soil features							
Moisture	Hydrologic and geochemical cycles	5 moisture levels	5 moisture levels	Microwave radiometer	1–10 km	2 days	20 cm ± 1 cm
Surface		5%	10%	Model	30–1000 m	1 wk	20 cm ± 1 cm
Root zone		5%	10%	Visible/SAR	30 m	Annual	20 nm/50 nm
Types — areal extent (peat, wetlands)	Geochemical cycle Agricultural and Forestry	10%	10%	Visible/SAR	30 m	Annual	20 nm/50 nm
Texture–color	Agriculture and Forestry	10%	10%	Visible/SAR	30 m	Annual	20 nm/50 nm
Erosion	Geochemical cycle	10%	10%	Visible/SAR	30 m	Annual	20 nm/50 nm
Elemental storage	Geochemical cycle						
Carbon		10%	10%	Visible/SAR	30 m	Monthly	20 nm/50 nm
Nitrogen		10%	10%	Visible/SAR	30 m	Monthly	20 nm/50 nm
Permafrost	Geochemical	10%	10%	Visible/SAR	30 m	Annual	20 nm/50 nm
Surface temperature							
Land	Primary production, soil moisture and respiration	0.5°C	±1°C	Thermal IR	1 km ± 0.5 km	12 hr	50 nm
Inland waters	Mass–energy flux	0.1°C	0.5°C	Thermal IR	30 m	12 hr	50 nm
Ocean	Mass–energy flux	0.1°C	0.5°C	Thermal IR, microwave	4 km (open ocean) 1 km (coastal ocean)	12 hr 12 hr	
Ice	Mass–energy flux	0.5°C	1°C	Microwave, thermal IR	1 km	1 day	
Vegetation							
Identification	Hydrologic cycle, biomass distributions	1%	5%	Visible, near IR, Thermal IR	1 km	7 days	10–20 nm

Parameter	Application			Measurement technique	Spatial resolution	Temporal resolution	Spectral resolution
Areal extent	and change, primary production, plant productivity, respiration, nutrient cycling, trace gas, source sinks, vegetation–climate interaction, microclimate	1%	10%	Visible, near IR, Thermal IR	30 m	30 days	30 nm
Condition (stress, morphology, phytomass)		10%	15%	Visible, near IR, Thermal IR, SAR	30 m	3 days	10–20 nm
Leaf area index canopy structure and density		10%	20%	Visible, near IR, Thermal IR, SAR	30 m	3 days	50 nm
Clouds							
Cover	Radiation balance, weather forecasting, hydrologic cycle, climatologic processes, tropospheric chemistry	2%	5%	Visible, thermal IR	1 × 1 km	6 hr	
Top height		±0.25 km	±0.5 km	Lidar	1 km	6 hr	
Emission temp.		±0.5°C	±1°C	Thermal IR	1 × 1 km	6 hr	
Albedo		±0.01	±0.02	Visible	50 × 50 km	6 hr	
Water content		0.05 kg/m²	0.1 kgm/m²	Microwave	50 × 50 km	6 hr	
Water vapor	Weather forecasting, hydrologic cycle, climatologic processes	0.001 ppm	0.002 ppm	Microwave, thermal IR, Lidar	100 × 100 km × 100 mbar (vertical)	12 hr	
Snow							
Areal extent	Hydrologic cycle	5%	10%	Visible/microwave	1 km	7 days	0.6–0.7 μm/1 cm
Thickness	Water equivalent	5%	10%	Microwave	1 km	7 days	3 bands, 0.8–3.0 cm
Radiation							
Short-wave	Surface energy budget	2%	5%	Visible	1 × 1 km	1 day	
Long-wave	Surface energy budget	2%	5%	Thermal IR	1 × 1 km	1 day	
Short- and long-wave	Hydrologic cycle	2%	5%	Visible, thermal IR	100 × 100 km	6 hr	Broadband

Table 7. Observational Needs — *Continued*

| Parameter | Application | Accuracy | | Approach | Spatial Resolution | Observation | |
		Desired	Required			Frequency	Spectral Resolution
Precipitation	Hydrologic cycle	5%	10%	Microwave or in situ	1 km	Daily	Several bands, 0.1–10 cm
	Climatologic cycle	5%	10%	In situ	1 km	Daily	
Evapotranspiration	Hydrologic cycle	5%	10%	Thermal IR, visible, Microwave combination (model)	1 km	Daily	Multiple sensors
Runoff	Hydrologic cycle	10%	10%	Thermal IR, visible, Microwave combination (model)	—	Daily	Multiple sensors
Wetland areal extent	Hydrologic cycle	2%	5%	Visible, thermal IR	30–100 m	Monthly	Multiple bands 20 nm/2 radar bands
	Biogeochemical cycle	10%	30%	Visible, near IR/SAR	30 m	3 days	
Phytoplankton Chlorophyll Open ocean/ coastal Ocean/inland waters	Biogeochemical cycle	10%	20%	Visible, near IR	4 km/1 km/30 m	2 days	10–30 nm
Fluorescence Open ocean/ coastal Ocean/inland waters		10%	20%	Visible, near IR	4 km/1 km/30 m	2 days	5–15 nm
Pigment groups Open ocean/ coastal Ocean/inland waters		10%	20%	Visible, near IR	4 km/1 km/30 m	2 days	10–20 nm

Parameter	Application	Accuracy	Accuracy	Technique	Spatial resolution	Frequency	Spectral
Turbidity	Biogeochemical cycle		20%				
Inland water/ coastal ocean	Erosion assessment	10%		Visible, near IR	30 m/1 km	2 days	10–30 nm
Bioluminescence	Ecological processes	Presence/absence	30%	Visible	4 km	Monthly	10–20 nm
Wetland areal extent	Biogeochemical cycle	10%		Visible, near IR, SAR	30 m	3 days	3 bands/20 nm
Surface elevation							
Land	Continental tectonics and surface processes	1 m	5 m	Laser or radar altimetry or stereophotogrammetry,	100 m IFOV	10 yr	
	Interpretation and modeling of gravity and magnetic field data		±3 m (from averaging within 3-km blocks)	SAR altimeter, 1-m, laser altimeter	300 m × 300 m for averaging into 3-km blocks	10 yr	
Ocean	Circulation	1 cm		Microwave altimeter	25 km	2 days	
Inland ice	Hydrologic cycle	0.1 m	1.0 m	Altimetry	30 m	5 yr	
Wave							
Height	Air–sea interactions	10%	10%	Scanning altimeter, SAR	50 km	3 days	
Spectrum		±10°	±20°	Scanning altimeter, SAR	50 km	3 days	
Inland Ice							
Thickness	Ice dynamics	1%	2%	Radar sounding (probably airborne)	1 km	50 yr	
Velocity field	Ice dynamics	5%	5%	SAR, ADCLS	1 per 100 × 100 km	10 yr	
Mass balance	Ice dynamics, Hydrologic cycle, climate	5%	10%		1 per 100 × 100 km	Annual total	
Temperature		1.0°C	1.0°C	Thermal IR, microwave, ADCLS		Annual mean	
Sea ice							
Areal extent	Hydrologic cycle	10 km	100 km	Microwave radiometer	5–20 km	Weekly	

Table 7. Observational Needs — *Continued*

Parameter	Application	Accuracy Desired	Accuracy Required	Approach	Spatial Resolution	Observation Frequency	Spectral Resolution
Concentration	Oceanic processes	1%	10%	Microwave radiometer	1 km	Biweekly	
Sea ice dynamics	Climatological processes	10 m	100 m	SAR, ADCLS	100 m	Daily	
Atmospheric constituents (ozone and compounds of carbon, nitrogen, hydrogen, chlorine, sulfur, etc.)	Tropospheric chemistry	5%	20%	Dial/correlation spectrum	10 × 10 × 1 km	1 day	
	Middle atmosphere	5%	10%	Thermal IR, UV, etc.	500 × 500 × 3.5 km	1 day	
	Upper atmosphere	10%	25%	Thermal IR, UV, etc.	500 × 500 × 3.5 km	1 day	
Aerosols	Tropospheric chemistry	5%	20%	Lidar	10 × 10 × 1 km	1 day	
	Stratospheric chemistry	25%	50%	Lidar/occultation	200 × 500 × 1 km	1 day	
Temperature	Troposphere	0.5°K	1°K	Thermal IR, microwave, lidar	100 × 100 × 5 km	1 day	
	Middle atmosphere	1°K	2°K	Thermal IR, microwave	500 × 500 × 3.5 km	1 day	
	Upper atmosphere	5°K	10°K	Thermal IR, microwave	500 × 500 × 3.5 km	1 day	
Winds	Troposphere		2 m/sec	Doppler lidar	100 × 100 × 3.5 km	12 hr	
	Middle atmosphere		3 m/sec	Visible, IR (interferometer)	500 × 500 × 3.5 km	1 day	
	Upper atmosphere		10 m/sec	Visible, IR (interferometer)	500 × 500 × 3.5 km	1 day	
	Surface	0.5 m/sec	1 m/sec	Scatterometry	50 km²	1 day	

Lightning (number of flashes, cloud to cloud, cloud to ground)	Tropospheric chemistry	Stroke count	Same	Visible to near IR	10 × 10 km	Continuously	10 nm; ~200 channels
Emission features	Atmospheric electricity			Electromagnetic spectrum from ground	1 × 1 km	Continuously	6 channels, 8–14 μm, 500 nm; 2 channels, 3–5 μm, 500 nm
Electric fields	Upper Atmosphere	10%	25%	Near IR	10 × 10 × 3.5 km	10 min	
	Global electric circuit		10%	In situ electric field probe			
Rock unit minerology	Continental rock types	1% absolute	1% relative	Visible, near IR—spectral reflectance	30-m pixel	10 yr	
	Continental soil and rock types and distribution	0.1°K (NEΔT)	.3°K (NEΔT)	Thermal IR—spectral emissivity	30-m pixel	Seasonal coverage once every 10 yr	
Surface structure	Tectonic history	7-dB SNR in image	5 db SNR in SAR image	SAR	30-m radar cell width (4 looks)	Yearly	Variable incidence, variable frequency, variable polarization
Gravity field	Mantle convection, oceanic lithosphere, continental lithosphere, sedimentary basins, passive margins, etc.	0.5 mgal	1 mgal	Gravity gradiometer tethered system, satellite tracking	<30 × 30 km	10 yr	
Surface stress	Weather forecasting, climate processes, oceanography	$u_* = 2.5$ cm/sec	$u_* = 5$ cm/sec	Radar scatterometer	50 × 50 km	12 hr	
Oceanic geoid	Mantle convention oceanic lithosphere	0.5 cm	1 cm	Altimeter	1 km	10 yr	

Table 7. Observational Needs—*Continued*

| Parameter | Application | Accuracy | | Approach | Spatial Resolution | Observation Frequency | Spectral Resolution |
		Desired	Required				
Magnetic field	Crust and upper mantle, composition and structure, lithospheric thermal structure, secular variation of main field (core problem) upper mantle conductivity	0.5 nT	1.0 nT	Magnetometer, magnetometer/gradiometer, tethered systems	<30 × 30 km	10 yr	
Plate motion	Plate tectonic theory, fault motion	0.5 cm in each component	1 cm in each component	Satellite tracking by radar laser, GPS, VLBI, ground transponder arrays in conjunction with satellite	Varies with problem, 1 km 1000 km	0.5 yr in most cases, more frequently in areas of very active deformation	

TABLE 8. Planned U.S. and Foreign Operational and Research Satellites for Observing the Earth

U.S. Operational Satellites

NOAA Weather Satellites, 1978–1990s
Objectives: operational weather data
Orbit: sun-synchronous, 833–870 km, 7:00 a.m. and 2:00 p.m. equator crossing times
Payload:

NOAA	E	F	G	H	I	J	K	L	M	N	O	P
Advances Very High Resolution Radiometer (AVHRR)	x	x	x	x	x	x	x	x	x	x	x	x
High Resolution IR Sounder (HIRS)	x	x	x	x	x	x	x	x	x	x	x	x
Stratospheric Sounding Unit (SSU)	x	x	x	x	x							
Microwave Sounding Unit (MSU)	x	x	x	x	x	x						
Data Collection System (DCS)	x	x	x	x	x	x	x	x	x	x	x	x
Space Environment Monitor (SEM)	x	x	x	x	x	x	x	x	x		x	x
Solar Backscatter UV Experiment (SBUV)		x		x		x		x		x		x
Earth Radiation Budget (ERB)		x	x									
Search and Rescue (SAR)	x	x	x	x	x	x	x	x	x	x	x	x
Advanced Microwave Sounder (AMSU)							x	x	x	?	x	x
Advanced Coastal Zone Color Scanner (ACZCS)								?		?		?
Planned or actual launch year	'83	'84	'85	'86	'87	'88	'89	'90	'91	'92	'93	'94
Equator crossing time (a.m. or p.m.)	a.m.	p.m.	a.m.	p.m.	a.m.	p.m.	a.m.	p.m.	a.m.	p.m.	a.m.	p.m.

Instrument description:
AVHRR: 5 bands, 0.58–12.5 μm, 1-km/4-km resolution, 1600-km swath, temperature of clouds, sea surface, and land
HIRS/2: 20 bands, atmospheric sounding, temperature and moisture profiles
SSU: 3 bands, atmospheric sounding, temperature profiles
MSU: 4 bands, 50.3–57.9 GHz, atmospheric sounding
DCS: random access from buoys, balloons, and platforms
SEM: 3 instruments, solar protons, alpha particles and "e" flux density
SBUV: 12 bands, 2550–3400 Å, solar spectrum, O_3 profiles, earth radiance spectrum

TABLE 8. Planned U.S. and Foreign Operational and Research Satellites for Observing the Earth—*Continued*

ERB: determine earth's radiation loss and gain

AMSU: 20 bands, 10–90 GHz, possibly 180 GHz, all-weather temperature profiles

ACZCS: 9 bands, 0.4–0.88 μm, 10.5–12.5 μm, ocean color, sea surface temperature

GOES—Geosynchronous Weather Satellite System, 1975–1990s

    Objectives: operational weather data, cloud cover, temperature profiles, real-time storm monitoring, severe storm warning

    Orbit: geostationary at east and west

    Payload: Visible and Infrared Spin Scan Radiometer (VISSR), VISSR Atmospheric Sounder (VAS), Data Collection System (DCS), Space Environment Monitor (SEM)

Instrument description:

    VISSR: 2 bands, 0.55–0.70 μm, 10.5–12.6 μm, 0.9-km resolution visible, 8-km resolution IR, sensitivity of 0.4–1.4°K, day/night cloud cover, earth/cloud radiance temperature measurements

    VAS: 12 bands, 0.55–0.70 μm, 3.9–14.7 μm, day/night cloud cover, atmospheric temperature and water content

    DCS: random access from buoys, balloons, and platforms

    SEM: solar protons, alpha particles, and "e" flux density

DMSP—Defense Meteorological Satellite Program, 1970s–1990s

    Objectives: operational weather data for DOD

    Orbit: sun-synchronous, 720 km, equator crossing time: as desired

    Payload: Operational Linescan System (OLS), Multispectral IR Radiometer (MIR), Microwave Temperature Sounder (MTS), Space Environment Sensor (SES) and Special Sensor Microwave Imager (SSM/I)

Instrument description:

    OLS: 0.4–1.1 μm, 10–13 μm, 0.56 km/2.78 km resolution, global cloud cover

    MIR: 9.8- and 12.0-μm bands, 13.4–15.0 μm (CO_2), 18.7–28.3 μm (H_2O), vertical temperature profiles

    MTS: 50–60 GHz, seven-band scanning microwave temperature sounder

    SES: charged particle monitor

    SSMI: 19.35-, 37.0-, 85.5-GHz, dual polarization, 22.23-GHz vertical polarization; 1400 km swath; precipitation, soil moisture, wind speed over ocean and sea ice morphology

Landsat, 1972–1984

    Objectives: operational and data: vegetation, crop, and land use inventory

    Orbit: sun-synchronous, 705 km, 9:30 a.m. node, 16-day repeat

Payload: Multispectral Scanner (MSS), Thematic Mapper (TM)
Instrument description:
MSS: 5 bands, 0.5–0.6 μm, 0.6–0.7 μm, 0.7–0.8 μm, 0.8–1.1 μm, 10.4–12.6 μm, 80-m resolution, 185-km swath
TM: 7 bands, 0.45–0.90 μm (4), 1.55–1.75 μm, 2.08–2.35 μm, 10.4–12.5 μm, 30 m/120 m (10.4-um) resolution, 185-km swath

N-ROSS—Naval Research Oceanographic Satellite System, 1988
Objectives: operational sea state data for DOD/NASA research
Orbit: sun-synchronous, 830 km, TBD node, TBD repeat
Payload: Special Sensor Microwave Imager (SSMI), Scatterometer (SCATT), Altimeter (ALT), Low Frequency Microwave Radiometer (LFMR)
Instrument Description:
SSMI: 4 frequencies, 19.3, 22.2, 37.0, and 85.5 GHz, 102° scan view, 1394-km swath, 70 × 45 to 16 × 17 km IFOV, sea surface wind precipitation, atmospheric moisture, soil moisture, and sea ice conditions
SCATT: 14.6 GHz, 6 antennas, dual/single pol, 50-km resolution, range 3–30 m/sec, accuracy 2 m/sec
ALT: Seasat class, sea surface topography
LFMR: 2 bands, 5, 10 GHz, 6-m antenna, sea surface temperature

Foreign Satellites

SPOT—Système Probatoire d'Observation de la Terre, 1985 launch with follow-on in 1987
Objectives: operational land use and inventory monitoring system
Orbit: sun-synchronous, 10:30 a.m. node, 2.5-day repeat
Payload: SPOT
Instrument Description:
SPOT: 3 bands, 0.5–0.6 μm, 0.6–0.7 μm, 0.78–0.9 μm, 20-m resolution color mode, 10-m resolution panchromatic mode, (0.51–0.73 μm), 60 km × 60 km viewing area, swath of 950 km centered around nadir, stereoscopic images

MOS/LOS—Marine Observation Satellite–Land Observation Satellite (Japanese Program)
MOS-1: 1986–1987; MOS-2-3 follow-ons
LOS-1: 1988–1989, LOS-2 follow-on
Objectives: MOS—color and temperature of sea surface; LOS—geological survey, land use, agriculture, forestry, disaster prevention
Orbit: sun-synchronous, 909 km, a.m. node, 17-day repeat
Payload:
MOS-1: Multispectral Electronic Self Scanning Radiometer (MESSR), Visible and Thermal IR Radiometer (VTIR); Microwave Scanning Radiometer (MSR)

TABLE 8. Planned U.S. and Foreign Operational and Research Satellites for Observing the Earth — *Continued*

MOS-2: altimeter, scatterometer

LOS-1: synthetic aperture radar (SAR), optical sensor

Instrument Description:

MESSR: 4 bands, 0.5–1.1 μm, 50-m resolution, 100-km swath

VTIR: 4 bands, 0.5–0.7 μm, 6.0–7.0 μm, 10.5–11.5 μm, 11.5–12.5 μm, 1-km/3-km resolution, 500-km swath

MSR: 2 bands, 23.8 and 31.4 GHz, 317 km swath

ALT: Seasat class

Scatterometer: 4- to 25-m/sec wind measurement, ±20° direction, 200- to 700-km swath

SAR: 1.2 GHz, 4 look angles, 25 m res

Optical sensor: TBD

ERS-1 — ESA First Remote Sensing Satellite, 1988–1989 launch. ERS-2 follow-on; ERS-3 planned. (European program)

Objectives: coastal ocean and ice studies, global weather, land use

Orbit: sun-synchronous, 777 km, a.m. equator crossing time, 3-day repeat cycle

Payload: Active Microwave Instrument (AMI), Along Track Scanning Radiometer (ATSR)

Instrument Description:

AMI: SAR: C band 5.3 GHz, 30 × 30 m res., 80- to 200-km swath

Scatterometer (wind mode): 3-beam C-band, VV polarization, 500-km swath, 50-km resolution, range 4- to 24-m/sec, accuracy 2 m/sec or 10%; scatterometer (wave mode): 5 km × 5 km image every 100 km; altimeter: KU band (12.5 GHz), 10-cm precision — land, 40-cm precision — ocean, 1.2-m-diameter antenna

ASTR: radiometer, 3.7-, 11- and 12-μm bands, 1 km × 1 km resolution, 0.1-km resolution, 50-km swath

PRARE: Precision Range And Range Rate Experiment; laser retroreflector

RADARSAT — Canadian Radar Program, 1990–2000, Three- to Four-Satellite Series

Objectives: high-resolution studies of arctic area; agriculture, forestry, and water resource management; ocean studies

Orbit: sun-synchronous, 1000-km altitude, 3-day repeat cycle

Payload: Synthetic Aperture Radar (SAR), Optical Sensor (TBD), Microwave Sensor (TBD)

Instrument Description:

SAR: C or L band, 150-km swath, 25- to 30-m resolution, 4- to 100-km look angles

U.S. Research Satellites

UARS—Upper Atmospheric Research Satellite, 1989 Launch

Objectives: coordinated measurement of major upper atmospheric parameters

Orbit: 57° inclination; 600-km altitude

Payload: Cryogenic Limb Array Etalon Spectrometer (CLAES), Halogen Occultation Experiment (HALOE), High Resolution Doppler Imager (HRDI), Improved Stratospheric and Mesospheric Sounder (ISAMS), Microwave Limb Sounder (MLS), Particle Environment Monitor (PEM), Solar Stellar Irradiance Comparison Experiment (SOLSTICE), Solar–UV Spectral Irradiance Monitor (SUSIM), Wind Measurement in the Mesosphere (WINTER), Active Cavity Radiometer Irradiance Monitor (ACRIM), Solar Backscatter UV Experiment (SBUV)

Instrument Description:

CLAES: global synoptic measurement of nitrogen and chlorine ozone destructive species, minor constituents temperature

HALOE: stratospheric species concentration

HRDI: middle atmospheric winds

ISAMS: atmospheric temperature and species concentration

MLS: vertical profiles of O_3 and O_2, wind measurements, inferred pressure

PEM: charged particle entry measurements for atmosphere

SOLSTICE: solar irradiance from 1.150 to 4000 Å

Chapter 1

Sensors

This book's first chapter describes sensors. Sections are devoted to imaging radar, the Coastal Zone Color Scanner (CZCS), the Thematic Mapper (TM), the Return Beam Vidicon (RBV), and a collection of future sensors including SPOT and the Earth Observing System. Certainly the list of instruments described is not intended to be exhaustive. The Landsat Multispectral Scanner (MSS), the Heat Capacity Mapping Mission (HCMM), the GOES Visible Infrared Spin Scan Radiometer (VISSR), and the Very High Resolution Radiometer (VHRR) are but a few candidates that were not included since descriptions of these instruments are more commonly available elsewhere. For a discussion of ESMR, see Section 3.1.1.

1.1 RADAR

When the Seasat satellite was launched into an 800-km orbit in 1978 a new method of remote sensing from space was demonstrated. Among the major instruments carried aboard Seasat was a Synthetic Aperture Radar (SAR) that operated at a frequency of 1.275 GHz with HH polarization to obtain images with 25-m ground resolved distance across a 100-km swath. The radar pointed 20° off vertical, which gave rise to a 23° angle of incidence at the surface of the earth. Unfortunately the Seasat SAR failed a few months after launch.

Using radar causes surface features such as mountains to stand out. Even small surface texture differences such as ocean wave patterns can be imaged directly. Ships provide strong returns and show up as bright dots.

SAR imagery from space continues to grow in importance. Shuttle flights have carried imaging radars, and future SAR flights are planned. One reason for this is that in dry areas SAR can penetrate vegetation and show underlying terrain. In desert areas, in fact, the radar can penetrate ground layers and provide a picture of what is beneath the surface. SIR-A (Shuttle Imaging Radar–A) demonstrated this in the Sahara, showing undiscovered river channels later confirmed by a team of scientists that visited the site.

SAR data is addressed in all three parts of this book. In this part radar imaging principles are described by an early but still valid and lucid discussion; this is followed by a description of the Seasat SAR instrument, including sample images. Part 2 includes an algorithm for processing SAR data into images. Part 3 includes a discussion with illustrations of how imaging radar can be applied to geology.

1.1.1 Radar Imaging*

Radar is unique among the commonly used sensors in that it provides its own illumination — radio waves. Because the properties of the source are accurately known, the time for a wave to travel to a remote object can be measured by comparing the time the wave left the radar with the time it returns. Using the known speed of electromagnetic waves in space (3×10^8 m/sec), this time measurement can be converted into a distance measurement. Furthermore, the wavelength of the returning wave can be compared with that transmitted to determine the Doppler shift, and from it the relative velocity of radar and object can be calculated. Both the distance-measurement and velocity-measurement properties of radar are used in imaging systems to improve the resolution over that possible with a passive system.

Radar's history apparently goes back to the 1920s, when observations were made of fluctuations in radio signals associated with passage of airplanes nearby. By the outbreak of World War II, German, British, French, American, and Japanese military groups had developed radars for detection and location of aircraft and, in some cases, ships. From these uses the acronym *radar* (for "radio detecting and ranging") was devised. Imaging radar became a reality early in World War II, but the quality of the images obtained by shipboard and airborne radars was unsuitable for observation of terrain properties other than land–water boundaries, major mountains, and cities. In the 1950s, imaging radars with much better pictorial abilities were developed for military reconnaissance, and the radars used today for studying earth resources are an outgrowth of these systems.

Radar's greatest advantage is that aerial surveys may be conducted on a preset time schedule without waiting for clear weather, without concern for time of day and without concern for problems of interpretation associated with differences between day and night thermal patterns. This advantage is due to the wavelength region of radar (between about 0.5 cm and 1 m), since waves of these lengths can penetrate clouds and smoke, and waves longer than 2 cm or so can penetrate fog and precipitation.

*Based on National Academy of Sciences, "Radar Imaging," *Remote Sensing, with Special Reference to Agriculture and Forestry,* Washington, D.C., 1970.

Imaging radar systems take many forms, but the one most likely to be useful for agricultural surveys is the side-looking airborne radar (SLAR). In this system, a continuous strip image is recorded on photographic film. The terrain imaged is to the side of the vehicle carrying the radar, and so a strip directly beneath the vehicle cannot be imaged readily. Such a system is illustrated in Figure 1. Strips may be imaged on both sides of the vehicle or only on one side. Such a strip may extend to the side from a distance about half the altitude out as far as 100 km or more. The vehicle carrying the radar may be an aircraft or a spacecraft.

Fundamental Principles of Radar Echoes

Radar returns depend upon reradiation (reflection) of energy supplied by the radar system. The strength of the reradiation depends both on the properties of the transmitted electromagnetic field and the properties of the irradiated object. These properties are summarized in the list below. They apply also to self-emission, except that temperature and other causes of self-emission must be added.

Parameters of source (or receiver)
Wavelength
Polarization
Direction
Parameters of surface
Dielectric and conducting properties, including quantum resonances
Surface roughness in wavelength units
Physical resonances
Surface slopes
Subsurface effects
Scattering area

The electromagnetic field generated by the radar transmitter is determined solely by the instrumentation. The wavelengths most commonly used for present-

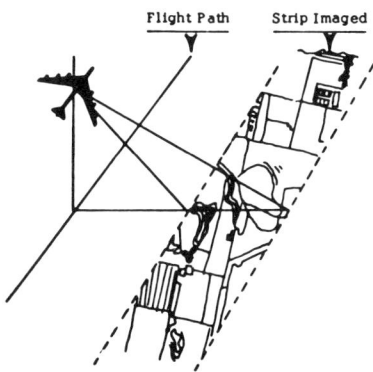

FIGURE 1. Diagram of application of side-looking airborne radar.

day operating imaging radar systems are between 0.86 and 3.3 cm, although experimental systems operating at wavelengths of up to 70 cm are producing synthetic aperture images. Present radars are monochromatic; that is, they use only a single wavelength of great spectral purity compared with the best obtainable passive systems using filters in the optical and infrared regions. Spectral purity of radar systems is comparable with that of lasers.

The direction at which the transmitted energy strikes depends on the position of the object being viewed with respect to the radar. This configuration is a matter of choice. The angle of incidence (measured with respect to the vertical) may range from zero (straight down) to 90° (horizontal). In practice, the vertical is avoided because geometric considerations make it very difficult to discriminate ground distance by time measurement near the vertical; near-horizontal incident angles are avoided because of shadowing. Normally, airborne imaging radars operate between incidence angles of 20° or 30° and 75° or 80°. The azimuth angle may be chosen anywhere from straight ahead to directly astern of the aircraft, but most high-resolution systems look directly to the side. The viewing angle is determined by the flight direction once the azimuth angle with respect to the aircraft has been fixed. For some ground objects, this angle can be important; that is, there is a difference at sea between observation upwind, downwind, and crosswind.

The reradiation for any particular surface depends on its dielectric and conducting properties. Since the dielectric constant of water is very high and plants often have high moisture content, many plants are good reflectors.

The degree of roughness of the surface is an even more vital factor in setting radar return strength. Numerous theoretical studies of radar scatter from statistically described surfaces are largely irrelevant for agriculture because of the complexities of the shape of growing plants. Only one mathematical description of growing crops has been attempted.[1]

The presence of resonant structures within surfaces can give rise to strong radar signals from apparently small objects if they are properly oriented, and it can also result in weaker signals with improper orientation. Surface slopes are, in a sense, measures of roughness, but they deserve separate listing because of their importance.

If the radar signal can penetrate a significant distance into the surface material, the signal observed will then be determined by the combination of the surface and the subsurface parameters. This is frequently true with crops, where the "surface" is the top of the vegetation and penetration to the soil is likely.

The scattering properties of the ground as a radar target are usually expressed in terms of the average differential scattering cross section σ^0. The signal received by the radar is proportional to σ^0 times the area contributing. All other factors in the radar power equation are parameters either of the radar system or of the geometric relations between radar-carrying vehicle and target area. The differential scattering cross section is a function both of the ground properties and of radar parameters such as wavelength and look angle.

As radar moves past a particular terrain element, its instantaneous cross section fluctuates over a wide range because of changes in the relative phase of signals from different parts of the element. This results in an "antenna pattern" for each

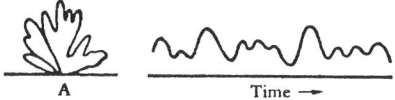

FIGURE 2. Antenna pattern and resulting fading. Left: Strength of signal from a target at A is proportional for each direction from A in the diagram. Right: Signal received in an aircraft going past a target at A but always pointing toward it.

surface element, as indicated in Figure 2. The signal fluctuates rapidly (fades) as the radar passes through this pattern. Thus, individual looks at targets can lead to erroneous conclusions, and averaging must be used to overcome the effects of fading on measurement precision. In optics, the same effect causes a laser-illuminated picture to be "speckled."

Rather than treating each patch of ground as having a single pattern, an approach based on superposition of the patterns of many flat facets is sometimes employed. The two methods give equivalent results. A rough surface is approximated by a collection of essentially flat (within a small fraction of a wavelength) facets of different sizes. "Rougher" parts of the ground require small facets, and smoother parts can permit larger facets. Backscatter patterns for facets of different sizes are indicated pictorially in Figure 3. The narrow pattern goes with a facet many wavelengths across, and the wide pattern goes with a facet that is a fraction of a wavelength across.

The direction of largest signal return is that having the most facets perpendicular to it. Returns come from the sea and from plowed fields at angles near grazing (the angle as it approaches 90°) for which no facets are perpendicular. Presumably, these are from the edges of the broad patterns for small facets.

Plants present a more complex situation than either plowed fields or oceans. Cylindrical scatterers, in most cases, behave to some extent like flat facets; that is, the maximum signal is returned if the axis of the cylinder is perpendicular to the direction of the incident wave (at least if the wave is polarized properly). Thus, radar return from plants with simple vertical stems should be highest near grazing incidence. Plants with complex structure, like corn stalks or apple trees, provide strong returns over a wider range of angles.

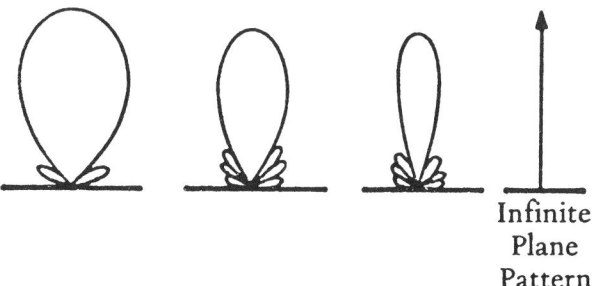

Infinite
Plane
Pattern

FIGURE 3. Facet patterns. Uniform normally incident illumination. Note that small facets are almost nondirective. Large facets give strong signals in a single direction only.

Return from plants is also wavelength-sensitive. At millimeter wavelengths, nearly all plant surfaces are "rough," except perhaps leaves. Leaves that are relatively flat can act as facets at these wavelengths, and a plant having most leaves almost horizontal should give strong returns for vertically incident radar signals. The leaves are only a fraction of a longer wavelength across, and the signals from them are almost nondirective. At still longer wavelengths (meter range), leaves are so small that they contribute very little to the signal return at all.

Moisture content of the plant can influence its radar return significantly, although a saturation level apparently exists beyond which additional moisture makes no difference. A completely dry (dead) plant, however, cannot return signals as well as a living, wet plant. Preliminary studies of this relationship have been reported,[2] and returns from wet and dry fields indicate this effect.[3]

The polarization of the transmitted energy is defined as the direction of its electric field vector, and it is determined by the antenna design. The receiving antenna may be designed to respond only to the same polarization transmitted, but it may also be designed to respond to electric field vectors in other directions. When separate receivers are used for different polarization directions, the shift in polarization by an echoing object may be observed and used as a discrimination feature.

Polarization is a significant factor in radar return strength. For straight-wire antennas, this direction is either parallel to the wire or at least in a plane passing through the wire. For other types of antennas, the direction of polarization is set by less obvious factors. Circular and other types of polarization can also be used.

Study of polarization influence on radar signals returned from vegetation has been reported recently.[4] Preliminary observations of centimeter-wavelength images show clearly that for some crops, returns are significantly different when signals received have the same polarization as those transmitted and when signals received have a polarization at right angles to that of the signals transmitted; these differences do not exist for other crops. Differences between signals with horizontal polarization, both transmitted and received, and signals with vertical polarization, both transmitted and received, have not been so clearly identified, but they are believed to exist.

A signal polarized in the direction of stalks and striking them at normal incidence should give a stronger return than a signal polarized at right angles to the stalks.

Cross-polarized returns (received polarization different from transmitted polarization) are usually stronger on surfaces having elements with small radii of curvature and on surfaces so rough that multiple scatter is easy. This should carry over to returns from plants as well.

Radar return from vegetation is more complicated than radar return from surfaces, and return from surfaces is not yet fully understood. Nevertheless, empirical determination of differences in return from vegetation as a function of wavelength, polarization, incident angle, and moisture indicates the sensitivity of radar to many significant plant differences. More research is needed to catalog this sensitivity adequately.

Atmospheric Effects on Radar

Serious atmospheric effects on radar are confined to the shorter wavelengths used for radar (usually wavelengths longer than 3 cm are relatively unaffected). The atmosphere only slightly absorbs radar signals at most of the wavelengths used. Radar echoes from precipitation and clouds present more of a problem than absorption, as they might obscure a desired echo. Cross-polarized returns from clouds and rain are so small that they never exceed ground echoes, so cross-polarization may be used if precipitation echoes are likely to be a problem.

Molecular absorption of radar signals is seldom a problem. Most radars operate at wavelengths longer than the 5-mm wavelength of the first oxygen absorption line. A water vapor absorption line that occurs at about 1.35 cm must be avoided, but no absorption lines occur at longer wavelengths.

Echoes from precipitation are proportional, for a single drop, to

$$\text{Diameter}^6/\text{wavelength}^4$$

Thus, for most wavelengths, clouds give little echo, but precipitation gives somewhat stronger echoes because of the larger particle diameter of raindrops. This property of moisture particle echoes is used in weather radar to distinguish regions of precipitation from regions of cloud.

Absorption due to liquid water droplets is also inversely proportional to wavelength,[5] but is only proportional to the third power of diameter, and hence is directly proportional to mass. Thus, total absorption is set by the mass density of water droplets, regardless of their size distribution. Some examples are given in Figure 4. In heavy rain this can have a significant effect on radars in the 1-cm-wavelength region, but the effect is negligible for wavelengths beyond about 8 cm. A more extensive treatment is given by Bean and Dutton.[6]

Fundamentals of Radar Operation

A radar system consists of the five basic elements shown in Figure 5. Since most imaging radars use pulse transmissions, a brief description of operation of a pulse system follows. The cycle starts with generation in the transmitter of a short burst of sinusoidally oscillating voltage at the carrier frequency (3×10^8/wavelength in meters). This burst, or pulse, varies in length from about 0.01 to 0.1 μsec. For most radars the power during the pulse is many kilowatts. The pulse is radiated from the transmitting antenna, and the waves travel through space to the ground. Waves reflected from different parts of the ground arrive at the receiving antenna at different times, depending on the distance from the radar. The synchronizing system that had initiated the transmitter pulse initiates action of the display or recording system either simultaneously with transmission or after a long enough delay that the first echoes are just arriving at the receiver. The receiver amplifies the returning signals (now at a power level often as small as 10^{-12} W) to a level suitable for operation of the display or recording systems.

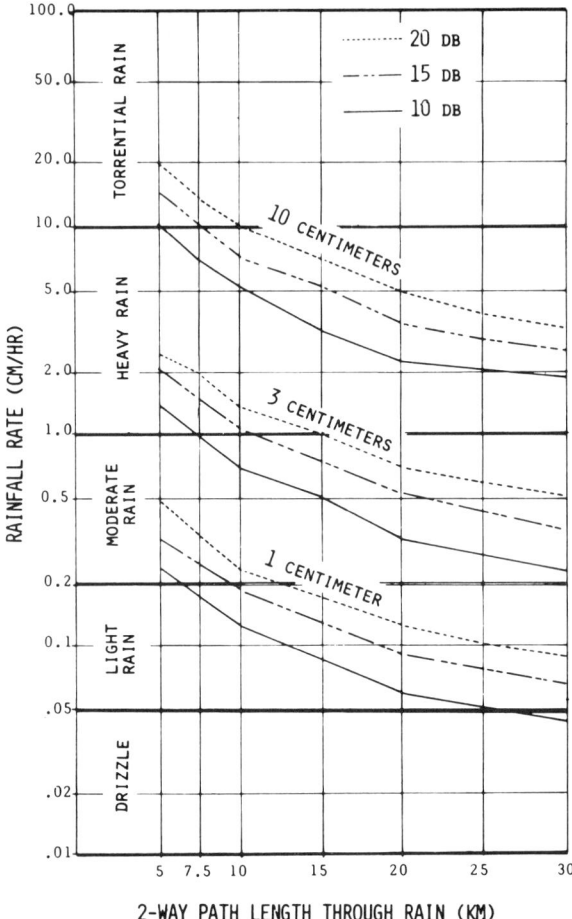

FIGURE 4. Attenuation of 1-, 3- and 10-cm radar by rainfall. The attenuations given are very conservative, being the most extreme determined by Medhurst.

Because of the many possible radar transmitting signal designs and the many ways one can relate transmitted and received signal, the design of a radar system is quite complex, and the problems are only touched on here. Because of these many possibilities, design compromises of many types must be made, and systems can be tailored to specific needs.

Limitations are placed on radar systems by available antenna space, available power, stability of the vehicle carrying the radar, available data storage or telemetry, and other factors. Transmitting and receiving systems themselves seldom limit performance within the wavelength regions commonly used. Thus, limitations on system performance are specified by the environment in which the radar operates, so design compromises are based on the environment. Probably the most severe restriction on radar systems is the limited space available for antennas. Ac-

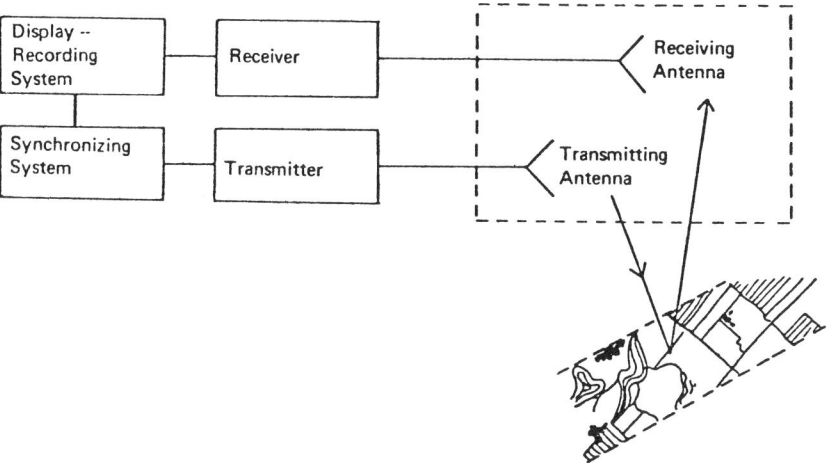

FIGURE 5. Diagram of five basic elements of a radar system. NOTE: The antenna system may use separate or common transmitting and receiving antennas. Common-antenna system contains switch to connect antenna to transmitter and receiver alternately.

cordingly, much of this discussion is devoted to techniques used to overcome this environmental limitation.

Information sensed by a radar is listed below. Numerous properties of the remote object may, of course, be inferred, but the information listed is, in fact, all that is sensed directly for a given wavelength and polarization. Because range and velocity measures are made independently of angle measurement, the radar system can have better resolution than a passive system that must determine position exclusively by angle measurements.

Signal strength ⎱
Angle to object ⎰ same as passive sensors

Distance to object, by time measurement ⎱
Relative velocity of object of Doppler measurement ⎰ unique to radar

Angles are measured by electromagnetic sensors by techniques depending on differences in phase shift along different paths. Lenses and antennas both depend on this principle. The precision of an angular measurement depends on the size of the aperture relative to the wavelength being used. Thus, an aperture of a given size permits better angular measurement at short wavelengths than at long ones.

Radar angular measurement ability is set by antenna patterns. The antenna pattern shows the strength in different directions of a signal transmitted from an antenna. It also shows the ability of the antenna to receive from those directions. An ideal antenna pattern might transmit (or receive) equal signals over an angular interval about its pointing direction and zero outside this interval. Figure 6A

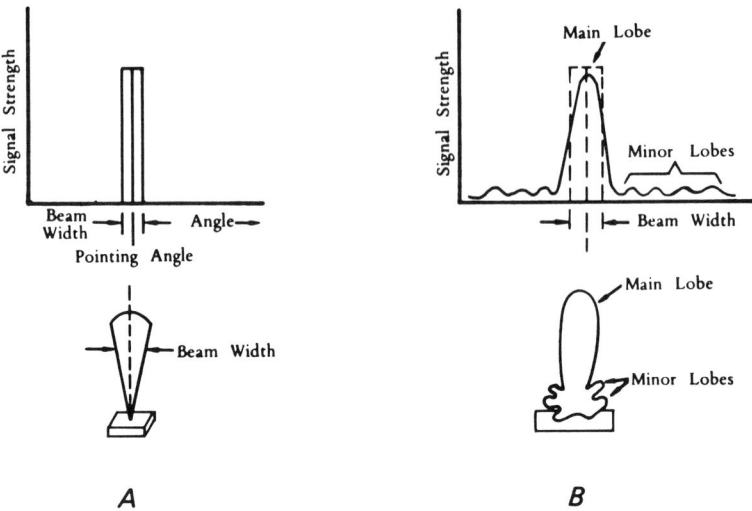

FIGURE 6. Diagrams of antenna patterns.

shows such an ideal pattern (in one angular dimension). In practice, antenna patterns look more like that of Figure 6B. The angular interval over which the pattern exceeds half its maximum value is defined as the *beamwidth*. The beamwidth (in radians) of an antenna (or a lens) is approximately

$$\text{Beamwidth} = \frac{\text{wavelength}}{\text{antenna length}}$$

Thus, the length of an antenna must be many wavelengths if the beam is to be narrow. In the centimeter, decimeter, and meter wavelength regions, this can be a very severe requirement; in the micrometer range it may not be as severe a limitation as equipment problems. For example, a 1-mrad beam can be achieved at 10 μm (infrared) with a 1-cm lens; for a 1-cm radar a 10-m antenna is required.

The radar measures range by comparing the received signal with a delayed sample of the signal transmitted earlier. The amount of delay required to obtain correlation is a measure of the range. Most radars used for imaging transmit short pulses of the microwave signal and achieve range resolution by separating the pulses returning from different parts of the ground by their differing times of arrival. Figure 7 illustrates the way a transmitted and received pulse would be displayed on an oscilloscope if the reflector were a single point. Since the velocity of the wave is known, the time t_d may be related directly to distance. Figure 8 shows the comparable oscilloscope picture for a signal returned from the ground. Here t_d is shown as the time delay for the signal from the closest point (directly under an aircraft radar); the signals extending to larger delays come from ground areas farther away (slant range). Hence, the total return shown is for an area from beneath the aircraft to a point some distance away, and the signals received from different

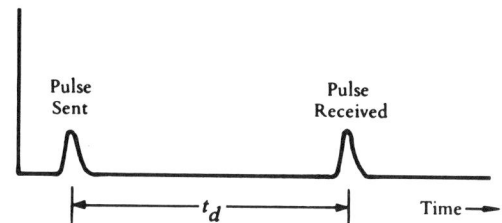

FIGURE 7. Diagram of the principle of radar range measurement (point target).

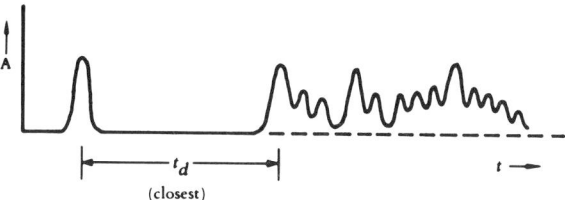

FIGURE 8. Diagram showing radar range measurement.

distant points may be identified simply by observing at the proper delay time. An image is produced by simultaneously looking at signals returning from many ranges and either moving the look direction by rotating an antenna or looking at new regions by moving forward while keeping an antenna fixed in a side-looking direction.

A radar measures relative speed, a quantity dependent on magnitude and direction of radar and target velocities. Multiple radars would be required to determine the velocity vector. Figure 9A illustrates what happens for a target coming toward the radar. The single frequency transmitted is received with a higher frequency. If the target were receding, the shift would be toward a lower frequency. Although only a single transmitted frequency is shown, the same shift occurs for all components of the transmitted signal.

For an airborne radar used against an area-extensive target, many Doppler frequencies are present. The reason is that the relative velocity between the radar and different points on the ground differs, depending upon the angles between the radar velocity vector and a line joining the point observed with the radar. Thus, the line spectrum indicated in Figure 9A for a single target becomes a distributed spectrum as in Figure 9B for the ground. The maximum relative velocity occurs along the flight track at the horizon. The relative velocity is zero on both sides and is negative behind the vehicle. Thus, the spectrum shown in Figure 9B corresponds to an antenna illuminating an area ahead of the radar since neither the zero frequency corresponding to the side nor the negative frequencies corresponding to the rear is present. This effect is used in the synthetic aperture system for high resolution.

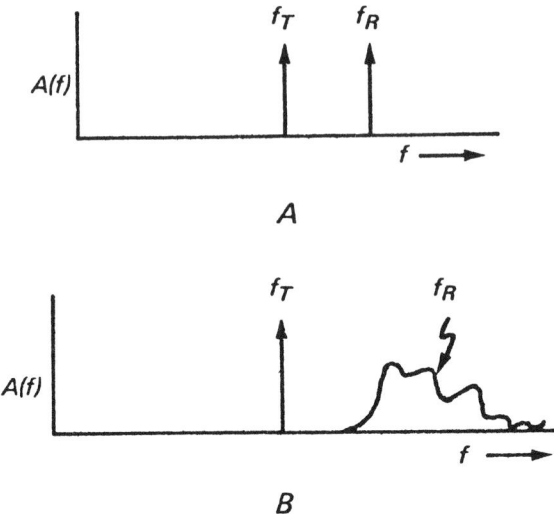

FIGURE 9. Diagrams of radar velocity measurement. *A:* single target; f = velocity; T = transmitted; R = received; $f_R - f_T$ = Doppler shift. Relative velocity toward radar makes $f_R > f_T$. *B:* Extended target. Signals simultaneously received from object with differing relative velocities toward radar.

Resolution

Figure 10 illustrates one definition of resolution. If two objects giving signals of equal intensity are sensed by a radar or other remote sensor as a single object, their spacing is less than the resolution distance. If the same two objects are sensed separately, their spacing is equal to or greater than the resolution distance. Usually, when two adjacent objects present signals of different intensity, they must be farther apart if the small one is to show up as distinct from the large one. Thus, the concept of resolution distance is not an easy one to apply to natural surfaces for which some pairs of objects may be of equal intensity but others may be of greatly different intensity.

In imaging radar, the term *resolution* is often loosely used to mean distinguishable spot size as calculated on some arbitrary basis. The size of the spot may be determined by the half-power angular width of the antenna pattern and the time width of the pulse. This quantity is certainly related to the resolution distance and is of the same order of magnitude, but it is not necessarily equal to the resolution distance defined in the preceding paragraph.

```
   O         O              O                      O
   |----D----|              |-----D----|
Appear as One Object        Appear as Two Objects
```

FIGURE 10. Diagram of one definition of resolution.

A distinction should be made between resolution, detectability, and precision. It may not be possible to resolve two objects 100 ft apart, yet one of the objects may be detectable even though it is only 1 ft across. Thus, the fact that an object is smaller than the resolution distance does not mean it goes undetected. For example, a metal fencepost may be resonant to the wavelength of the radar. If so, and if the illumination is at the correct angle, the fencepost will show up clearly, but one will not be able to distinguish between fenceposts spaced more closely than the resolution distance.

With an infinite signal-to-noise ratio, range can, in theory, be measured to any desired degree of precision, regardless of bandwidth and regardless of resolution. Thus, a radar with a resolution distance of 100 ft could, in theory at least, measure the range from the radar to the target with a precision of 1 ft.

Passive sensors must achieve all their resolution by angular measurements. Radar can improve on this resolution by using its range and velocity measurement capabilities. Figure 11 illustrates this. A passive system or a radar system using continuous transmission with no modulation and no relative velocity depends on angular resolution set by the antenna beam. If only angular resolution were used, the radar beam would have to be scanned as with the beam of a microwave radiometer or optical scanner. Since the radar can discriminate in range by other means, scanning is unnecessary. Figure 11A illustrates this point for a pulsed SLAR. Instead of a beam narrow in both dimensions, as for a radiometer, the beam is narrow in the direction along the flight path but wide perpendicular to the flight path. Such a "fan beam" is achieved by an antenna long in the flight direction but short in the vertical direction. The antenna is fixed in position on the aircraft, and the effect of scanning in range is achieved by the different delay times for the signals returning from different ranges.

Where the required beamwidth calls for an antenna that is too large to be carried conveniently, its effective length is increased by the "synthetic aperture" technique illustrated in Figure 11B. Here the resolution cell is not established by the antenna at all. Instead, range resolution is determined by pulse length, and azimuth resolution (along the flightpath) is determined by a measurement of Doppler shift. If the radar looks directly to the side, the center of the resolved area is on the zero relative-velocity line having zero Doppler shift. By filtering so that only those signals experiencing a small positive and a small negative Doppler shift are detected, the resolution indicated can be achieved.

The term *synthetic aperture* arises from a different way of looking at the technique.[7] The operation involved in Doppler filtering may be thought of in different terms, and, indeed, it can be performed by an entirely different approach. Thus, if one stores the signals returned from a large number of pulses in some sort of memory, being careful to retain phase as well as amplitude information, he can process the signals in the memory in the same way that a large antenna processes them. For example, all the pulses transmitted during 300 m of flight may be combined to give a resolution cell of the same size as that achievable with an antenna 600 m long. Thus, although the real antenna might be only 1 or 2 m long, the synthetic antenna is 600 m long.

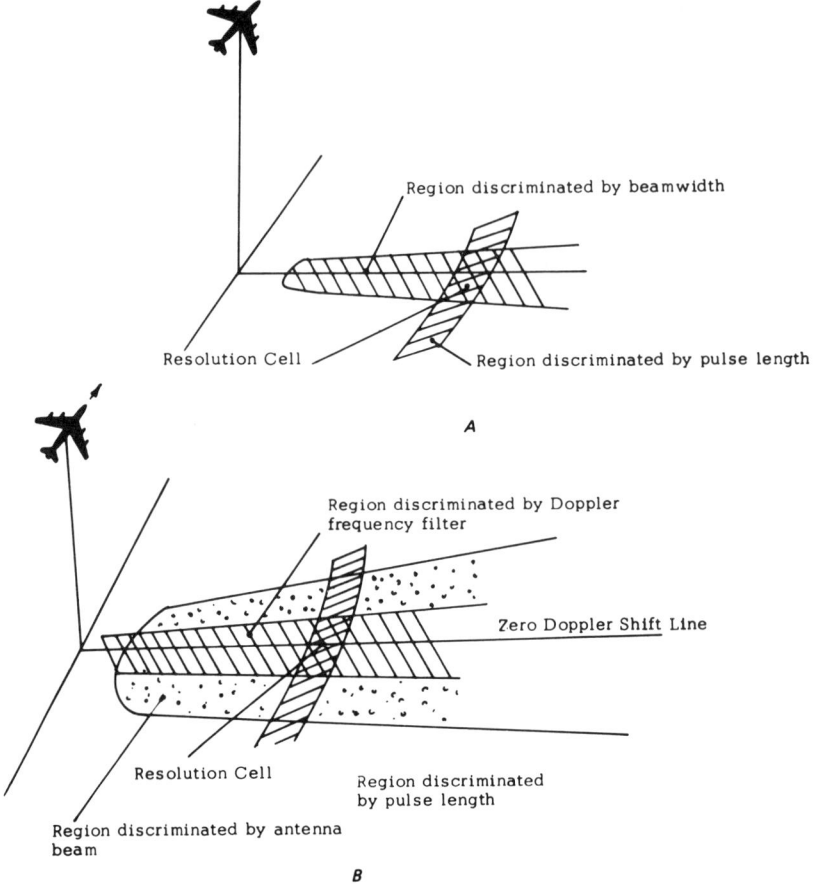

FIGURE 11. Diagrams of radar resolution. *A:* Real-aperture system; *B:* Synthetic-aperture system.

For remote sensing uses, antennas and lenses with small apertures are ordinarily focused at infinity. If, however, the resolution desired is as small as or smaller than the aperture, the antenna or lens must be focused for the distance to the region imaged. Many times synthetic apertures are much longer than the desired resolution, so they must be focused. This can be achieved in the processing, although the simple Doppler filter associated with Figure 11B is insufficient to do the focusing. Focused synthetic aperture radars can, in theory, achieve resolutions equal to half the length of the physical antenna used; unfocused systems are simpler but achieve resolutions only down to half the square root of wavelength times range.

Radar Presentations

Many types of presentations have been developed for radar systems. Some of these are particularly appropriate to special purposes such as airport landing systems or mapping radars. The highest resolution airborne systems use a fixed side-looking antenna or a synthetic aperture (equivalent to a longer side-looking physical antenna).

The presentation commonly used for these, at least for recording and later viewing, involves intensity modulation of a swept CRT beam by the signal. This is illustrated in Figure 12. The beam is swept in one direction only, so it always appears in the same line on the tube; thus, a single line is scanned rather than an area, as with television. The other dimension in the map is achieved by moving a film past this line in synchronism with the motion of the vehicle. Thus, the first sweep appears as a line on the film. When the second sweep comes along, the film has advanced so the second sweep appears as a different parallel line on the film. In this way a maplike image is produced. Similar techniques are used to produce images with infrared scanners.

Radar System Choices

Most of the radar system design choices of interest to the user relate to resolution. This in turn relates to the space available on the vehicle for an antenna, to the range to the farthest target, to the choice of wavelength, and to the complexity permitted by cost and reliability constraints. A side-looking radar with a "physical aperture" (no synthetic aperture techniques used) is intrinsically much simpler than a synthetic aperture system. An unfocused synthetic aperture system is simpler than a focused system.

Wavelength choices depend on the application planned, although it is clear that multispectral systems are preferable to single-wavelength systems. Physical aperture systems must operate at the shortest possible wavelengths to achieve good

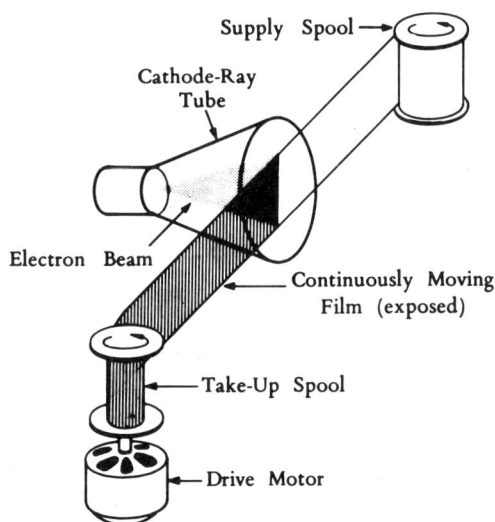

FIGURE 12. Diagram of recording technique for SLAR. Electron beam scans vertically with position proportional to distance from flight track.

resolution, but as a result they pay the penalty of susceptibility to attenuation by and echoes from atmospheric liquid moisture. The optimum wavelengths for different agricultural applications have not been determined, but it is fairly clear that wavelengths should be in the centimeter region for crop distinction, whereas it appears the meter wavelengths may be superior for soil studies. Longer wavelengths (decimeter to meter) appear desirable for study of stream patterns and similar patterns under forests because they penetrate the tree cover better. Presumably, some intermediate wavelength may be shown to be best for timber studies.

Figures 13 and 14 have been prepared to relate these choices to the above considerations. Resolutions with two different antenna lengths (3 and 10 m) have been plotted for each of three types as systems versus wavelength and versus maximum slant range. Theoretical values have been used for these diagrams; one should not necessarily infer from this that these resolutions are actually attainable in the present state of the art. Furthermore, although the curves for focused synthetic aperture indicate, like the theory, that resolution is independent of range, the length of the synthetic aperture required is proportional to range, so vehicle stability and equipment stability problems are obviously more severe at the longer ranges. Thus, realistic focused synthetic aperture curves would show degradation in resolution, especially at longer ranges.

For many agricultural purposes, resolutions of 15–30 m should be adequate. Clearly, these cannot be achieved with any physical aperture system with a wavelength of more than 1 cm at a 50-km range, and the aperture required at that range would be so large that it would be difficult to carry the antenna on existing aircraft. Hence, if a swath this wide is desired, some sort of synthetic aperture system is called for unless a considerable relaxation in resolution is possible. A 3-cm

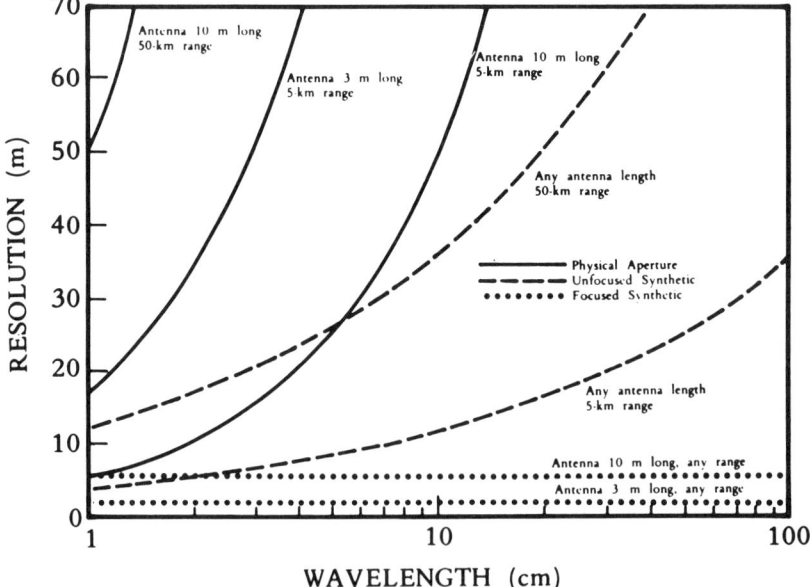

FIGURE 13. Design trade-off information for side-looking radars, I.

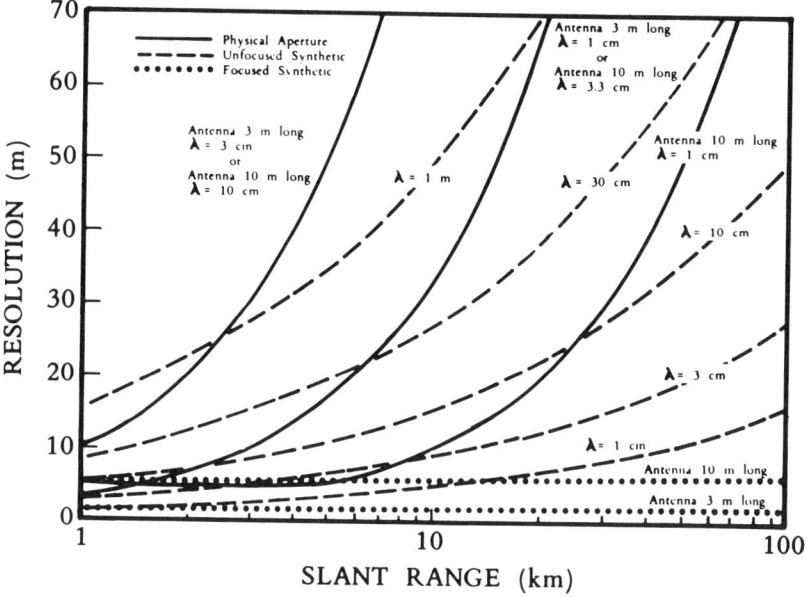

FIGURE 14. Design trade-off information for side-looking radars, II.

wavelength unfocused system should be adequate to meet this resolution swath requirement, but at 10 cm a focused system is required. On the other hand, if a 5-km total slant range is sufficient (probably a 4-km swath at about 1-km altitude), a physical aperture 3 m long would permit a 1-cm system, and an aperture 10 m long would permit a 3.3-cm system. An unfocused synthetic system could meet this requirement at 30-cm wavelength. For longer wavelengths, a focused synthetic aperture system is required.

Many other examples could be developed from these curves, and the curves themselves could be expanded. Relations used in plotting the curves are as follows:[7]

Imaging with Nonphotographic Sensors

Physical aperture system: $\quad r = \lambda R / l$

Unfocused synthetic aperture system: $\quad r = \sqrt{\lambda R / 2}$

Focused synthetic aperture system: $\quad r = l / 2$

where r = resolution distance

R = slant range

l = antenna physical length

λ = wavelength

To achieve reasonable resolution from space, a synthetic aperture system is obviously required. Since, in theory at least, synthetic aperture focused systems have resolution independent of range, the resolutions indicated might be achieved. A system proposed to NASA would have 15-m resolution at wavelengths of 60, 15, and 3.75 cm from orbital altitudes. For various complex reasons, the swath width of such systems is limited. The system proposed would have a swath width of 40 km, based on an antenna 8 m long. A shorter antenna could theoretically permit better resolution, but it would also reduce the permissible swath width. Techniques have been devised for overcoming this swath width limitation, but they are costly in power and in complexity. The vast quantity of information that can be gleaned from a spaceborne system sweeping out an area 8 × 40 km/sec makes it appear that the additional complexity is not warranted for some time, as analysis of images obtained with the 40-km width will be a staggering task.

The Unique Role of Radar in Agriculture

Radar has certain characteristics important for agricultural surveys. Some of these are due to the nature of the radar system (usually an active device that operates at single frequency employing a coherent radiation technique), and others due to the spectral region commonly used.

Values Inherent in Radar Systems

1. Images are produced under the same conditions day and night, regardless of temperature history. This is true for soils and variations in illumination conditions. However, plant leaves do exhibit changes in configuration with variations in temperature, illumination, and wind. The influences of these adaptive plant changes on radar returns have not yet been explored.
2. Wide areas may be imaged simultaneously with essentially uniform resolution.
3. Polarization of illumination may be controlled.

Values Due to Use of Radar System in Centimeter–Meter Wavelengths*

1. Images may be produced in any weather.
2. Crop cover may be penetrated to give soil response (longer wavelengths).
3. Tree cover (orchard and forest) may be penetrated to show underlying cultivation, drainage, and geologic patterns (longer wavelengths).

*Because of technical and economic considerations, radar development has tended to cluster at and around a few frequencies. These bands are designated by the following letters:

Band	Nominal Range (MHz)	Wavelength (cm)
P	220–390	133–77
L	390–1,550	77–19
S	1,550–5,200	19–5.8
C	3,900–6,200	7.7–4.8
X	5,200–10,900	5.8–2.7
K	10,900–36,000	2.7–0.83

4. Additional multispectral discriminations between crops, crop states, soils, and so on, may be added to those in other wavelengths.

For quick surveys with aircraft, item 2 of the first list above is significant, since the useful swath width from aircraft using radar is greater than for other sensors. If the sensors are carried in spacecraft, however, the swath width advantage of high-resolution radar is lost. The other advantages remain.

Radars operating in the decimeter (and perhaps meter) range give a composite image in which the soil or other surface plays a larger role, for the waves are attenuated only slightly in passing nearly vertically through vegetation. With the shorter radar wavelengths, the plants themselves provide more attenuation, so the signal is primarily due to vegetative cover where there is any. Thus, combinations of radars operating at short and long wavelengths should provide information on both soil and crop characteristics.

Item 4 in the list above is, of course, quite significant in multispectral analysis to permit identification of terrain properties strictly on the basis of remote sensing, for each portion of the wavelength spectrum (from ultraviolet to long waves) is sensitive to different properties of the surface.

REFERENCES

1. Beckman, P., and A. Spizzichino, *The scattering of electromagnetic waves from rough surfaces,* Macmillan, New York, 1963.
2. Cosgriff, R. L., W. H. Peake, and R. C. Taylor, *Terrain scattering properties for sensor system design* (Terrain Handbook II), Engineering Experimental Station, Ohio State University, Columbus, 1960, pp. 16–18.
3. Carlson, N. L., *Dielectric constant of vegetation at 8.5 Ghz,* Technical Report 1903-5, Electroscience Laboratory, Ohio State University, Columbus, 1967.
4. Morain, S. A., and D. S. Simonett, *K-band radar in vegetation mapping,* CRES Report 6123, Center for Research in Engineering Sciences, University of Kansas, Lawrence, 1967.
5. Medhurst, R. G., Rainfall attenuation of centimeter waves: comparison of theory and measurement, *IEEE Trans.* AP-13, 1965: 550–564.
6. Bean, C. R., and E. J. Dutton, *Radio meteorology,* U.S. National Bureau of Standards Monograph 92, U.S. Government Printing Office, Washington, D.C., 1966, chap. 7.
7. Cutrona, L. J., and G. O. Hall, A comparison of techniques for achieving fine azimuth resolution, *IRE Trans.* MIL-6, 1962: 119–121.

1.1.2 Synthetic Aperture Radar (SAR) Specifications*

IDENTIFICATION

Discipline: Environmental Observations

Status: Flight Missions

Acronym: SAR

Instrument Type: Radar

Spacecraft: Seasat-A

Contractor: Jet Propulsion Laboratory

*Based on *Handbook of Sensor Technical Characteristics,* NASA Headquarters, Office of Space Technology Applications, Communications Information Division, February 1981.

OBJECTIVES

1. To obtain microwave imagery at 1275 MHz of the sea surface
2. To discern the length and direction of ocean waves
3. To determine the size, location, and speed of sea ice, oil spill, and coastal features

DESCRIPTION

Summary: The Seasat-A satellite bearing the SAR was launched to attain a near-polar orbit of 108° inclination and altitude of 790 km. The period will be approximately 100 minutes, resulting in $14\frac{1}{2}$ orbits per day. This yields complete ocean coverage in 12 hours. Using an antenna measuring 10 m (azimuth) $\times$ 2 m (elevation), the SAR receives reflected radiation from a swath of 100 km width, starting 150 km off nadir. Data are transmitted directly to ground via S-band link; video pass band is 2–21 MHz. No on-board storage is provided due to high data rates. The average on-time is 10.85 minutes over each of the following STDNs: Alaska, Godstone, Rosman, Madrid, and Orroral. The swath is illuminated by 1275-MHz microwave pulses of 32 microsecond or μsec duration occurring at 1400 pps. The transmitter is operated at a peak power level of 800 W. Azimuth compression ratio is 155:1. Range compression ratio is 576:1. Transmitter operates in four modes: standby, on, operate, calibrate. Mode is selectable by ground command.

Heritage/Derivation: The only satellite-borne SAR flown prior to 1976 was the Apollo Lunar Sounder Experiment. It produced profiles on the lunar subsurface at 5, 15, and 150 MHz.

DATA

Data Products: Four negative 70-mm rolls, each measuring 25$\times$ up to 4000 km in length, with calibration wedges and time; location marks for up to 500 passes will be available.

Data Archives Location: Jet Propulsion Laboratory.

Period of Operation: July 7 through October 10, 1978.

BIBLIOGRAPHY

NASA, *Seasat-A program summary,* August 1974.

SYSTEM CHARACTERISTICS

Range compression ratio: 776
Range swath width: 100 km

Total pulses in beam: 2915
Pulses coherently integrated/look: 909
Coherent integration time/look: 0.552 second
Number of looks in azimuth: 3 or 4
Range migration correction capability: 128 bins
Nominal pixel spacing (range and azimuth): 20 m
Integrated side-lobe ratio: -17 dB
Radiometric accuracy: ±0.5 dB
Geometric accuracy due to processing: 0.5 pixel
Throughout rate capability: real time
Time and range/azimuth reference marks: 1-sec intervals/10-km spacing
Number of wavelength bins: 20
Number of angular divisions: 30
Angular resolution: 6°
Wavelength resolution: 50 m
No true image available
Volume: 0.01 m^3
Weight: 10 kg
Power consumption: 30 W
Data rate: 110 Mbs, 20 kbs

MEASUREMENTS

Nominal range resolution (3-dB width): 25 m
Nominal azimuth resolution (3-dB width): 25 m
Dynamic range for point target: 50 dB
Dynamic range for distributed target: 27 dB

POTENTIAL DERIVED PARAMETERS

1. Ocean surface imagery capable of yielding directional wave spectra in the open ocean
2. Monitoring of coastal processes
3. Charting ice fields
4. Land imagery useful in geological, hydrological, and glaceological studies
5. Iceberg detection
6. Fishing vessel surveillance

Figures 15 and 16 are from Lee-Lueng Fu and Benjamin Holt, *Seasat Views Oceans and Sea Ice with Synthetic Aperture Radar,* Publication 81-120, Jet Propulsion Laboratory, Pasadena, Calif., February 15, 1982.

FIGURE 15. Seasat—artist's conception.

1.2 COASTAL ZONE COLOR SCANNER

The Coastal Zone Color Scanner (CZCS) was launched on Nimbus-7 in October 1978 as a research instrument to determine whether the biological and nonbiological content of the ocean could be determined by remote sensing to a degree of accuracy useful for oceanographers. By the end of the first year of the sensor's life, the algorithms for calculating such things as pigment concentration and diffuse attenuation coefficient had been developed, and ship measurements showed that the agreement with surface truth was better than the original goal set for the sensor data product in the open ocean. Near shore, however, where high sediment levels prevented the atmospheric correction algorithm from working properly, gradients could be observed, but quantitative accuracy was not as good as in offshore waters. Moreover, the calibration has degraded with time.

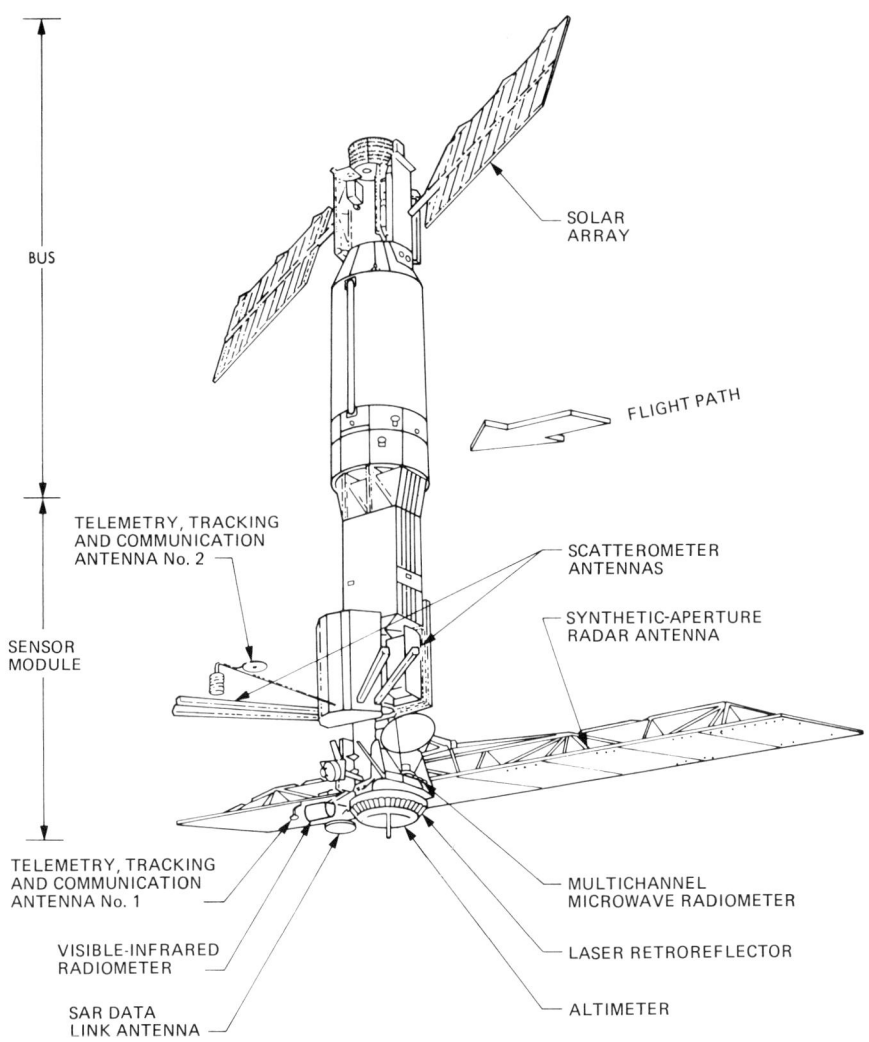

Seasat bus and payload configuration.

FIGURE 16. Seasat bus and payload configuration.

1.2.1 The Coastal Zone Color Scanner Experiment*

Introduction

The Coastal Zone Color Scanner is the first instrument devoted to the measurement of ocean color and flown on a spacecraft. Although instruments on other

*Based on Warren Hovis, *The Coastal Zone Color Scanner Experiment,* National Oceanic and Atmospheric Administration, National Environmental Satellite Service, Washington, D.C.

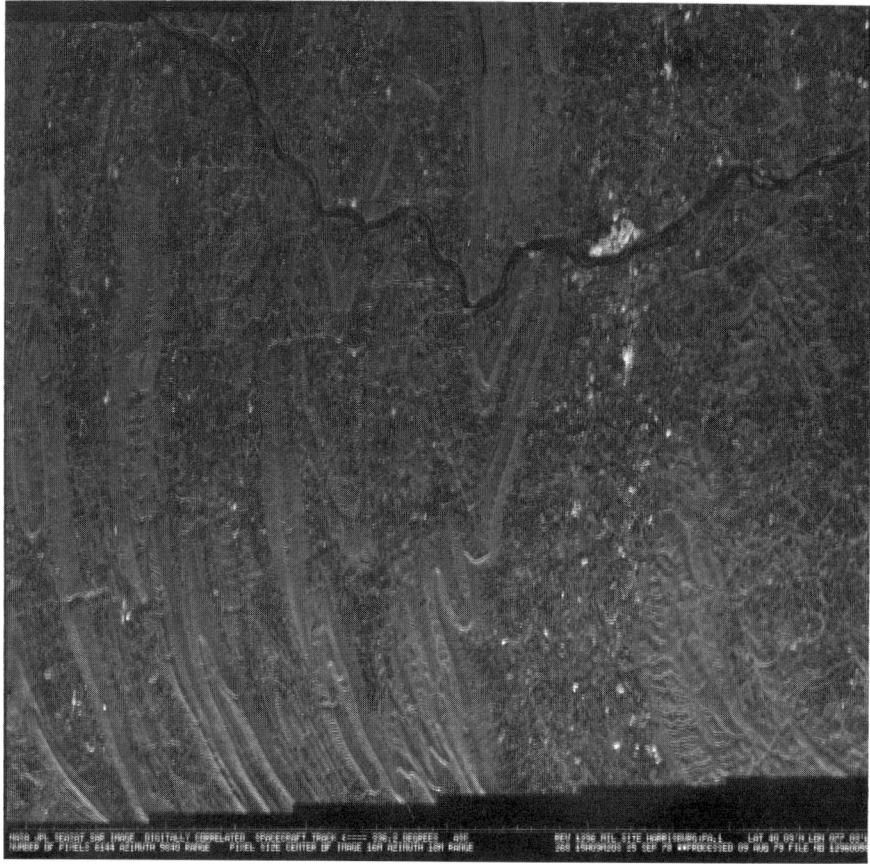

FIGURE 17. Harrisburg, Pa. This scene was acquired by Seasat SAR on 25 September 1978 at 15:9:20 GMT at coordinates latitude 40°9'N, longitude 77°3'W, rev 1296 and shows clearly the area's geologic appearance.

satellites have sensed ocean color, their spectral bands, spatial resolution, and dynamic range were optimized for land or meteorological use. In the CZCS, every parameter is optimized for use over water to the exclusion of any other type of sensing. The signal-to-noise ratios in the spectral channels sensing reflected solar radiance are higher than those required in the past. These ratios need to be high because the ocean is such a poor reflecting surface that the majority of the signal seen by the reflected energy channels at spacecraft altitudes is backscattered solar radiation from the atmosphere rather than reflected solar energy from the ocean. The CZCS thermal channel utilizes the 10.5- to 12.5-μm region used on many other thermal mappers. This CZCS channel is unique, however, since it is registered with the reflected solar energy bands and has the same spatial resolution.

The data-processing techniques for the CZCS are also unique in that offsetting is used to enhance contrasts over the ocean and remove much of the effect of the

NASA JPL SEASAT SAR IMAGE DIGITALLY CORRELATED SPACECRAFT TRACK ←—— 200.9 DEGREES DES   REV 0788 MIL SITE NEW ORLEANS,LA.1   LAT 30 06'N LON 089 41'W
NUMBER OF PIXELS 6144 AZIMUTH 5840 RANGE   PIXEL SIZE CENTER OF IMAGE 16M AZIMUTH 18M RANGE   233 02H47M39S 21 AUG 78 **PROCESSED 01 MAY 79 FILE NO 07880012

FIGURE 18. New Orleans, La. Seasat acquired this SAR image on 21 August 1978 at 2:47:39 GMT at latitude 30°6'N, longitude 89°41'W, rev 788. It shows 6144 pixels (azimuth) × 5840 pixels (range). The size of a pixel at the center of the image is 16 × 18 m.

backscattered atmosphere. Attempts will be made to process the data into derived products such as pigment concentration and diffuse attenuation coefficient prior to distribution to users. The archived magnetic tapes contain both calibrated radiances and equivalent blackbody temperatures, plus the derived products, so a user with a large computer facility would be able to utilize a more complicated algorithm than that used in production and processing at Goddard Space Flight Center (GSFC). A user without such a facility can utilize the derived products provided by NASA.

The CZCS is a conventional multichannel scanning radiometer utilizing a rotating-plane mirror at a 45° angle to the optic axis of a Cassegrain telescope. The rotating mirror scans 360°; however, only ±40° of data centered on the spacecraft nadir is collected for ocean color measurements. During the rest of the scan, the instrument acquires a view of deep space and of internal instrument sources for

FIGURE 19. Canyonlands, Utah. This scene was acquired on 18 September 1978 at 16:53:17 GMT. The center of the frame has coordinates latitude 38°13′N, longitude 109°32′W, rev 1197. (Seasat SAR)

calibration of the various channels. The radiation collected by the telescope is divided into two portions by a dichroic beam splitter. One portion is transmitted to a field stop that is also the entrance aperture of a small polychromator. The radiant energy entering the polychromator is disbursed and reimaged in five wavelengths on five silicon detectors in the focal plane of the polychromator. The portion of the beam reflected off of the dichroic mirror is directed to a cooled mercury cadmium telluride detector sensing in the 10.5- to 12.5-μm region. The CZCS utilize a radiative cooler that cools the mercury cadmium telluride detector to approximately 120 K during spacecraft flight.

The CZCS is intended primarily as a tool for determining the content of water. It is well known that the content of water, be it organic or inorganic, particulate matter or dissolved substances, affects its color. Ocean water, containing very little particulate matter, scatters as a Rayleigh scatterer with the well-known deep purple or bluish color of the ocean. As particulate matter is added to the water, the

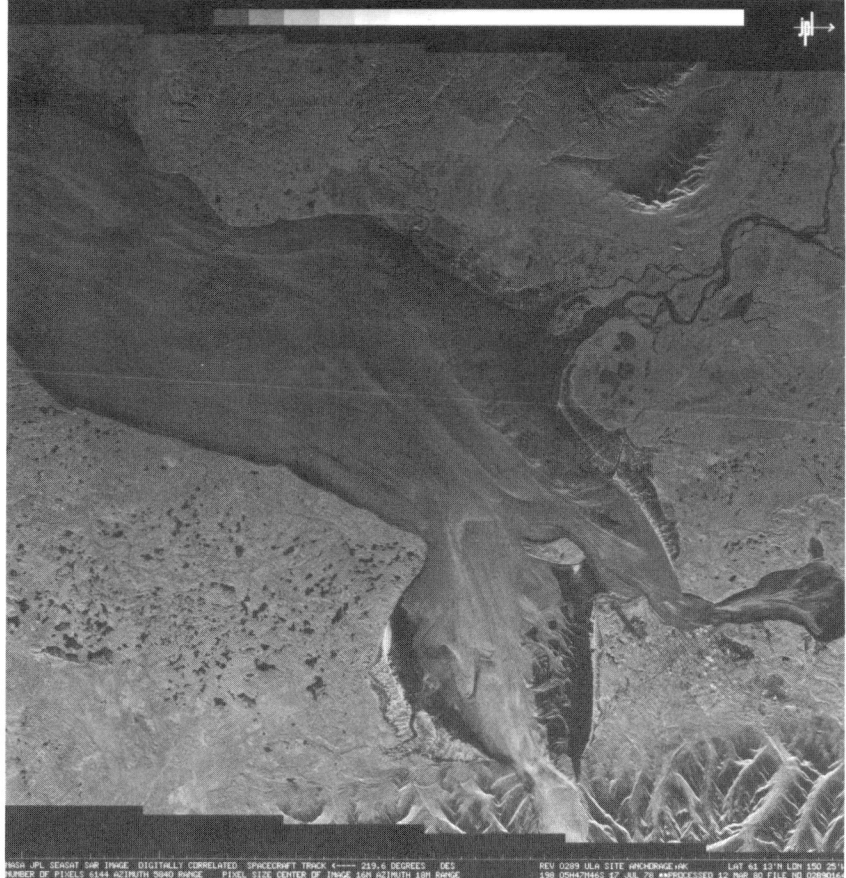

FIGURE 20. Anchorage, Alaska. On 17 July 1978 at 5:47:46 GMT Seasat acquired this scene at 61°13'N, 150°25'W on rev 289. (SAR)

scattering characteristics are changed and the color is changed. Phytoplankton, for instance, have specific absorption characteristics and normally change the water to a more greenish hue, although some phytoplankton, such as the various red tides, can change the water to colors such as red, yellow, blue–green, or mahogany. By sensing the color with very high signal-to-noise ratios, the CZCS provides a mechanism for analyzing that color for the content of the water. Inorganic particulate matter in water, such as the terrigenous outflow from rivers, has a different color from organic material typically brownish in color, but sometimes varying with red.

Scientific and Technical Objectives

Scientific Objectives

The scientific objective of the CZCS is to determine the specific nature of the contents of water as quantitatively as possible and to carry out such measurements over large areas in short periods of time in a way not possible with other techniques such as surface ship investigations. Specifically, the CZCS experiment attempts to discriminate between organic and inorganic materials in the water, determine the quantity of these materials in the water sample to the best degree possible, and, in certain instances, attempt identification of organic particulates such as discriminating between various types of red tide organisms.

In one 2-minute data segment, the CZCS covers approximately 1.3 million km^2 of the ocean surface, allowing examination, nearly simultaneously, on a scale never before accomplished. Measurements on this scale allow oceanographers to determine such things as the standing stock of phytoplankton and its distribution in various fishing areas and potentially, allows them to assess the ability of that area to support a standing stock of fish. In addition to examining the existing fisheries, the CZCS will be used to look for new areas of potential fish production around the globe.

Technical Objectives

The technical objective of the CZCS program is to determine if remote sensing of color can be used to identify and quantify material suspended or dissolved in water. If ocean color measurements can be used to derive such products as chlorophyll and sediment concentration, they will guide further development of the ocean color discipline and help to determine if such an instrument is a candidate for operational satellite use in the future.

The algorithms being developed for the derived products from CZCS are the result of the most extensive ocean color measurements ever made and are a considerable step forward from those available in the past. Corrections for such things as atmospheric backscatter and limb brightening are included in the CZCS processing algorithms. The processing goal is to take the observed radiance, determine the radiance that would be seen directly above the ocean surface, and then derive from that radiance the content of the water below the ocean surface.

Instrument Description

Operation

The CZCS has considerable flexibility built into it to accommodate a wide range of conditions. The first four spectral bands, for instance, have four separate gains that change, on command, to accommodate the range of sun angles observed during a complete orbit and throughout the various seasons. The gains are changed to utilize the best dynamic range possible without saturating over water targets. Normally, the gain used in the first four channels is determined by the solar elevation angle of the target to be acquired. When a special circumstance is expected, such as particularly bright material in the water, the gain can be changed to accommodate the special circumstances.

In addition to gain change, the CZCS scan mirror can be tilted from nadir to look either forward or behind the spacecraft line of flight. It can tilt in 2° increments up to 20° in either direction. This feature was built into the instrument to avoid the glint caused by capillary waves on the ocean that would obscure any scattering from below the surface. The angle of tilt of the scan mirror is determined by the solar elevation angle. It is normally tilted to avoid sunlight and would only be commanded to look into the glint for a special sunglint study.

Viewing Geometry

The CZCS is a scanning multispectral radiometer with a recorded scan width of 1566 km centered on spacecraft nadir. The scanner actually scans through 360°, but the electronics limit the high-data-rate sampling to 39.34° about nadir. The ground resolution of the IFOV is 0.825 km at nadir and degrades somewhat as the instrument scans away from nadir on either side. The viewing geometry of the instrument is illustrated in Figure 21.

Channel Characteristics

The CZCS has six spectral bands, five sensing backscattered solar radiance and one sensing emitted thermal radiance. Figure 22 illustrates the method by which discrimination of the spectral bands is achieved. The beam is split by a dichroic

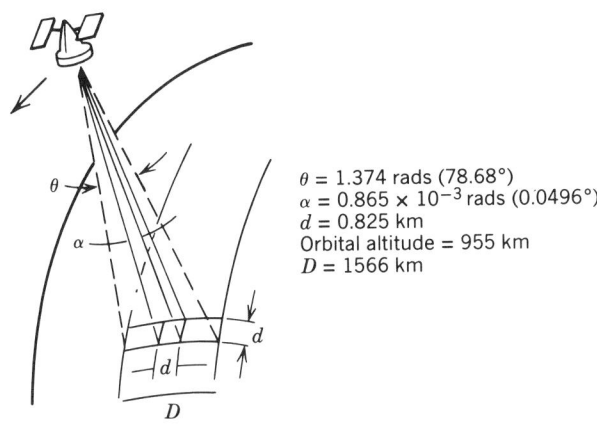

$\theta = 1.374$ rads (78.68°)
$\alpha = 0.865 \times 10^{-3}$ rads (0.0496°)
$d = 0.825$ km
Orbital altitude = 955 km
$D = 1566$ km

CZCS scanning arrangement

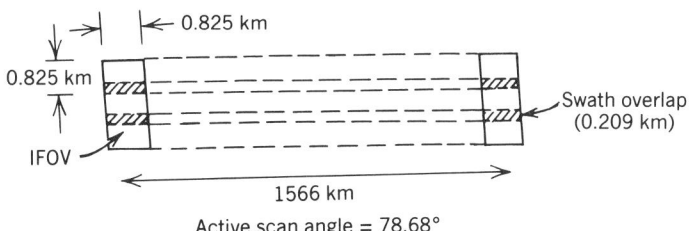

FIGURE 21. CZCS viewing geometry and earth scan pattern.

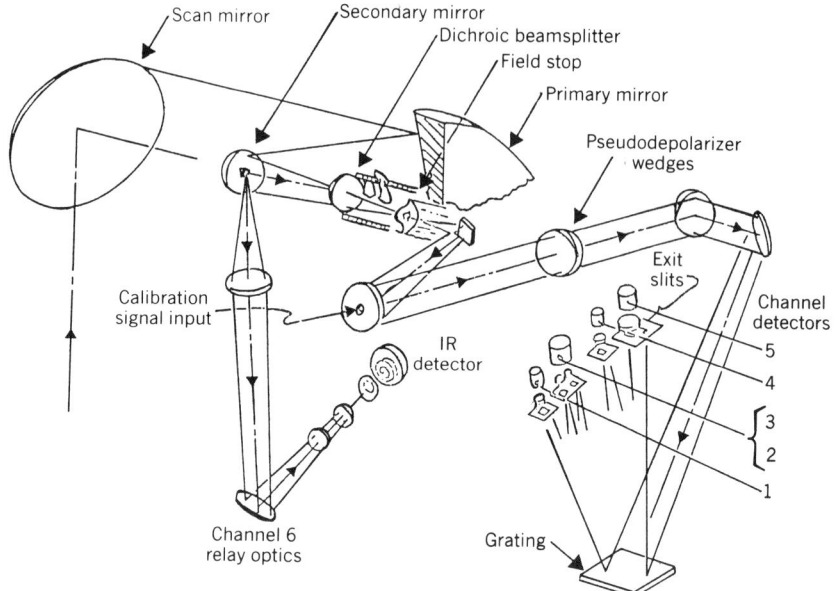

FIGURE 22. CZCS optical arrangement.

beam splitter, one portion of the beam going through a set of depolarizing wedges to a small polychromator where the radiance is dispersed and detected by five silicon diode detectors in the focal plane of the polychromator. Radiance in the 10.5- to 12.5-μm spectral band is reflected off the dichroic and then imaged onto an infrared detector of mercury cadmium telluride cooled to approximately 120 K. Table 9 shows the center wavelengths, the spectral bandwidths, and the minimum signal-to-noise ratio specified for the instrument at the most sensitive gain setting, that is, the gain setting that would be used for the darkest targets. (Prelaunch tests show the instrument exceeded the specification for signal-to-noise in every channel.) The first four channels were selected to cover specific absorption bands and the so-called hinge point. These channels are meant to look at water only and saturate when the field of view is over most land surfaces and clouds. The spectral response of channels 1 through 5 is illustrated in Figure 23.

Channel 5 has the same spectral response as channel 6 of the Landsat Multispectral Scanner series. The gain of channel 5 is fixed and set to produce the same percentage of maximum signal over land targets as the Landsat channel 6. However, the actual radiance for saturation is higher since the Nimbus-7 spacecraft crosses the equator at high noon, whereas Landsat crosses the equator at 9:30 a.m. local time.

Operational Modes

Since Nimbus-7 flies from south to north in daylight, the scan mirror is positioned to look behind the satellite when the spacecraft is south of the subsolar point and ahead of the spacecraft when it is north of the subsolar point. Tilt and gain setting

TABLE 9. CZCS Performance Parameters

Performance Parameters	Channel					
	1	2	3	4	5	6
Scientific observation	Chlorophyll absorption	Chlorophyll correlation	Yellow stuff	Chlorophyll absorption	Surface vegetation	Surface temperature
Center wavelength λ (μm)	0.443 (blue)	0.520 (green)	0.550 (yellow)	0.670 (red)	0.750 (far red)	11.5 (infrared)
Spectral bandwidth $\Delta\lambda$ (μm)	0.433– 0.453	0.510– 0.530	0.540– 0.560	0.660– 0.680	0.700– 0.800	10.5– 12.5
Instantaneous field of view (IFOV)			0.865 × 0.865 mrad (0.825 × 0.825 km at sea level)			
Coregistration at nadir			<0.15 mrad			
Accuracy of viewing position information at nadir			<2.0 mrad			
Signal-to-noise ratio (min.) at radiance input $N <$ (mW/ cm^2 · stera-dian · μm)	>150 at 5.41	>140 at 3.50	>125 at 2.86	>100 at 1.34	>100 at 10.8	NETD of 0.220°K at 270°K
Consecutive scan overlap			25%			
Modulation transfer function (MTF)		1 at 150-km target size, 0.35 min. at 0.825-km target size				

information is transmitted with the CZCS data and is part of the data product records.

The CZCS data is transmitted from the spacecraft to ground receiving stations at a rate of 800 kbs either in real time or in playback of the tape recorder. Whenever possible, the data are recorded in real time. However, when the satellite is out of the range of tracking stations, the data are recorded on an on-board tape recorder. The tape-recorded data will normally be played back at the Alaska tracking station. Nine other STDNs also have the capability to receive these playbacks.

To improve the instrument response to ocean color, a DC offset can be inserted into the on-board processing of the radiance measured in the first four bands. In

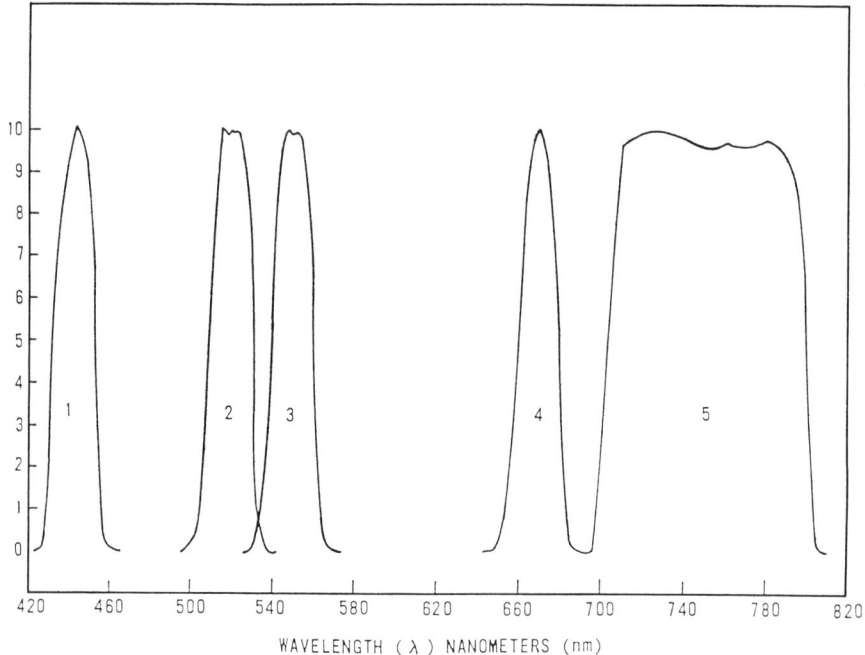

FIGURE 23. CZCS spectral response for channels 1–5.

this DC offset mode, the entire digital capability of the on-board digitizer is utilized to cover approximately the top 30% of the signal, which contains modulation due to change in ocean color. Since the exact amount of the offset eliminated in the on-board processing is always known, it can be reinserted where needed for processing on the ground.

The sensor is turned on in sufficient time prior to collection of data to allow for instrument warmup and sensor stabilization. Since all channels are calibrated continuously during flight, any effect of turn-on transient should be noticed immediately, but none is expected.

The most important aspect to be understood about the CZCS operation is that the operation is limited by spacecraft power constraints to approximately 2 hours per day. Because of the requirement to operate the sensor 2 hours per day, data must be taken in carefully preselected locations. Minimum on–off data-taking time is a 2-minute segment. Frequently, longer segments are taken — up to a maximum of 10 minutes of continuous data.

Interested users are reminded that if they wish to acquire CZCS data over a particular site, they should contact a member of OSTA and inform that member of the location of the site and the dates on which coverage is most highly desired. Even though all of the data are placed in the public archive, there is no guarantee that all areas of the world will be covered. A special effort will be made, however, to cover major oceanographic expeditions where surface truth is being collected by a ship.

The prime operational areas for the CZCS are the coastlines of the United States and the Gulf of Mexico. Other areas of coverage are the coastline of Europe, including the Mediterranean Sea, the Baltic Sea, the North Sea, the channels between England and the Continent, the Irish Sea, and the test sites designated by the EURASEP Group of the Commission of European Communities. South African NET participation has requested coverage around the southern tip of Africa on both the east and west coasts and extending toward the Antarctic. In the Antarctic summer following the launch of Nimbus-7, the Scientific Committee on Antarctic Research (SCAR) conducted a large expedition in Antarctic water. This also was a prime target area for the CZCS. Other areas, such as the Deep Ocean Mining Experiment Stations (DOMES), and the Antarctic coverage are limited by the extent of the activities such as the time of the DOMES action or the availability of sunlight, as in the Antarctic, where the sensor can only operate usefully during the Antarctic summer. Requests for coverage of other areas have been received from a number of institutions around the world, and every attempt will be made to accommodate requests from other oceanographic institutions, especially when surface truth is being measured by institutions.

All channels of the CZCS instrument operate simultaneously. During daytime operations all six channels provide useful information. If the sensor operates at night, only data from channel 6 are usable.

Data Processing, Formats, and Availability

Data Processing

The data from the CZCS are transmitted to the ground either in real time or from tape recorder playback at a rate of 800 kbs. The data are recorded on magnetic tapes and sent to the Image Processing Division (IPD) at GSFC. These tapes contain both radiometric information from the imagery and CZCS housekeeping information. IPD uses these data, plus image location tape (ILT) data, to produce the user tapes and the images. After making sufficient tape and film copies for Nimbus Experiment Team (NET) users, IPD forwards the tapes and film to the Environmental Data Information Service (EDIS) of NOAA for archiving and reproducing copies.

At IPD the data are converted from voltages to radiances for bands 1 through 5 and to equivalent blackbody temperature for band 6. This is accomplished by using the calibration curves derived before launch and applying in-flight calibration sources. After calibration, the data are processed using algorithms developed by the CZCS NET to derive products of suspended and dissolved material in the water. As knowledge is gained from the experiment, the algorithms may need to be changed. All algorithms are available to users, and those used to process each tape and image are identified on those products.

Tape Products

The following tape products are produced by IPD and are sent to the EDIS at NOAA for archiving:

CRCST (CALIBRATED RADIANCE, PIGMENT, DIFFUSE ATTENUATION COEFFICIENT, AND TEMPERATURE TAPE). These tapes contain calibrated and located CZCS data from all six channels scan line by scan line with the channels separated, plus derived pigment and diffuse attenuation coefficient parameters, where computed. There are a maximum of three 2-minute blocks (files) of data per tape. Statistical and calibration summaries are at the end of each file.

CAT (CATALOG TAPE). These tapes contain cataloged information on all images (2-minute files on the CRCST's). Entries are organized chronologically by target area (location).

The form and content of each of these tapes are described in a tape specification document for each tape type. The appropriate document will accompany a tape shipment to a user.

The CZCS is expected to operate a maximum of 2 hours per day. If the sensors operate for 2 hours per day for 1 year, approximately 22,000 images and 7000 magnetic tapes would be generated. Because of weather conditions, principally cloud cover, the CZCS probably will not operate its scheduled 2-hour period per day as planned. Thus, a more reasonable estimate of total output per year is approximately 12,000 images and 4000 magnetic tapes.

Film Format

Figure 24 is an example of the format for all CZCS images. Each display is produced on 241 mm × 241 mm (9.5-in.) black-and-white image stock. The title and reference information at the top of each display includes the gain and tilt angle in effect during the scene and whether the threshold mode of data enhancement was on or off.

Each of the 10 chips on a single display has the same latitude and longitude grid around the chip boundaries. Channel displays 1–6 show radiances as shades of gray (referenced to the gray scale at the bottom of the display). The physical parameters of pigment concentration and diffuse attenuation coefficient, if calculated for the scene, are shown as shades of gray (two chips) and as contoured plots (two chips). The top displays (channels 5 and 6) represent the maximum amount of data from a single channel: ±39.36° from nadir for each scan line, and 2 min of data along the orbit track. The bottom eight displays (channels 1–4 on the left, plus the four scenes of pigment and diffuse attenuation coefficient parameters) show only the region of best spatial resolution and least geometric distortion, which is within ±20° of nadir. Each scene (chip) is rectified along each scan line and from scan line to scan line, so there is an approximately equal scale over each scene.

The crosses at the four corners of each of the lower eight chips and along the top and bottom of the top two chips are for reference if two or more channels are to be photographically color-composited. Cutting apart the chips and aligning two or more sets of crosses provides chip-to-chip registration.

The 16-step gray scale beneath the chips is calibrated in radiances for reference to the images. The values for the gray scale versus radiances for each display are

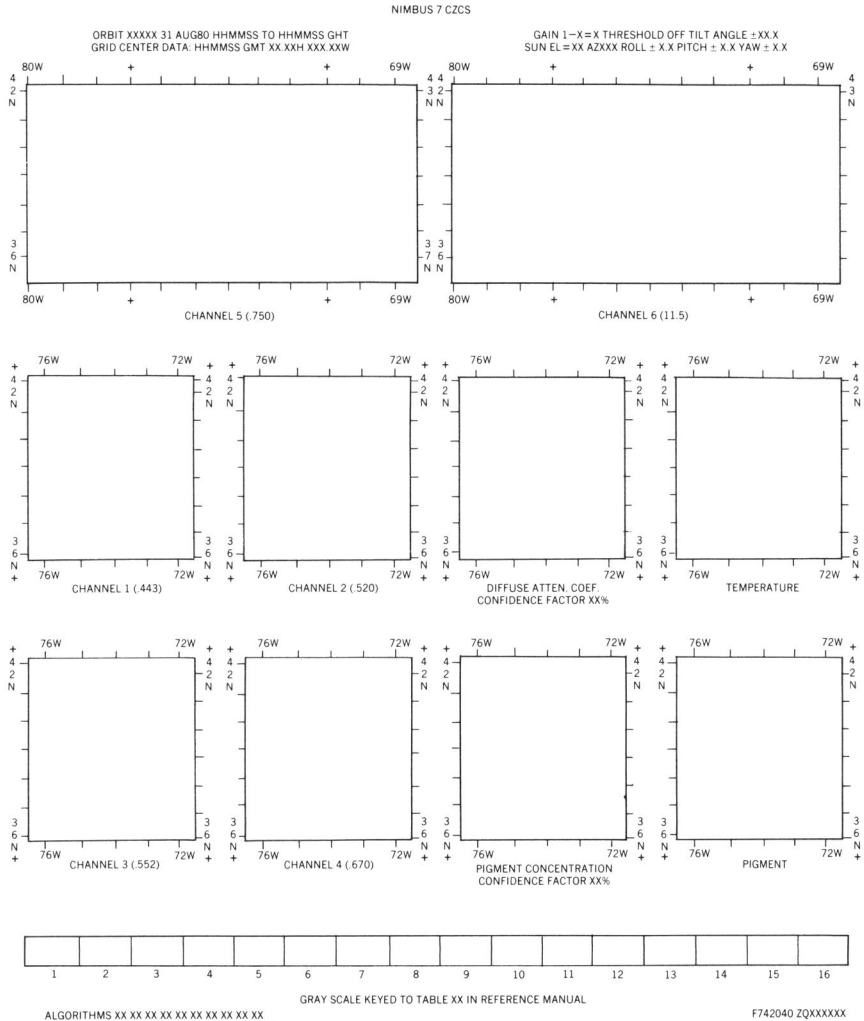

FIGURE 24. CZCS image display format.

provided in the EDIS catalogs. The appropriate table for each display is identified beneath each gray scale.

The 10 algorithms used to generate the 10 chips are identified in the lower left corner. The last line in the lower left lists the number of scan lines processed (maximum is 960) and used as input for the chips and the number of these scan lines containing errors.

The reference data in the lower right corner are the film specification number (F742040), the projection data format code (ZQ), and the film frame number (XXXXXX).

Data Availability

All CZCS data are archived with the Satellite Data Services Branch of the Environmental Data Information Service of NOAA at the following address: World Weather Building, Room 606, Camp Springs, Maryland 20233. A catalog is planned that will show the orbital track of the Nimbus-7 spacecraft on a day-by-day basis with the areas where CZCS was operated indicated on the orbital tracks. In addition, there will be a short description of the imagery, giving such parameters as cloud cover for each image. The catalogs will be sent to an initial mailing list and will then be available through the Environmental Data Information Service, Satellite Data Services Branch.

The cost of the CZCS data product has not been established, but it is estimated at approximately $3.50 for a photographic transparency and $60.00 for the magnetic tape. All data are available to any user who wishes to purchase them. Data will normally be ordered from the CZCS catalog by specifying the orbit and GMT of the data desired. The first validated data sets should be available to users between 3 and 6 months after launch.

Planned NET Experiment Investigations and Data Applications

The Nimbus Experiment Team for the Coastal Zone Color Scanner presently plans two major expeditions after launch of Nimbus-7 for validation of the derived product of the CZCS. One expedition will be carried out of the Gulf of Mexico utilizing the research vessel *Gyre* from Texas A&M University. This expedition will cover various water mass types in the Gulf of Mexico.

The other NET surface validation expedition will be carried out in the Pacific off Southern California and in the Gulf of California utilizing a research vessel from the Scripps Institute of Oceanography.

Foreign experiment team members will carry out validation investigations in European waters and off South Africa. The Joint Research Center of the Commission of European Communities will coordinate the activities of the EURASEP group in carrying out surface truth validations in waters around Europe. Information on their planned activity can be obtained from Dr. Bruno Sturm of the Joint Research Center, Ispra, Italy.

The South African experiment team member, Dr. Frank Anderson, director of the National Research Institute for Oceanology, will coordinate South Africa's efforts in validation measurements made in conjunction with Nimbus-7 overpasses.

BIBLIOGRAPHY

Clarke, G. L., The significance of spectral changes in light scattered by the sea, in *Remote Sensing in Ecology*. University of Georgia Press, Athens, 1969, chap. 11.

Clarke, G. L., G. C. Ewing, and C. J. Lorenzen, *Science* 167, 1970: 1119.

Hovis, W. A., M. L. Forman, and L. R. Blaine, Detection of ocean color changes from high altitude, NASA Publication X-652-73-371, Nov. 1973.

Morel, A., Analysis of variation of ocean color, limnology, and oceanography, *Limnology and Oceanography,* 22(4), July 1977.

Smith, R. C., and K. S. Baker, *The bio-optical state of ocean waters and remote sensing,* Ref. 77-2, Scripps Institute of Oceanography, 1977, La Jolla, CA.

Smith, R. C., and K. S. Baker, Optical classifications of natural waters, Ref. 77-4, Scripps Institute of Oceanography, 1977, La Jolla, CA.

Yentsch, C. S., The absorption and fluorescence characteristics of biochemical substances in natural waters, *Proc. symp. on remote sensing in marine biology and fisheries, March 1971.* Texas A&M University, College Station, 1971, pp. 75–97.

1.2.2 Coastal Zone Color Scanner Specifications*

IDENTIFICATION

Discipline: Environmental Observations **Acronym:** CZCS
Status: Flight Missions **Instrument Type:** VIS/IR Spectroradiometer
Spacecraft: NIMBUS-7
Contractor: Ball Brothers

OBJECTIVES

1. To map chlorophyll concentration in water, sediment distribution, and Gelbstoffe (yellow stuff) concentrations as a salinity indicator
2. Temperature of coastal waters and ocean currents

DESCRIPTION

Summary: The CZCS is a conventional six-channel scanning radiometer utilizing a rotating plane mirror at a 45° angle to the optic axis of a Cassegrain telescope. The rotating mirror scans 360°; however, only ±40° of data centered on the spacecraft Nadir are collected for ocean color measurements. During the rest of the scan, the instrument acquires a view of deep space and of internal instrument sources for calibration of the various channels. The radiation collected by the telescope is divided into two portions by a dichroic beam splitter. One portion is transmitted to a field stop that is also the entrance aperture of a small polychromator. The radiant energy entering the polychromator is disbursed and reimaged in five wavelengths on five silicon detectors in the focal plane of the polychromator. The portion of the beam reflected off the dichroic mirror is directed to a cooled mercury cadmium telluride detector sensing in the 10.5- to 12.5-μm region. The CZCS utilizes a radiative cooler that cools the mercury cadmium telluride detector to approximately 120°K during spacecraft flight.

Heritage/Derivation: Similar to Nimbus-5 Surface Composition Mapping Radiometer (SCMR).

---

*Based on *Handbook of Sensor Technical Characteristics,* NASA Headquarters, Office of Space Technology Applications, Communications Information Division, February 1981.

DATA

Data Products

1. CRCST (Calibrated Radiance, Pigment, Diffuse Attenuation Coefficient, and Temperature Tape)
2. CAT (Catalog Tape)
3. Film (241 mm × 341 mm (9.5-in.) black-and-white image stock)
4. Data available from *CZCS Data Catalog* and the *Satellite Data User's Bulletin,* June 1979

Data Archives Location: Satellite Data Services Branch, Environmental Data Information Services, National Oceanographic and Atmospheric Administration, Washington, DC 20025.

BIBLIOGRAPHY

National Space Science Data Center Computer Printout, 1980.
NASA/GSFC, *The NIMBUS-7 User's Guide,* August 1978.

POTENTIAL DERIVED PARAMETERS

1. Determine the specific nature of the contents of water as quantitatively as possible and carry out such measurements over large areas in short periods of time.
2. Discriminate between organic and inorganic materials in the water.
3. Determine the quantity of these materials in the water sample to the best degree possible.
4. Attempt identification of organic particulates such as discriminating between various types of red tide organisms.

The specific observations include:

a. Chlorophyll concentration
b. Sediment distribution
c. Gelbstoffe concentrations as a salinity indicator
d. Temperature of coastal waters and the open ocean

1.3 THEMATIC MAPPER*

The Thematic Mapper (TM) is a remote sensor with seven spectral bands covering the visible, near infrared, and thermal infrared regions of the spectrum.

*Based on O. Weinstein, L. Linstrom, and J. Bremer, *Development of the Thematic Mapper,* NASA Goddard Spaceflight Center, Greenbelt, Md.

The Thematic Mapper is designed to satisfy more demanding performance specifications than have previously been applied to an instrument of its type. In response to these requirements, the design incorporates state-of-the-art materials, structures, control techniques, calibration mechanisms, data handling, and electronics. Examples of structural uses of materials include the beryllium "eggcrate" scan mirror, the ultralow-expansion lightweight primary mirror, and the graphite epoxy telescope tube. The mirror is mounted on flex pivots and is magnetically compensated to linearize its scan motion. A microprocessor-controlled magnetic torquer is used to maintain scan repeatability. Each of the 100 detectors in the two focal plane arrays undergoes a two-reference calibration every half-cycle, and an on-board sun calibration is implemented once per orbit to correct for long-term degradations. The multiplexer handles a data rate of 84.9 Mbps by using emitter-coupled logic to perform high-speed operations.

The Thematic Mapper is carried by Landsat-D. The sensing scenario is shown in Figure 25.

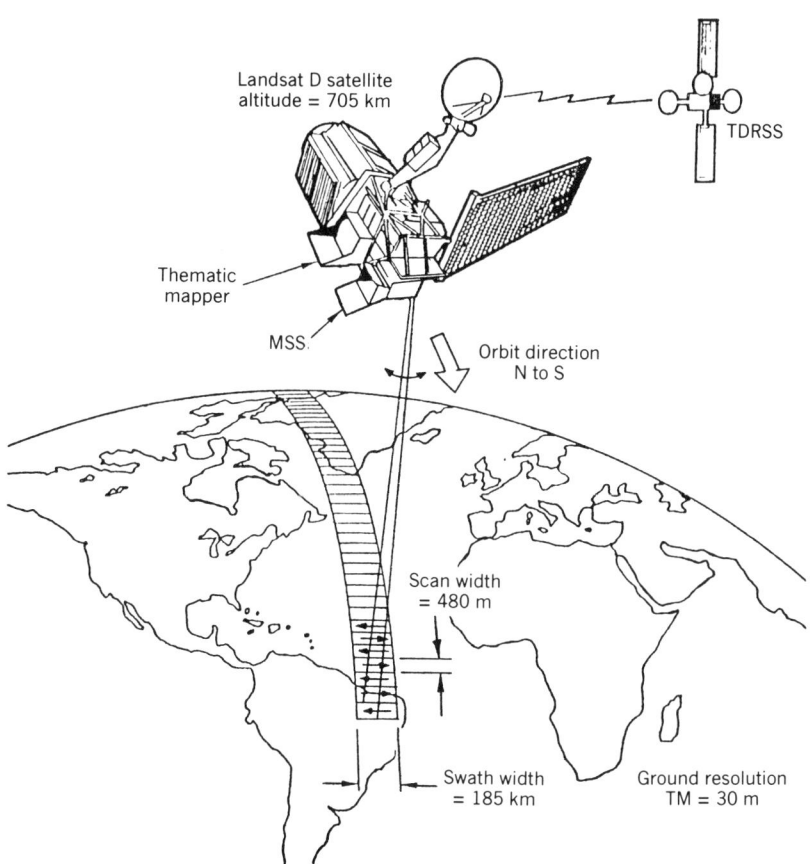

FIGURE 25. Landsat-D mission.

1.3.1 Thematic Mapper Design Description

The performance parameters for the Thematic Mapper have evolved from experience gained in the operation of the MSS, from technical working group efforts to define user requirements, and from trade-off analyses of performance goals in terms of technological feasibility. Table 10 represents the major TM requirements which have been established as a result of this effort. The TM pixel size is 30 m^2 as compared to 80 m^2 for the MSS. At this ground resolution, most small agricultural fields may be accurately characterized.

The seven spectral bands which are specified in Table 11 have been selected for the Thematic Mapper. Their passbands and radiometric resolutions have been defined to satisfy user requirements which represent the current state of the art in remote sensing. For example, the short wavelength band of the MSS, whose spectral passband is 0.5–0.6 μm, has been able to map underwater features to a far greater extent than was anticipated. Band 1 of the Thematic Mapper coincides with the maximum transmissivity of water and should therefore demonstrate coastal water mapping capabilities superior to those of the MSS. It also has beneficial features for the differentiation of coniferous and deciduous vegetation. Bands 2–4 cover the spectral region which is most significant for the characterization of vegetation. Vegetation moisture may be estimated from band 5 readings, and plant transpiration rates may be estimated from the thermal mappings in

TABLE 10. Major TM Performance Requirements

Square-wave response	
Bands 1–5, 7	0.35 at 30 m
Band 6	0.35 at 120 m
Band-to-band registration	<6 m
Scan profile repeatability	<6 m
Along-track overlap/underlap	<6 m
Swath width	185 km
Radiometric resolution	
Bands 1–5, 7	0.5–2.4% NEρ (noise-equivalent reflectance)
Band 6	0.5°K NETD (noise-equivalent temperature difference)
Absolute radiometric accuracy	10%
Band-to-band radiometric precision	2%
Channel-to-channel radiometric precision	$< \dfrac{\text{rms noise}}{4}$
Spectral coverage	0.45–12.5 μm
Signal quantization levels	256
Data rate	84.9 Mbps
Weight	<243 kg
Power	<300 W
Envelope	0.6 m × 1.1 m × 2.0 m

TABLE 11. TM Spectral Passbands

Band	Spectral Range (μm)	Radiometric Resolution	Principal Applications
1	0.45–0.52	0.8% NEρ	Coastal water mapping Soil–vegetation differentiation Deciduous–coniferous differentiation
2	0.52–0.60	0.5% NEρ	Green reflectance by healthy vegetation
3	0.63–0.69	0.5% NEρ	Chlorophyll absorption for plant species differentiation
4	0.76–0.90	0.5% NEρ	Biomass surveys Water body delineation
5	1.55–1.75	1.0% NEρ	Vegetation moisture measurement Snow–cloud differentiation
6	10.4–12.5	0.5°K NETD	Plant heat stress measurement Other thermal mapping
7	2.08–2.35	2.4% NEρ	Hydrothermal mapping

band 6. Band 7 is primarily motivated by geological applications, including the identification of hydrothermally altered rocks. The band profiles which are narrower than those of the MSS are specified with stringent tolerances, including steep slopes in spectral response and minimal out-of-band sensitivity. The measured spectral response of the band 3 filters is shown in Figure 26. All of the other filters exhibit similar qualities.

TM Subsystem Development
A complete system analysis was conducted in order to translate the above TM requirements into a sensor design with quantitative performance parameters. The remainder of the section will address some of the more interesting and challenging aspects of this development program.

Basic Optical Configuration
The basic optical design is driven by the need for a compact, lightweight optical system capable of satisfying the simultaneous requirements for broad spectral coverage, high spectral and radiometric resolution, and high image quality over a 185-km-wide ground track. The principal feature of the design is the use of an object space scan mirror which simplifies the performance requirements for the rest of the optical system by requiring the telescope to operate only at very small field angles. Further, the same zone of each element is used at all scan angles. The scan mirror moves synchronously with a Scan Line Corrector (SLC) to provide bidirectional scanning.

A telescope of the Ritchey–Chretien configuration is used as the primary energy collector. Here 40.6-cm-diameter optics are employed based on analysis of signal-to-noise requirements, and an $f/6$ design is adopted to provide a reasonable image size at the focal plane, Uncooled silicon photodiode detectors are positioned

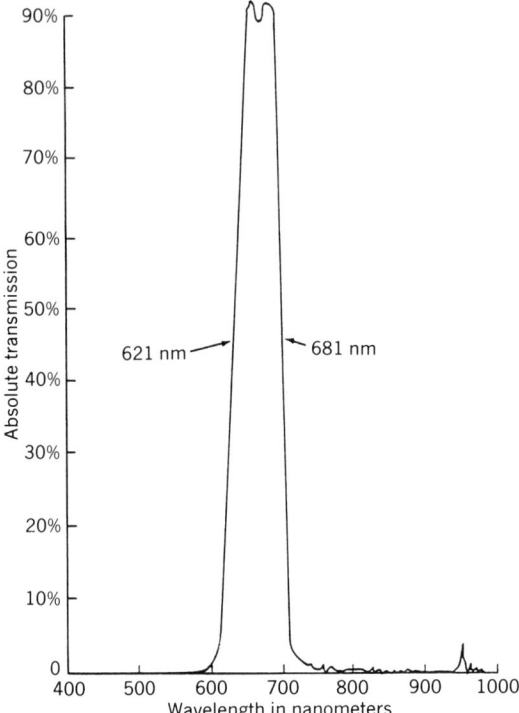

FIGURE 26. Band 3 filter spectral response.

at the focus of the telescope, while energy for long-wavelength detectors is relayed to a focal plane in the radiative cooler.

The SLC is used to shift the optical line of sight by approximately the length of the detector array at the end of each scan. The SLC employs two small mirrors that are in parallel but are rotating on a common axis to displace the optical axis. This assembly is highly insensitive to alignment errors. Image displacement errors due to the scan geometry are less than 0.1 μrad over the small angle of travel.

Relay optics of reflective design are used to transfer energy to the cooled focal plane. The folded configuration is adopted as a convenient way to obtain an optical axis parallel to the axis of the radiative cooler. The image of the relay is excellent over the required field. The object space scanning concept frees the infrared detectors from scan modulation of internal element radiances. The basic optical configuration is illustrated in Figure 27.

The spectral bands are defined by bandpass filters. The filters are positioned close to the detector elements to reduce the effects of optical crosstalk. In the cooled spectral bands (bands 5, 6, and 7), it is important to keep the filters cool to reduce the background radiation from the filter.

The internal calibrator uses a tungsten lamp source for bands 1–5 and 7. A small blackbody radiator is used to calibrate band 6. Energy for calibration is introduced at the end of each scan line by the action of a moving shutter. Radi-

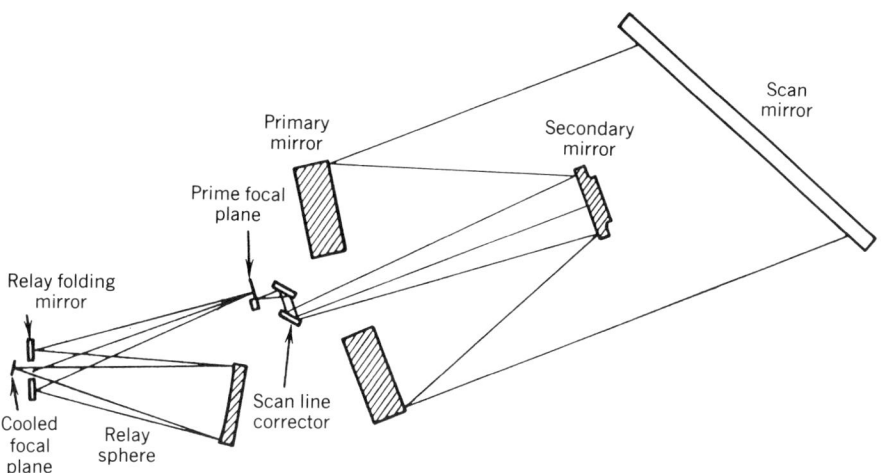

FIGURE 27. Optical system elements.

ance changes are imposed to establish the gain and zero offset for each detector channel.

TM Scanning Subsystems

The scanning subsystems for the TM consist of the Scan Mirror Assembly (SMA) and the Scan Line Corrector (SLC). The motion of the scan mirror causes the telescope field of view to scan the 185-km ground swath bidirectionally, while motion of the SLC provides the required along-track scan contiguity.

The critical SMA requirements comprise three groups: (1) operational requirements, (2) requirements pertaining to geometric accuracy, and (3) optical requirements. The operational group has importance in matching the distance between sets of 16 lines scanned with the nominal orbit parameters (effective nadir rate) so that consecutive sweeps are contiguous. The tolerances have also been selected to assure synchronization with the other parts of the TM, such as Scan Line Corrector, DC restore, and calibration. The precision required for this group is much less severe than for the geometric group.

The geometric requirements bear upon mapping accuracy, radiometric accuracy, the ability to register spectral bands, and the modulation transfer function. These influences vary according to the random or systematic nature of the inaccuracy.

Requirements for a stable along-scan profile and continuous cross-scan accuracy have received major attention in the scan mechanism design.

The optical group has been specified to support radiometric performance and modulation transfer function (MTF). Reflectivity influences signal-to-noise ratio, while both static and dynamic flatness affect MTF. Polarization and scattering influence radiometric accuracy.

The schematic representation of the SMA is given in Figure 28. The scan mirror is mounted on flex pivots to eliminate bearing wear and lubrication problems.

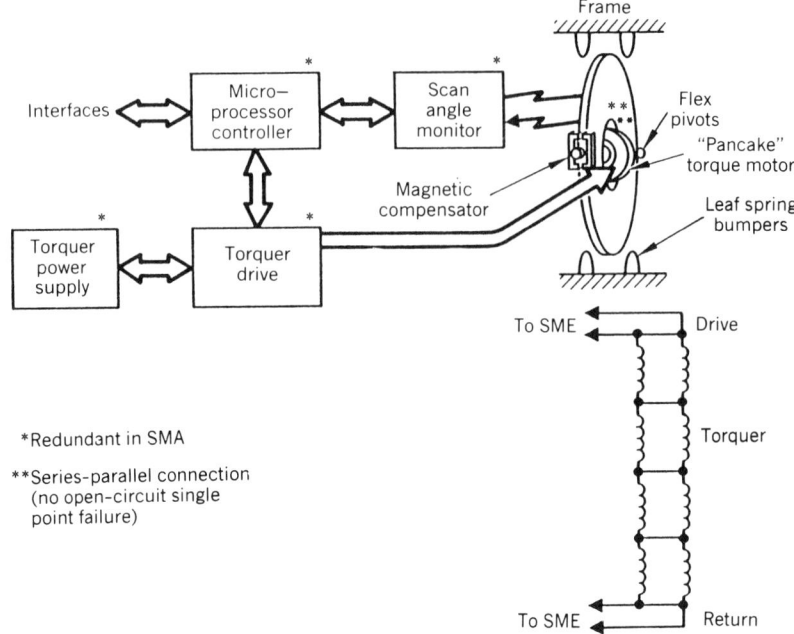

FIGURE 28. Functional components of scan mirror assembly.

To linearize the scan, it is necessary to minimize forces on the mirror during the active portion of the scan. Magnetic compensators are used to cancel the flex pivot restoring torque, reducing it by three orders of magnitude relative to its uncompensated value. A preloaded leaf spring mechanism, illustrated in Figure 29, is used to reverse the angular momentum during the turnaround. The bumper mass is minimized to minimize the energy dissipation upon impact, and the bumper surface rolls on the striker plates to minimize sliding friction.

A pancake torque motor is used to apply torque to the mirror during the turnaround portion of its motion, as shown in Figure 30. The duration of the half-scan period is measured by the scan angle monitor, an electrooptical device which produces reference signals at the beginning, midpoint, and end of the active scan angle. These signals are used as inputs by a microprocessor, which controls the length of the pancake torquer's operation and thereby provides active control of the scan duration. The entire digital control system has been designed and tested and has demonstrated a scan line length repeatability of 1 part in 10^5, significantly better than specified.

The pancake torquer and the magnetic compensator both introduce eddy current effects which retard the mirror rotation, resulting in residual nonlinearities. The possibility of using the pancake torquer to compensate for the above eddy current effects is now being investigated.

The moment of inertia of the scan mirror must be kept to an absolute minimum while its structural rigidity is maximized. The mirror is therefore fabricated with a

CONTROLLED BY BUMPER ASSEMBLY
- **SPRING CONSTANT**
- **BUMPER POSITION**

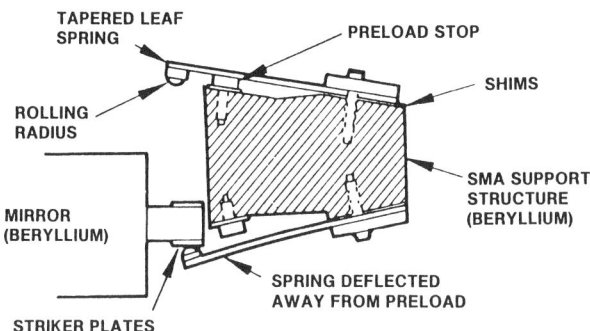

FIGURE 29. Bumper assembly.

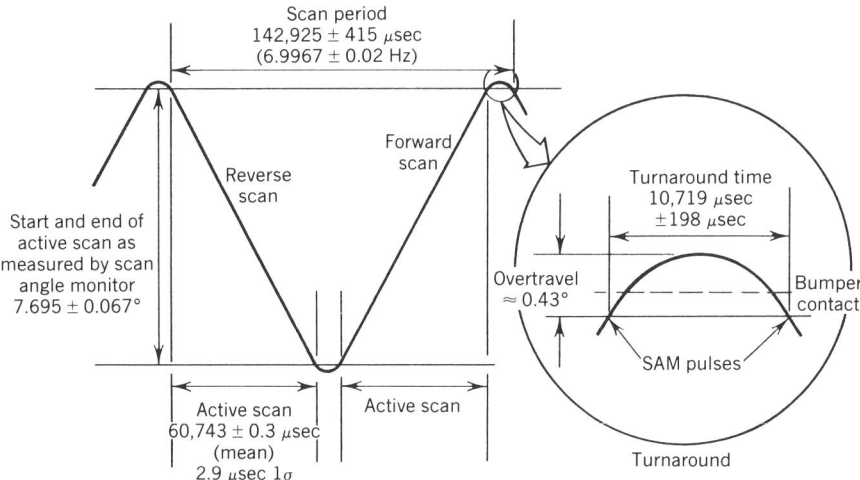

FIGURE 30. Scan mirror assembly operational requirements.

lightweight beryllium eggcrate structure. Due to the critical role of the scan mirror, two alternative design efforts were pursued in parallel. In one approach, two solid slabs of beryllium, each half as thick as the final mirror thickness, were hollowed out on one face with a series of weight-reducing holes. The holes in the respective faces of the two slabs were aligned and the slabs were brazed together. This machined lightweight configuration is shown in Figure 31. The alternative lightweight configuration consists of an eggcrate grid to which two sheets of beryllium are attached to form the front and backs, as shown in Figure 32.

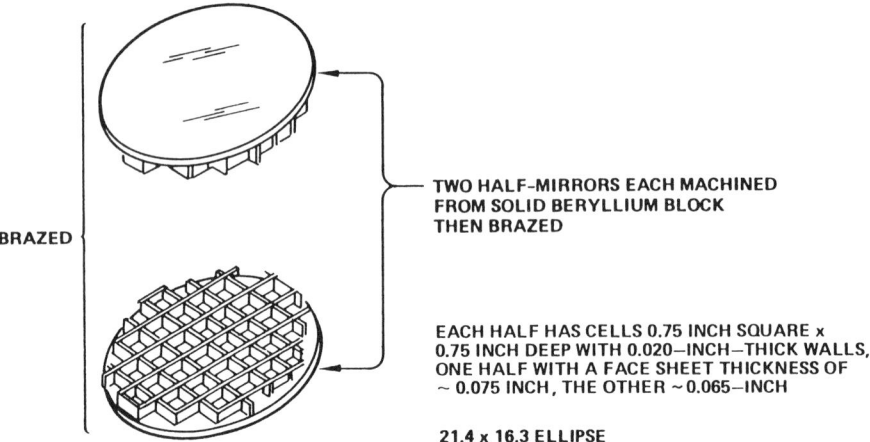

TWO HALF-MIRRORS EACH MACHINED
FROM SOLID BERYLLIUM BLOCK
THEN BRAZED

BRAZED

EACH HALF HAS CELLS 0.75 INCH SQUARE x
0.75 INCH DEEP WITH 0.020—INCH—THICK WALLS,
ONE HALF WITH A FACE SHEET THICKNESS OF
~ 0.075 INCH, THE OTHER ~0.065—INCH

21.4 x 16.3 ELLIPSE

FIGURE 31. SM-machined eggcrate approach.

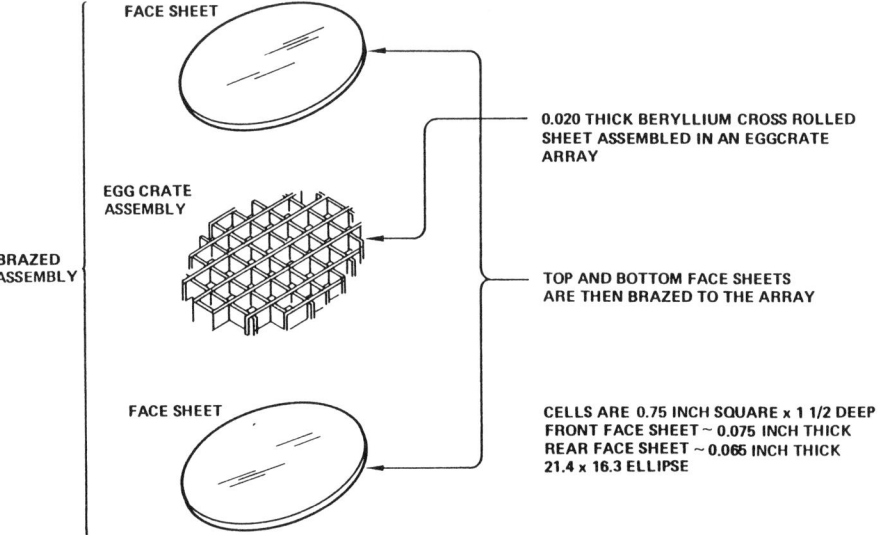

FACE SHEET

0.020 THICK BERYLLIUM CROSS ROLLED
SHEET ASSEMBLED IN AN EGGCRATE
ARRAY

EGG CRATE
ASSEMBLY

BRAZED
ASSEMBLY

TOP AND BOTTOM FACE SHEETS
ARE THEN BRAZED TO THE ARRAY

FACE SHEET

CELLS ARE 0.75 INCH SQUARE x 1 1/2 DEEP
FRONT FACE SHEET ~ 0.075 INCH THICK
REAR FACE SHEET ~ 0.065 INCH THICK
21.4 x 16.3 ELLIPSE

FIGURE 32. SM-brazed eggcrate approach.

Full-sized mirrors (40.6 cm × 58.4 cm ellipses) of both configurations have been constructed and polished. Each mirror weighs approximately 2.5 kg.

The SLC (Figure 33) is the second component of the scanning subsystem; it aligns the traces of the scan mirror to produce parallel, nonoverlapped scans. Because of the forward motion of the spacecraft during each scan, the successive traces form angles with each other and alternately overlap and underlap the preceding and following scan lines. (See Fig. 34.) The SLC rotates perpendicular to

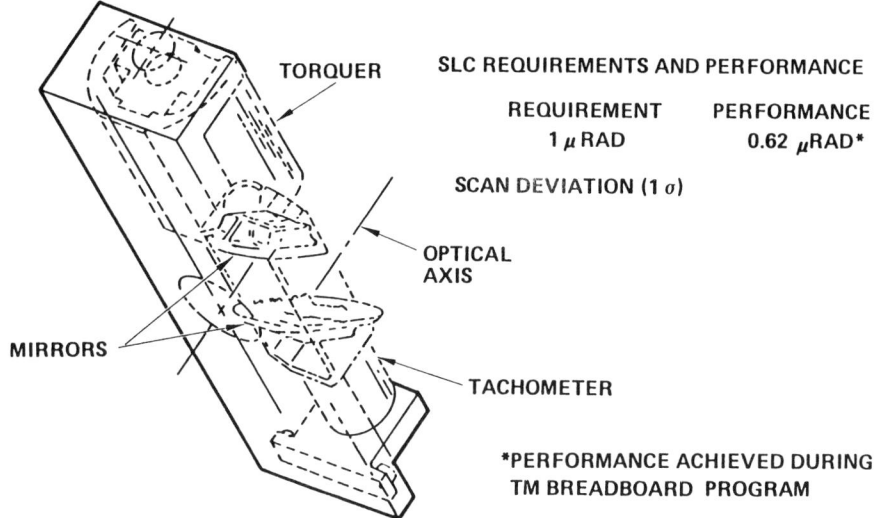

FIGURE 33. Scan line corrector layout.

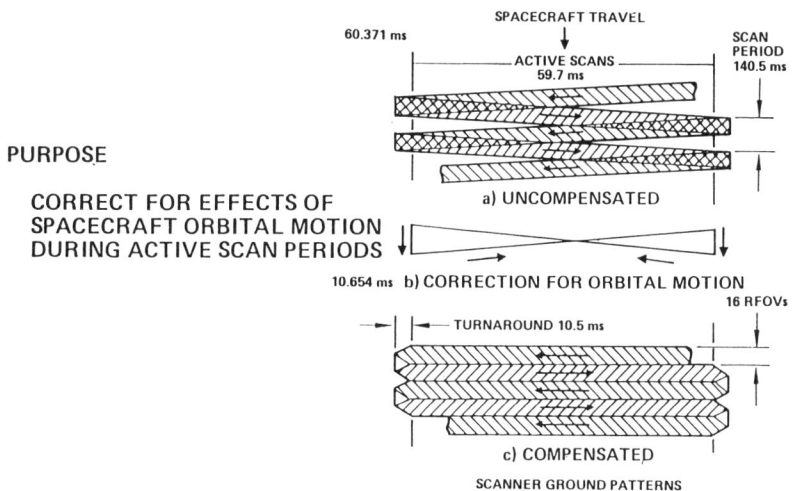

FIGURE 34. Compensation for spacecraft motion by the scan line corrector.

the scan mirror at a rate which cancels out the orbital rate. This produces effectively parallel scan lines with ideally no overlap or underlap.

Figure 33 shows the SLC configuration developed during the breadboard program. The two beryllium mirrors are bonded to the beryllium box structure.

One end of the box structure is attached to a motor coil housing, and the rotor of a tachometer is clamped to the opposite end. The resultant assembly is supported by a pair of flex pivots, completing the galvo-like mechanism. The mirror assembly is driven by a moving coil DC torque motor with velocity feedback provided by the tachometer.

Telescope Assembly

In order to maintain proper focus and alignment (both short-term and long-term), the telescope design requires techniques which will maintain the relative position of the primary and secondary mirrors to less than 2 μm. The choice was made to use a passive system which relies on the extremely low coefficients of thermal expansion of the mirrors and metering structure to stabilize the primary–secondary mirror intervertex distance.

The basic telescope structure is illustrated in Figure 35. It consists of three major subsystems: the primary and secondary mirror assemblies and the telescope housing. The primary mirror is rigidly attached to the housing, and the secondary is adjusted to produce the proper mirror positioning.

A graphite epoxy composite cylinder is the main structure of the telescope housing. The graphite epoxy was chosen because of its extremely low coefficient

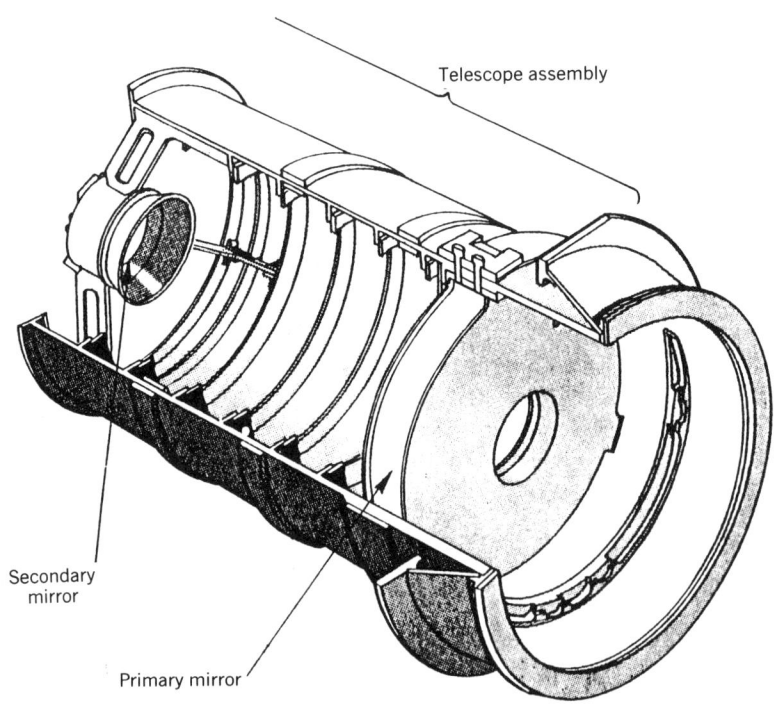

FIGURE 35. Basic telescope structure.

of thermal expansion and its high rigidity-to-weight ratio. It is a hygroscopic material, however, and its contact with humid air must therefore be minimized. It has been fabricated in a humidity-controlled environment, and precautions were implemented throughout the assembly, testing, storage, shipping, and spacecraft integration phases of the TM program to minimize the absorption of water vapor by the graphite epoxy structure.

A three-legged "spider" support for the secondary mirror is attached at one end of the telescope housing. The aft optics bulkhead is attached behind the primary mirror. An invar ring mounted on this end of the tube provides the interface between the telescope metering structure and the radiation cooler.

Internal light baffles are bonded to the inside wall of the cylinder and function to stiffen its cross section as well as to intercept unwanted light.

Both primary and secondary mirrors are fabricated from ULE, an optical glass whose coefficient of thermal expansion is virtually zero. The primary mirror, illustrated in Figure 36, is of lightweight configuration. An eggcrate core, a faceplate, and a backplate are fused together to form the mirror. Three points join it to the telescope housing.

Detector Array Configuration and Sampling Technique

Each band, with the exception of band 6, has a pixel dimension of $(30 \text{ m})^2$ on the ground. Silicon detectors for bands 1–4 are located in the prime focal plane and

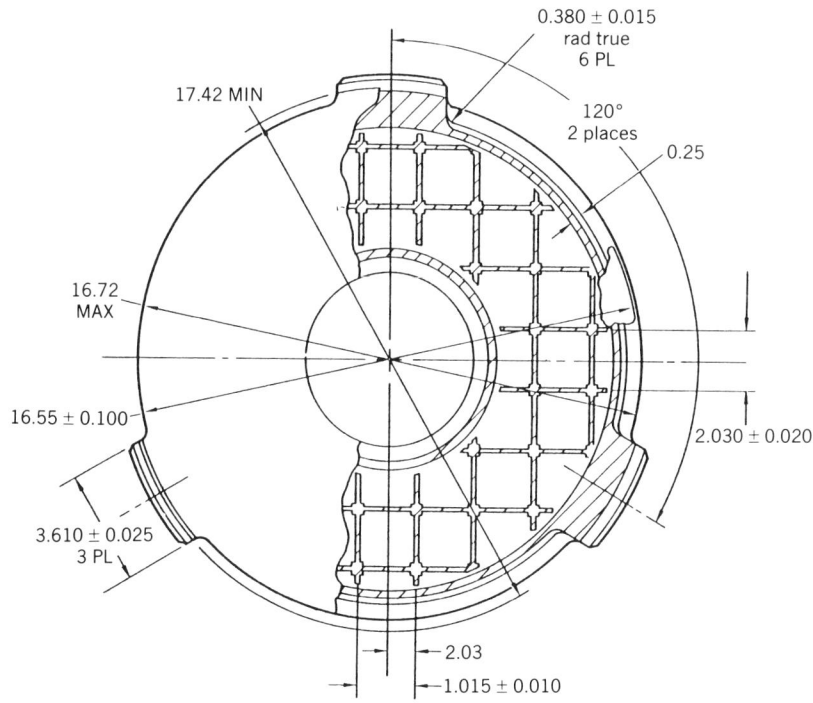

FIGURE 36. Primary mirror blank design.

have dimensions of $(0.1035 \text{ mm})^2$, corresponding to an instantaneous field of view (IFOV) of 42.5 μrad. In the cooled focal plane, indium antimonide detectors for bands 5 and 7 have dimensions of $(0.0533 \text{ mm})^2$ corresponding to an IFOV of 43.75 μrad. This slightly oversized angular dimension is specified because the linear dimension of the indium antimonide detectors is half as great as that of the silicon detectors and therefore fabrication tolerance, in angular dimensions, is greater. Band 6 has a pixel dimension of approximately $(120 \text{ m})^2$ on the ground, corresponding to an IFOV of 170 rad. The band 6 detectors are 0.207-mm squares of mercury cadmium telluride operated in the photoconductive mode. Four detector channels are required for band 6 and 16 detector channels are required for each of the other spectral bands.

There are 100 detector channels in the Thematic Mapper. Figure 37 represents the detector array as projected at the prime focal plane. The band 1–4 detectors are actually located at this plane; the cooled focal plane is inverted and magnified by two by the relay mirrors. The image is scanned vertically across the detector plane, as illustrated. Each swath is 16 pixels wide and 6320 pixels long. The dwell time per pixel is 9.611 μsec for all bands except band 6, for which it is four times longer. Figure 38 shows the detector array geometries. The detectors of

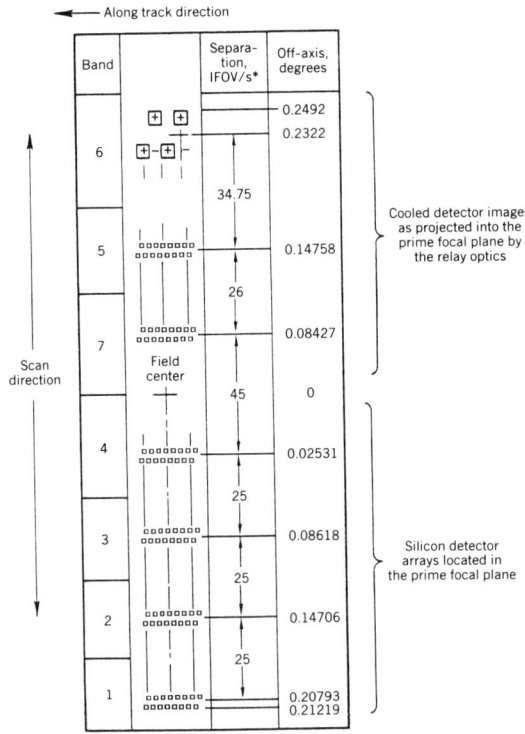

*High resolution IFOV/s (band 6 IFOV/s are 4 times greater).

FIGURE 37. Detector projection at prime focal plane.

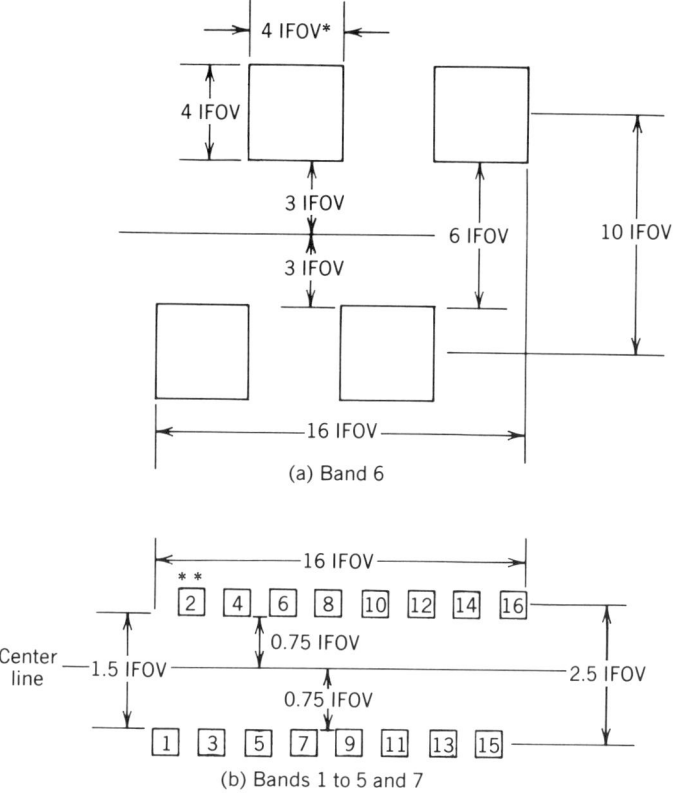

(a) Band 6

(b) Bands 1 to 5 and 7

*Typical for all indicated spacings, high resolution IFOV/s (band 6 IFOV/s are 4 times greater).
**Typical channel number, as indicated.

FIGURE 38. Detailed optical layouts.

bands 1–5 and 7 are configured in 16 element arrays of identical geometric layout. Each array is displaced from the others by an integral number of IFOVs in the scan direction. To obtain continuous coverage in the along-track direction, the detectors of each band are staggered into two rows. These rows are displaced from each other by 1 IFOV in the along-track direction and by 2.5 IFOVs in the scan direction. The four detectors of band 6 are also staggered and separated by 2.5 band 6 pixels (10 IFOVs) in the scan direction.

The multiplexer samples all odd detector channels, then all even detector channels, then all odd channels, and so on, with sampling intervals occurring each time the scan has traversed one-half IFOV (every 4.806 μsec). This technique, with stringent tolerances on detector plane geometry and sampling time budget, makes it possible to measure a given pixel sequentially in the six high-resolution bands and properly coregister the IFOVs. The detector layout and sampling scheme is further designed to coregister each band 6 pixel with 4×4 blocks of pixels from the other six spectral bands.

Inchworm Focus/Alignment Drive

The on-orbit coalignment of the cooled and uncooled focal planes is also necessary for proper coregistration of channels. Also, the relay optics require reposi-
tioning after launch and cooldown in order to meet the specification of 0.3 IFOV
for alignment of the two detector planes. The spherical relay mirror, which is the
last element in the optical train, is attached to movable shafts at three equally
spaced points on its periphery. The mirror can be tilted to change alignment by
moving the shafts differentially; it can be translated to change the focus by displacing all three shafts the same distance in the same direction.

Each of these shafts is driven by a piezoceramic device known as an *inchworm*.
An inchworm is composed of segments which can be made to change dimension,
clamping or releasing the rod, stretching or contracting, depending upon the
orientation of the materials and the presence or absence of a high-voltage control
signal.

The sequence of movements in Figure 39 would translate the shaft to the left by
1.2 μm, displacing the image by 0.1 IFOV on the cooled detector plane.

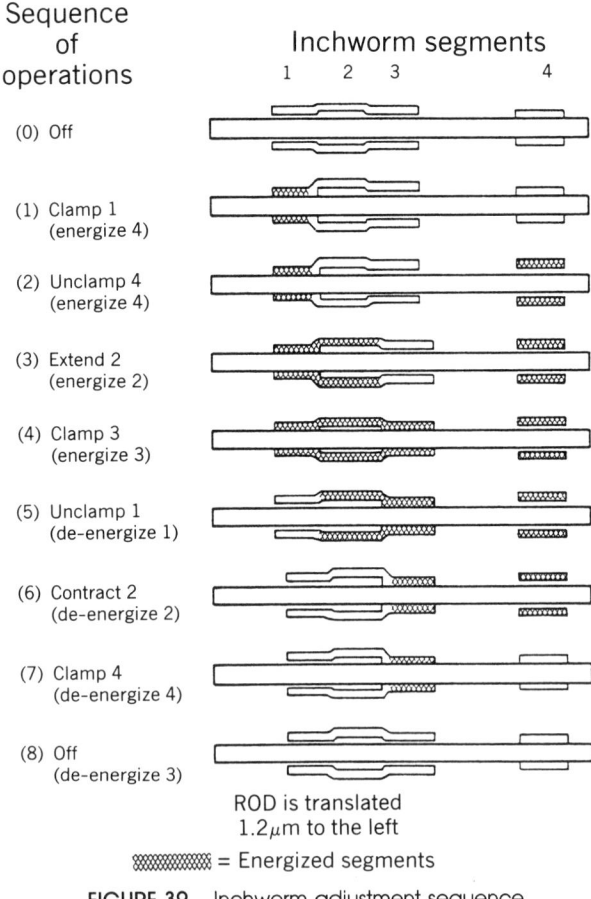

FIGURE 39. Inchworm adjustment sequence.

The piezoceramic-driven shafts eliminate the need for the motors, motor-driven parts, and lubrication which would otherwise be required. They do require high voltage, however, and the support circuitry must be designed so that it will not be damaged by the reverse effect, that is, the production of high voltages across the inchworms due to changes in dimensions produced by acceleration or vibration.

Radiative Cooler

The TM uses a two-stage passive radiative cooler to refrigerate the focal plane arrays for bands 5–7. In particular, the band 6 HgCdTe detector array requires a temperature of 95°K for proper operation. The TM cooler design is based on the proven VISSR design with modifications to increase the cooling power.

In order to achieve the required low temperature, the cooler design requires a two-stage approach. An intermediate stage is provided in tandem with the cold stage to reduce the heat losses from the cold stage. The design of the cooler is shown in Figure 40. The principal parts of the cooler are a radiative shield which causes the cold patch to view outer space in the direction away from the sun and earth, a highly reflective door which prevents radiant energy from the earth hitting the cold stage, and honeycomb radiation panels for high emissivity. The cooler has been designed so that the detector focal plane can be cooled by external bench test equipment. This feature allows the TM cooled detector channels to be readily tested in the laboratory without a complicated space background simulation.

An adequate number of vent paths has been provided to reduce the possibility of cooler contamination. Worst-case thermal analyses have shown that this design

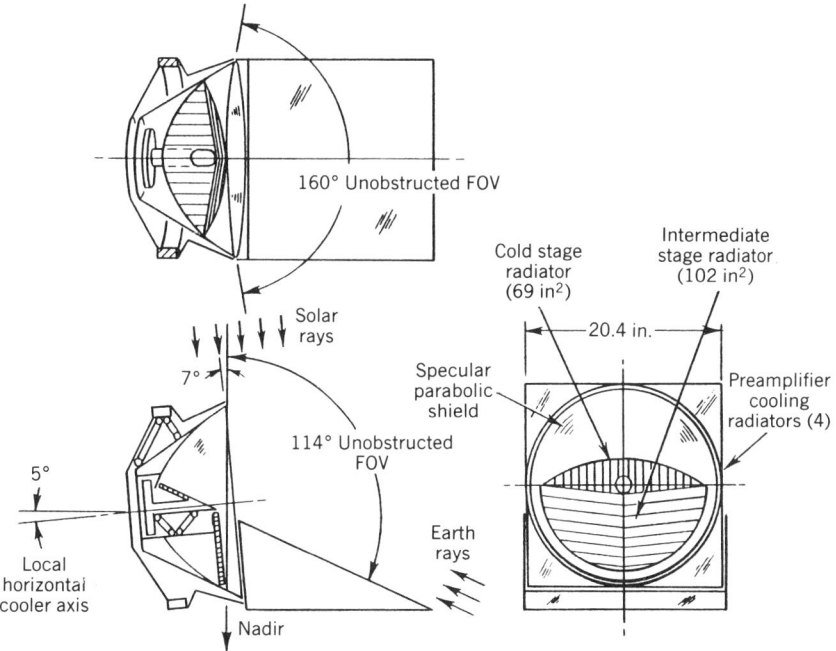

FIGURE 40. Radiative cooler schematic.

can cool the 36 detectors mounted in the cold patch to 84°K, providing an 11° margin from the operating temperature of 95°K. The operating temperature is controlled to better than 0.1°K via proportional heaters. This tight thermal control stabilizes the detector responsivity, thus improving the system calibration.

On-Board Calibration

The TM must satisfy exacting requirements for absolute radiometric accuracy, band-to-band and channel-to-channel precision, and signal drift compensation. A sophisticated on-board calibration system was constructed to satisfy these requirements. The system will perform a two-reference calibration of each detector channel every half-scan.

The two inputs are presented to each detector by a calibration shutter which oscillates like a metronome in synchronization with the scan mirror motion. The shutter blocks the scene radiation twice per cycle (during the turnaround periods of the scan mirror) and sequentially provides DC restore level and a calibration input to each detector. This mechanism is illustrated in Figure 41.

The calibration signals for the reflected radiation channels originate from three tungsten filament lightbulbs, as shown in Figure 42. Colored filters and fiber optic bundles of various sizes are used to produce the proper intensity for each spectral band. The radiant intensity of each lightbulb is controlled by a separate narrow-band silicon photometer and an aperature mask to produce relative radiometric intensities of 2:3:4 for the three bulbs.

This multibulb source permits the calibration input level to be varied in known discrete steps for all channels by turning the lamps on and off in sequence. Lamp

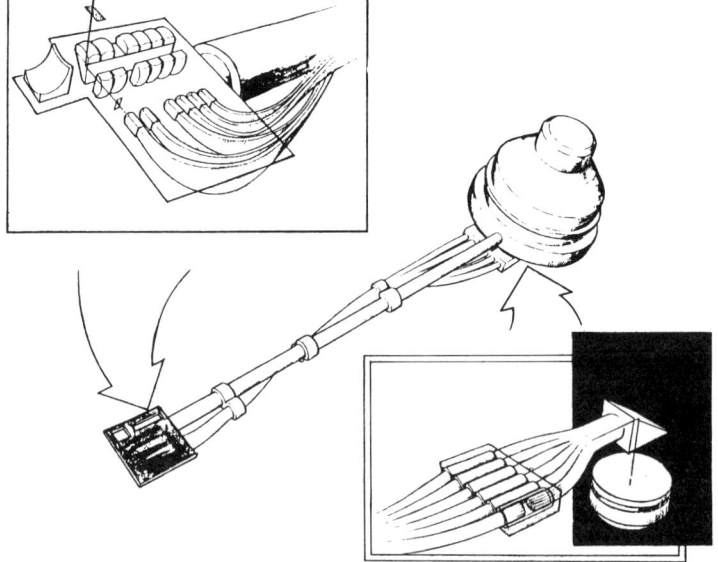

FIGURE 41. Internal calibrator flag rotation.

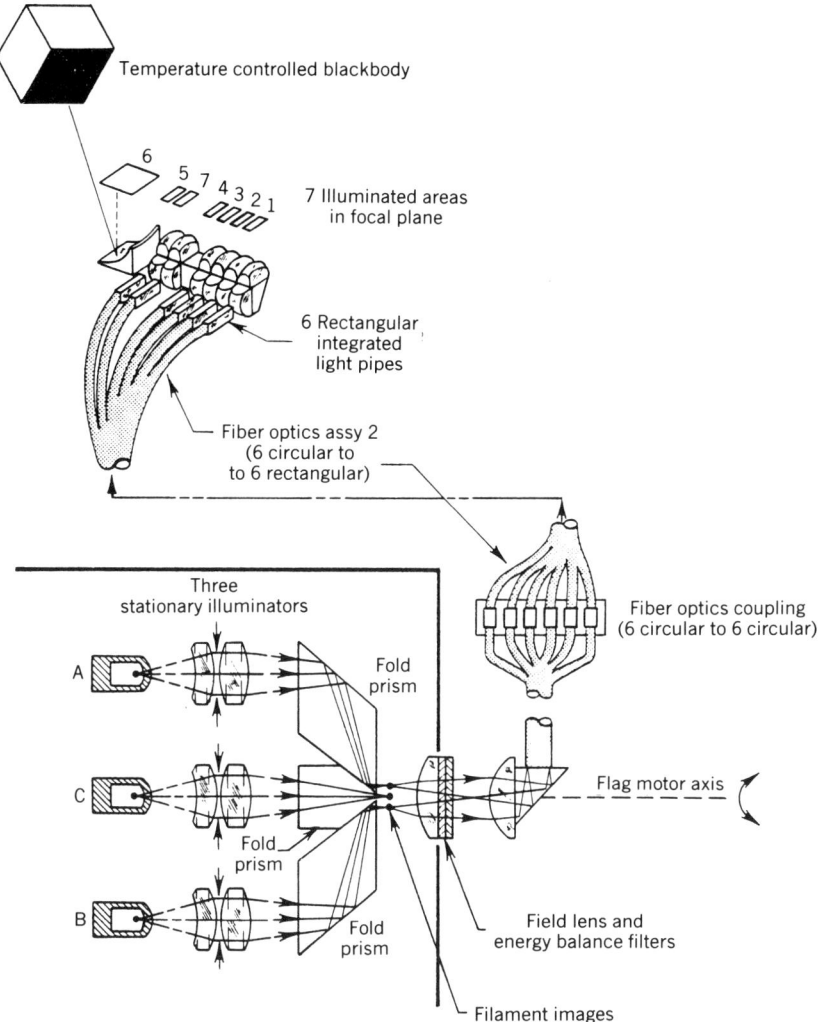

Temperature controlled blackbody

7 Illuminated areas
in focal plane

6 Rectangular
integrated
light pipes

Fiber optics assy 2
(6 circular to
to 6 rectangular)

Three
stationary illuminators

Fold
prism

Fiber optics coupling
(6 circular to 6 circular)

Flag motor axis

Fold
prism

Field lens and
energy balance filters

Fold
prism

Filament images

FIGURE 42. Internal calibrator scheme.

characteristics may be cross-calibrated by observing the channel outputs as they are turned on and off. The lamp calibration level is changed after every 40 scan lines by turning one lamp on or off. The relative lamp intensities and radiance level sequence are shown in Table 12. A temperature-controlled blackbody source is imaged onto the band 6 array by a concave mirror attached to the shutter. A second reference is provided by the shutter, whose temperature is measured by a thermistor.

A sun calibration is implemented once per orbit, shortly after the spacecraft passes from local night into local day. This calibration is performed to test for long-term changes in the tungsten lamp internal calibrator. It has the additional

TABLE 12. Lamp Command Sequence and Calibration Lamp Intensity

Command	Lamps On[a]	Relative Intensity
All lamps off	—	—
Lamp C off	None	0
Lamp A on	A	2
Lamp B on	A + B	5
Lamp A off	B	3
Lamp C on	B + C	7
Lamp A on	A + B + C	9
Lamp B off	A + C	6
Lamp A off	C	4

[a]Relative intensities: lamp A, 2; lamp B, 3; lamp C, 4.

advantage of using a light path which includes reflections from the five fore-optic mirrors: the scan mirror and the primary, secondary, and two scan line corrector mirrors. Degradations of these optical surfaces cannot be detected by the internal calibrator, but may be compensated by use of data from the solar calibration measurement.

Multiplexer

The primary task of the multiplexer is the encoding and formatting of the signals from the 100 detector channels. The detector signals are amplified and filtered in the radiometer electronics and then provided to the multiplexer. The multiplexer performs a DC restoration function to compensate for offset in the detectors and preamplifiers, samples the signals, and converts these samples into eight-bit digital words. These digital words are then formatted with synchronization words and spacecraft telemetry data. This results in a composite 84.9-Mbps data stream which is provided to the spacecraft communications subsystem for transmission to earth. A simplified block diagram of the multiplexer is shown in Figure 43. All detector signals are sampled at a rate determined by the field-of-view requirements of the instrument and the mirror scan rate. Each reflected radiation detector is sampled once every 9.61 μsec, while the four band 6 detectors are sampled at one-fourth this rate.

In addition to encoding and formatting detector information, the multiplexer provides a time reference and signals for timing and coordinating the operation of the TM instrument. For example, the multiplexer provides a 10.61-MHz clock signal to the Scan Mirror Assembly to be used as a time reference for controlling the highly precise scan mirror motion, and a 208.1-kHz clock to the TM power supply. The power supply DC-to-DC converter synchronizes itself to this clock, which in turn is synchronous with the multiplexer's sample rate. This precaution reduces the likelihood of power supply noise appearing as coherent patterns in the final image products.

The 84.9-MHz output data from the TM represents a substantial increase in raw data rate over that generated by other multispectral imagers. For example, the TM

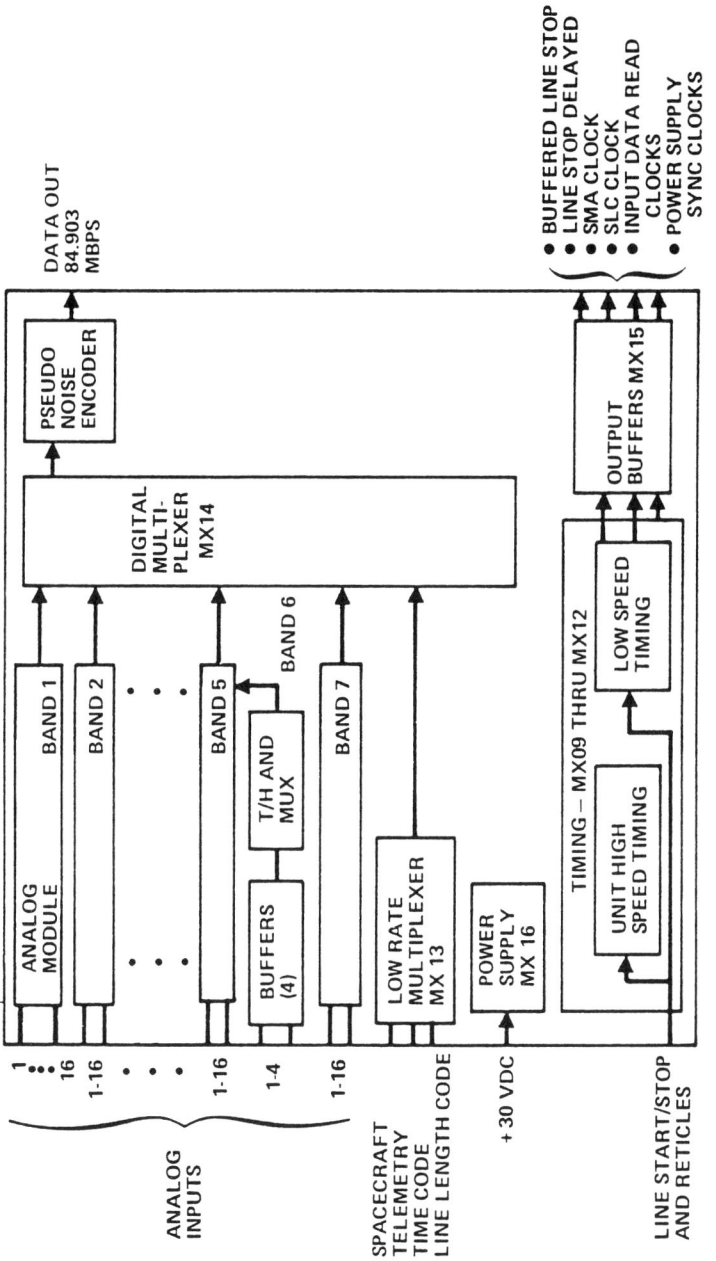

FIGURE 43. TM multiplexer.

83

data rate is nearly six times that of the MSS. This high data rate has substantial implications for the selection of electronic components and the layout of the circuitry. Emitter-coupled logic is used to perform all high-speed functions.

The multiplexer design is highly modular and hybridized. Rather than utilizing a single analog-to-digital (A/D) converter capable of converting all 100 channels, the multiplexer provides dedicated buffering, multiplexing, and A/D conversion circuitry for each full band. This circuitry is packaged on a single module, which is replicated in the unit six times, one for each of the six full-rate bands. The low-rate band 6 inputs are converted by the band 5 A/D converter. Most of the circuitry which is replicated for each band or each detector is packaged in hybrid microcircuits. These hybrids are of three types: an analog input buffer with DC restoration, an analog multiplexer and track/hold function, and an A/D converter. The nonreplicated portions of the multiplexer, that is, the timing and formatting and digital buffering functions, are built of standard flat-pack integrated circuits on multilayer printed circuit boards. The timing and output formatting circuitry is common for all bands.

1.3.2 Thematic Mapper Specifications*

IDENTIFICATION

Discipline: Resource Observations
Status: Future Flight Missions

Acronym: TM
Instrument Type: Multispectral Scanner
Spacecraft: Landsat-D
Contractor: Goddard Space Flight Center

OBJECTIVES

To improve land use, water resources, and food supply/distribution/management by imaging filtering and detecting reflected solar radiation from the surface of the earth in several spectral bands simultaneously through the same optical system.

DESCRIPTION

Summary: The Thematic Mapper (TM) is a seven-band, earth-looking, scanning radiometer with a 30-m ground element resolution covering a 185-km ground swath from a 705-km altitude. The instrument consists of primary imaging optics, scanning mechanism, spectral band discrimination optics, detector arrays, radiative cooler, in-flight calibrator, and required operating and processing electronics. The scanning mechanism provides the cross-track scan, while the progress of the spacecraft provides the scan along the track. The optical system images the earth's surface on a field stop or a detector sized to define an area on the earth's surface of 30 m². Several lines are scanned simultaneously to permit suitable dwell time for each resolution element. The variation in radiant flux passing through the

*Based on *Handbook of Sensor Technical Characteristics,* NASA Headquarters, Office of Space Technology Applications, Communications Information Division, February 1981.

field stop onto the photo and thermal detectors creates an electrical output that represents the radiant history of the line. Seven spectral bands are used to provide the spectral signature capability of the instrument. The information outputs from the detector channels are processed in the TM multiplexer for transmission via the tracking and data relay satellites (TDRSS) and/or direct readout to local receiving stations.

Heritage/Derivation: Landsat MSS.

DATA

Data Products: $9\frac{1}{2} \times 9\frac{1}{2}$ latent film made from band sequential format tapes, fully radiometrically and geometrically correct.

Data Archives Location: EROS Data Center (EDC), National Space Science Data Center.

Period of Operation: 1982–1986.

BIBLIOGRAPHY

Landsat-D Project Plan, April 1978, Goddard Space Flight Center.
National Space Science Data Center computer printout, 1980.

SYSTEM CHARACTERISTICS

Size: 1.1 m $\times$ 0.7 m $\times$ 2.0 m $= 1.54$ m^3
Mass: 239 kg
Power requirement: 320 W
Data rate: 85 Mbps
Quantization levels: 256
Interband registration: 0.1 IFOV
Long-term scan stability: 0.5 IFOV

MEASUREMENTS

Spectral Band	Width (μm)	Sensitivity
1	0.45–0.52	0.8%
2	0.52–0.60	0.5%
3	0.63–0.69	0.5%
4	0.76–0.90	0.5%
5	1.55–1.75	1.0%
6	10.40–12.50	0.5°K
7	2.08–2.35	2.4%

Ground IFOV: 30 m (bands 1–5 and 7)
120 m (band 6)

POTENTIAL DERIVED PARAMETERS

1. Measuring crop acreages and the associated errors with differing field sizes
2. Mapping of floodplains, flooded areas, and inland wetlands
3. Mapping of geomorphic features and structural, geologic features
4. Mapping of land use categories in urban and suburban areas
5. Determining shoreline changes
6. Attaining national map accuracy standards at various scales
7. Delineating areas of crop stress, including moisture, salinity, disease, insect damage, and nutrient deficiency stress
8. Eliminating confusion in land use categories, that is, urban and nonurban categories
9. Objectively discriminating between cloud-covered and snow-covered areas
10. Defining intrusives of different iron–mineral content from surrounding rock and hydrothermally altered rocks from unaltered rocks.

1.4 RETURN BEAM VIDICON

The three-channel Return Beam Vidicon (RBV) is not in operation today. A fair-sized library of relatively unexploited imagery was accumulated that may be accessed through the EROS Data Center, Sioux Falls, South Dakota. The instrument was originally flown in the interest of the mapping community, since it offered better geometric accuracy than was available from the Multispectral Scanner, with which the RBV shared space on Landsats-1, -2 and -3.

1.4.1 Return Beam Vidicon Camera*

The Return Beam Vidicon (RBV) camera subsystem contains three individual cameras that operate in different nominal spectral bands. The measured spectral response of the three cameras is shown in Figure 44.

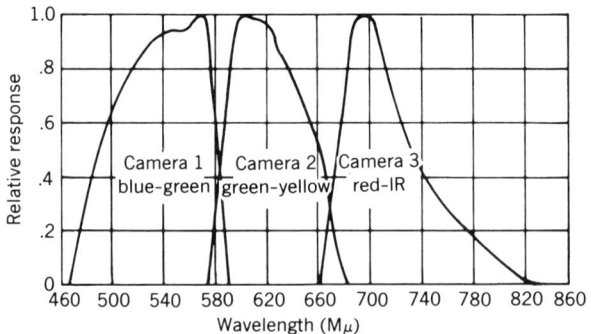

FIGURE 44. Spectral response RBV camera system.

*From the *ERTS Data Users Handbook,* NASA Goddard Spaceflight Center, Greenbelt, Md., May 4, 1972.

Each camera contains an optical lens, a shutter, the RBV sensor, a thermo-electric cooler, deflection and focus coils, erase lamps, and the sensor electronics. The cameras are similar except for the spectral filters contained in the lens assemblies to provide separate spectral viewing regions. The sensor electronics contain the logic circuits to program and coordinate the operation of the three cameras as a complete integrated system and provide the interface with the other spacecraft subsystems. Table 13 shows the major camera parameters and their expected performance.

Operation

The three RBV cameras are aligned in the spacecraft to view the same nominal 185-km (100-nautical-mile) square ground scene as depicted in Figure 45. When the cameras are shuttered, the images are stored on the RBV photosensitive surfaces, then scanned to produce video outputs. As shown in the RBV timing relationships illustrated in Figure 46, the three cameras are scanned in sequence during the last 10.5 seconds of the basic 25-second picture time cycle. The video from each is serially combined with injected horizontal and vertical sync. The readout sequence is camera 3, then camera 2, then camera 1.

The video data interval for each camera lasts for 3.3 seconds, lines 251 through 4375 of the composite video output. The format of the video data is presented in Figure 47. The 720 μsec of active video in each of the lines is replaced with 1.6-MHz sine wave when a camera is turned off and the camera controller–combiner is still operating.

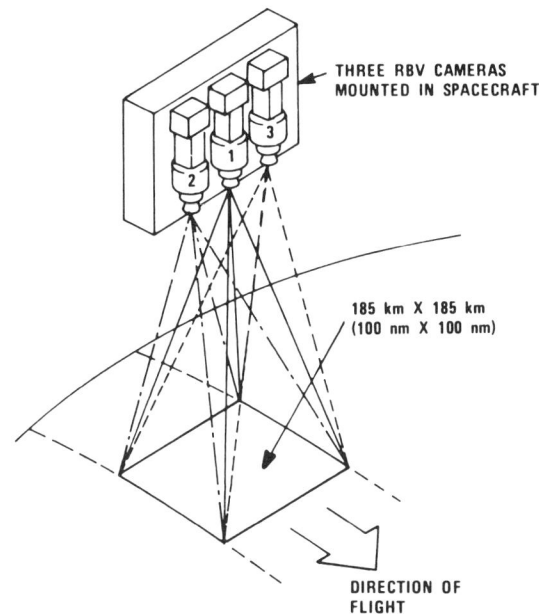

THREE RBV CAMERAS
MOUNTED IN SPACECRAFT

185 km X 185 km
(100 nm X 100 nm)

DIRECTION OF
FLIGHT

FIGURE 45.   RBV scanning pattern.

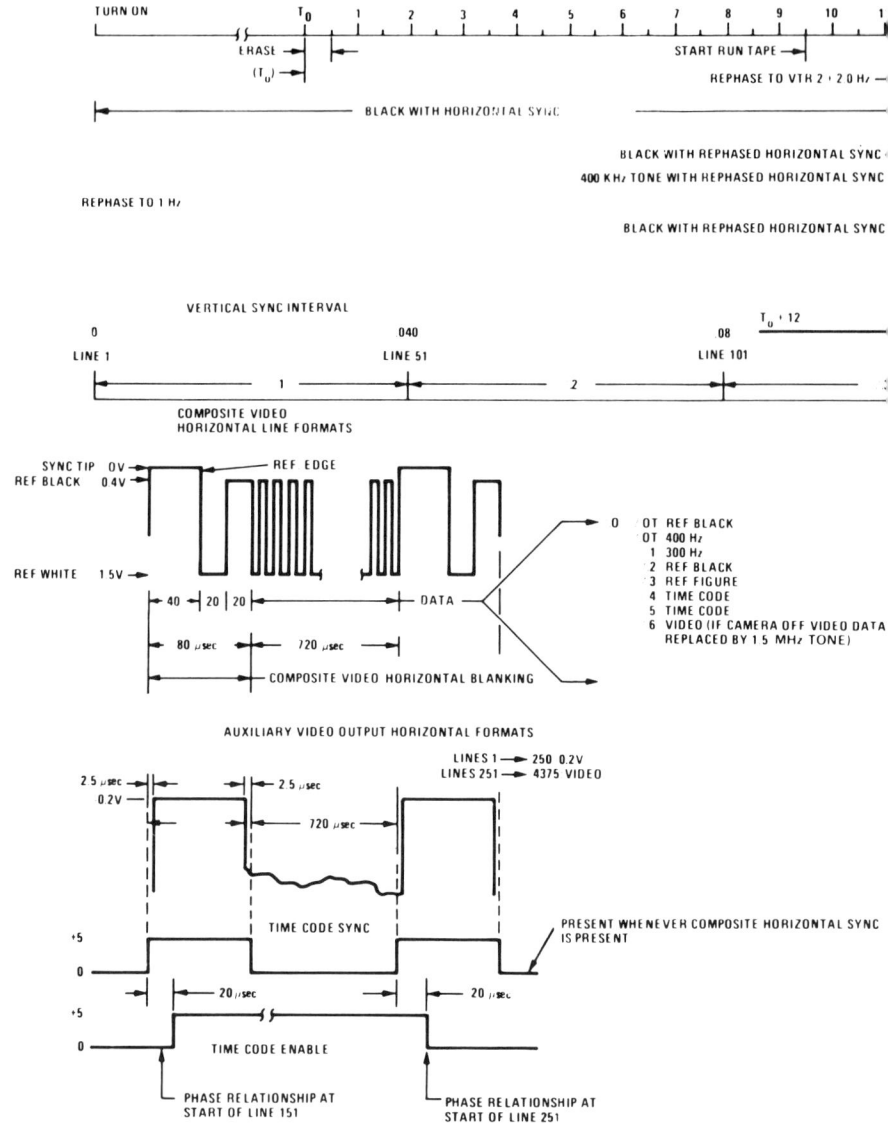

FIGURE 46. RBV camera subsystem timing relationship.

Two modes of operation are possible and are selectable by ground command. Normally the continuous cycle mode is used.

1. *Continuous cycle.* This mode is the normal operating mode of the three-camera system. The system continues to take pictures every 25 seconds, the three cameras operating by one command, until the system is commanded off.

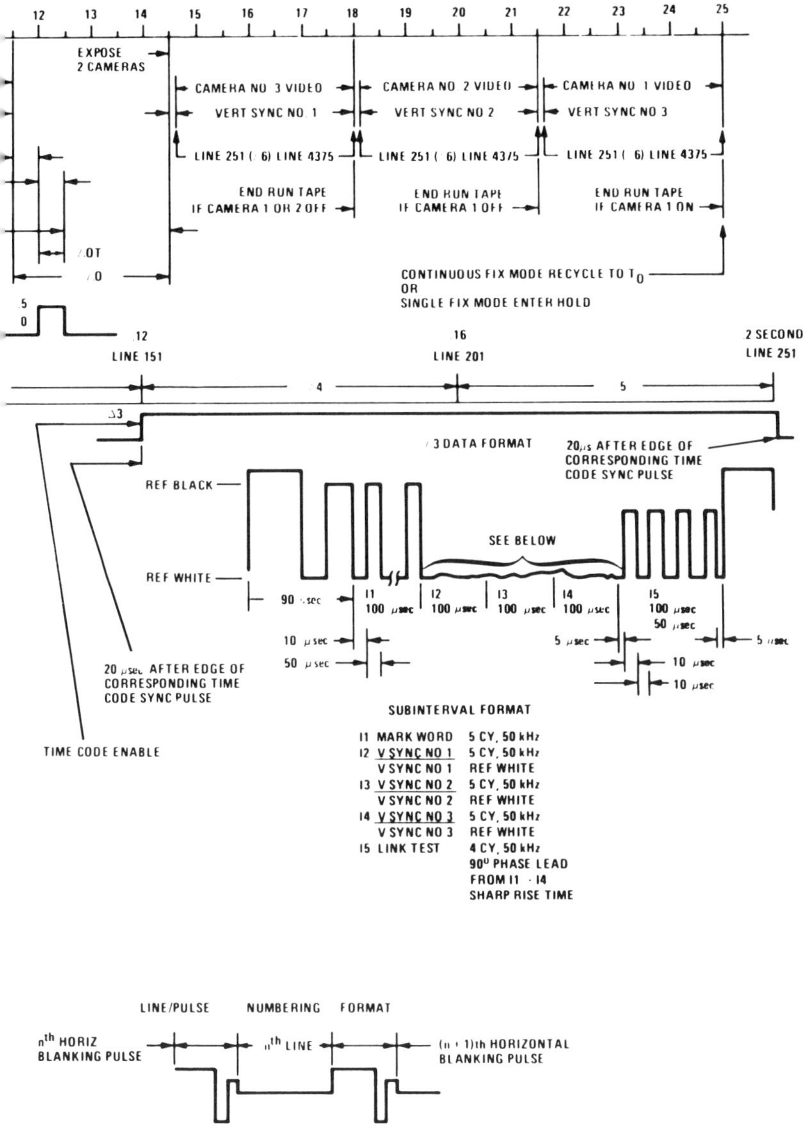

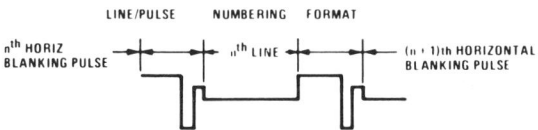

12   13   14   15   16   17   18   19   20   21   22   23   24   25

EXPOSE
2 CAMERAS

← CAMERA NO 3 VIDEO →   ← CAMERA NO 2 VIDEO →   ← CAMERA NO 1 VIDEO →

VERT SYNC NO 1   VERT SYNC NO 2   VERT SYNC NO 3

LINE 251 (6) LINE 4375   LINE 251 (6) LINE 4375   LINE 251 (6) LINE 4375

END RUN TAPE   END RUN TAPE   END RUN TAPE
IF CAMERA 1 OR 2 OFF → IF CAMERA 1 OFF →   IF CAMERA 1 ON →

T0
T0

CONTINUOUS FIX MODE RECYCLE TO T0
OR
SINGLE FIX MODE ENTER HOLD

12   16   2 SECOND
LINE 151   LINE 201   LINE 251

4   5

REF BLACK

SEE BELOW

REF WHITE

90 sec   I1 100 μsec I2 100 μsec I3 100 μsec I4 100 μsec I5 100 μsec
                                                                    50 μsec

10 μsec   5 μsec   5 sec
50 μsec   10 μsec

20 μsec AFTER EDGE OF   10 μsec
CORRESPONDING TIME
CODE SYNC PULSE

TIME CODE ENABLE

20 μs AFTER EDGE OF
CORRESPONDING TIME
CODE SYNC PULSE

/3 DATA FORMAT   20μs AFTER EDGE OF
                     CORRESPONDING TIME
                     CODE SYNC PULSE

SUBINTERVAL FORMAT

I1 MARK WORD   5 CY, 50 kHz
I2 V SYNC NO 1   5 CY, 50 kHz
   V SYNC NO 1   REF WHITE
I3 V SYNC NO 2   5 CY, 50 kHz
   V SYNC NO 2   REF WHITE
I4 V SYNC NO 3   5 CY, 50 kHz
   V SYNC NO 3   REF WHITE
I5 LINK TEST   4 CY, 50 kHz
                  90° PHASE LEAD
                  FROM I1 - I4
                  SHARP RISE TIME

LINE/PULSE   NUMBERING   FORMAT

nth HORIZ   nth LINE   (n + 1)th HORIZONTAL
BLANKING PULSE   BLANKING PULSE

2. *Single cycle*. The camera will take one picture and then revert back to hold mode until a "start prepare" command is received. This mode allows a single 25-second picture cycle to be taken of selected areas with the enabled cameras.

In addition, a calibration mode is provided and is exercised by ground command. In this mode the erase lamps provide three different exposures to each

TABLE 13. RBV Camera Parameters

Item	Camera 1	Camera 2	Camera 3
Nominal spectral band (μm)	0.475–0.575	0.580–0.680	0.698–0.830
	Blue–Green	Green–Yellow	Red–IR
Abbreviated band reference	Blue	Yellow	Red
Edge resolution (% of center)	80%	80%	80%
Video bandwidth (MHz) without aperture correction	3.2 (−20 dB)	3.2 (−20 dB)	3.2 (−20 dB)
Signal-to-noise ratio (at 100% high light) aperture correction out	33 dB	33 dB	31 dB
Horizontal scan rate (lines/second)	1250	1250	1250
Number of scan lines (active video)	4125	4125	4125
Readout time (seconds of active video)	3.5[a]	3.5[a]	3.5[a]
Readout sequence	3	2	1
Focal length of lens (mm)	125.865	125.824	125.979
Exposure set time (msec)			
No. 1	4.0	4.8	6.4
No. 2	5.6	6.4	7.2
No. 3	8.0	8.8	8.8
No. 4	12.0	12.0	12.0
No. 5	16.0	16.0	16.0

[a]Readout time includes 3.3 seconds ground video and 0.2 second sync and time code information.

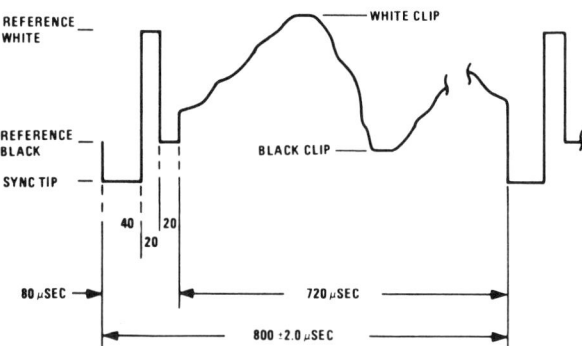

FIGURE 47. Video data format for one horizontal line.

camera which are nominally 0%, 15%, and 100% of the maximum specified input radiance for each camera (designated as Cal 0, 1, and 2, respectively).

The calibration command exercises the sequence depicted in Figure 48. The shutters of each camera are inhibited and the cameras then proceed through three 25-second picture cycles, producing nine images corresponding to three illumination levels for each of the three cameras.

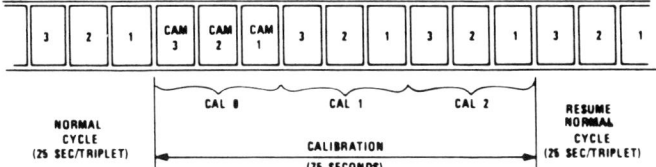

FIGURE 48. Calibrate mode operation.

Reseau Marks and Scan Orientation

A reseau pattern is inscribed on the photoconductive surface of the RBV tube. Figure 49 shows the reseau pattern as it projects into the scene being viewed by the camera. The orientation of the pattern is indicated by using unique anchor marks in the pattern. These reseaus and anchor marks are detailed in Figure 50. All dimensions shown in the figure are in millimeters measured on the faceplate of the RBV camera (multiplier for 70 mm, 2.165; for 242 mm: 7.362). The arrows in Figure 49 marked "H" and "V" (upper left-hand corner) indicate the direction of the line and frame scan. The two-digit numbers are assigned to identify each cross in the reseau pattern; the first digit is a row number and the second digit is a column number.

The orientation of the whole camera with respect to the projection of the reseau pattern into the scene is given by the "camera feet" indication in Figure 51. The camera lens reverses and inverts the scene, so that the actual orientation of the reseau pattern on the vidicon in the camera is also inverted and reversed. The orbit track direction and shutter motion direction are also shown. The shutter mechanism in each RBV camera consists of two adjacent blades with offset cutouts

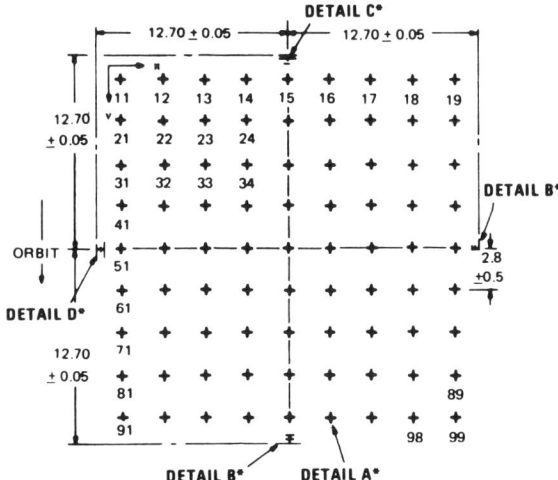

FIGURE 49. Reseau marks on scene.

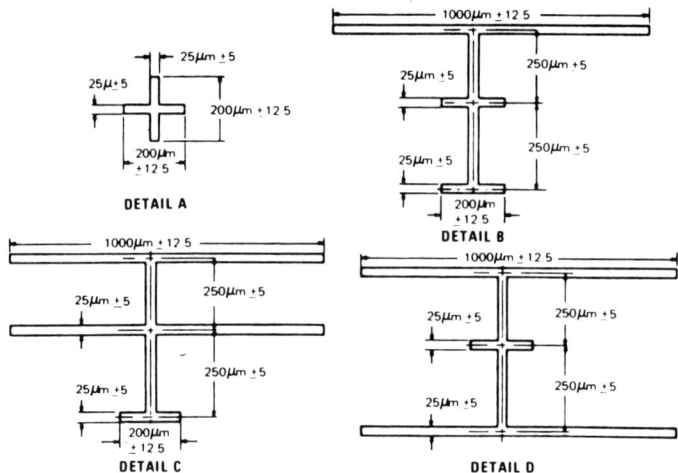

FIGURE 50. Details of reference marks.

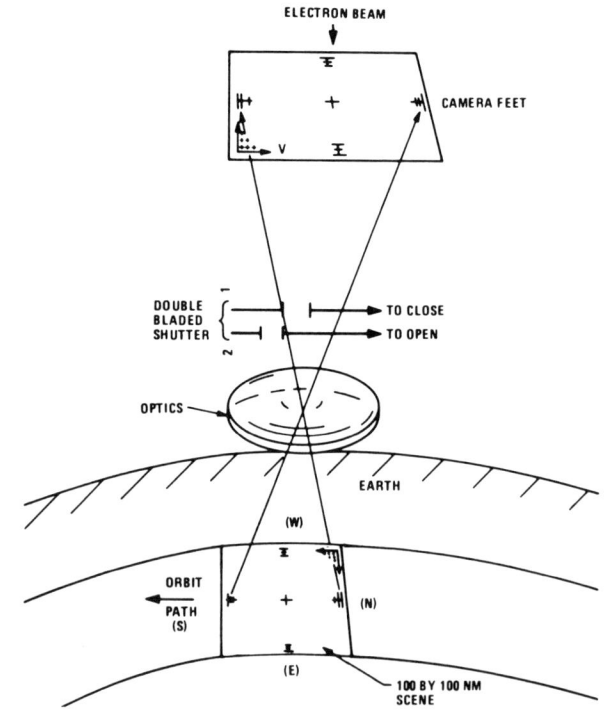

FIGURE 51. Camera–scene orientation.

which sweep across the vidicon aperture to provide the precommanded exposure time to each portion of the photoconductor. The shutter provides uniform exposure over the photoconductor within a maximum variation of ±5%.

The unique anchor marks are located at the (nominal) edges of the scans. The edges will drift somewhat because of circuit tolerances (the overall size-centering tolerance is ±2%); however, the starting point of the scan is somewhat tighter. The reseau locations have been mapped on the vidicon faceplate with approximately 3-μm accuracy and are used during image generation to remove geometric distortions.

Resolution

Measured square wave response for the RBV (lens, vidicon, and amplifier) are shown in Figure 52. An improvement in response, with a corresponding decrease in signal-to-noise ratio, is possible by utilizing the aperture compensation command. With this command each RBV camera employs a secondary amplifier system for the raw video which incorporates specific frequency response shaping networks. It is important to note that this improvement applies to the cross-track direction only and cannot compensate for smear degradations occurring in the along-track direction. Annotation on each image will state if aperture compensation was "in" or "out." . . .

Geometric Fidelity

Table 14 shows the raw internal RBV errors observed during test and includes, for reference only, the positional effect of these errors on the output image. All errors are effects associated with the electromagnetic characteristics of the vidicon camera.

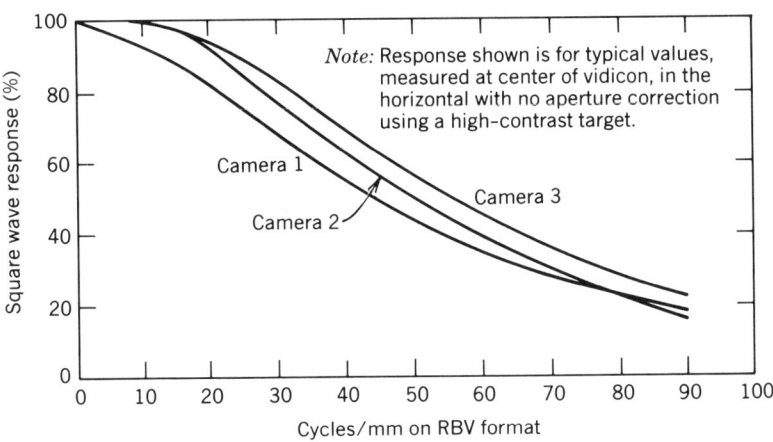

FIGURE 52. Typical sine wave response—RBV camera.

TABLE 14. Positional Effects of Raw Internal RBV Errors

Item	Name of Error	Illustration of Error Type	Observed Value	Image Positional Effect (m) 1σ
1	Magnetic lens distortion		1% of maximum	432
2	S curve		0.200 mm at corners	418
3	Scale		<0.1%	<432
4	Centering		<0.74% each axis	<982
5	Nonlinearity		0.1% maximum each axis	518
6	Skew		<0.26°	<210
7	Raster rotation		<0.1°	<75

RBV Exposure Capabilities

The capability of the RBV cameras to recognize specific scene radiance is a function of the light transfer characteristics (LTCs) and time of exposure of each camera. The LTC relates voltage output to radiance for mean levels or levels in large areas (near zero spatial frequencies). Figure 53 is the measured LTC for the three cameras for the on-axis (center) location of the vidicon. The radiance is the equivalent spectrally flat radiance in front of the lens within the bandpass of each camera.

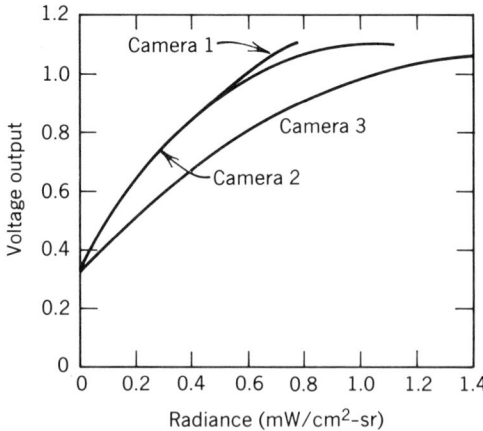

FIGURE 53. RBV light transfer characteristics.

The equivalent spectrally flat radiance is obtained by integrating the scene radiance and camera spectral responses.

$$N = \frac{\int R(\lambda) N_S(\lambda)}{\int R(\lambda)}$$

where $R(\lambda)$ = camera spectral response
$N_S(\lambda)$ = scene spectral radiance

The camera spectral response is shown in Figure 44.

The exposures for the various spectral bands corresponding to 1-V video output (white reference, defined as saturation exposure) are:

Band	Exposure
1	0.552 $\mu J/cm^2$
2	0.598 $\mu J/cm^2$
3	0.985 $\mu J/cm^2$

The maximum mean radiance of a scene at the vidicon faceplate is related to the saturation exposures and exposure time by

$$N = \frac{4T^2 Ex}{\pi t} \quad (W/cm^2\text{-s})$$

where N = mean radiance of scene at vidicon faceplate
T = effective f number of lens
t = exposure time
Ex = saturation exposure

Based on this equation, Table 15 delineates the exposure time settings along with the value of scene radiance at saturation of the vidicon.

TABLE 15. Scene Radiance at Saturation for Various Exposure Times

	Band 1		Band 2		Band 3	
Exposure Set	1 (ms)	N_{SAT} (mW/cm²-sr)	1 (ms)	N_{SAT} (mW/cm²-sr)	1 (ms)	N_{SAT} (mW/cm²-sr)
A	4	1.80	4.8	1.62	6.4	2.01
B	5.6	1.29	6.4	1.22	7.2	1.78
C	8	0.90	8.8	0.89	8.8	1.46
D	12	0.60	12	0.65	12	1.07
E	16	0.45	16	0.49	16	0.80

TABLE 16. Total Scene Radiance N (mW/cm²-sr)

Typical Scene	Band 1 (Zenith Angle)				Band 2 (Zenith Angle)				Band 3 (Zenith Angle)			
	0°	30°	45°	60°	0°	30°	45°	60°	0°	30°	45°	60°
Specular snow	4.21	3.83	3.01	2.21	3.39	3.38	2.79	2.02	3.04	2.63	2.16	1.54
Fresh snow	—	2.91	2.32	1.61	—	2.56	2.12	1.54	—	1.85	1.52	1.09
Icy snow	—	2.82	2.43	1.78	—	2.69	2.22	1.62	—	2.07	1.70	1.22
Clay	—	2.23	2.16	1.67	—	2.37	1.97	1.44	—	1.91	1.58	1.13
Sand	—	1.02	0.88	1.08	—	1.07	0.90	0.68	—	1.16	0.96	0.69
+1σ plants	—	0.70	0.62	0.99	—	0.53	0.46	0.37	—	1.04	0.86	0.62
−1σ plants	—	0.47	0.43	0.64	—	0.31	0.28	0.25	—	0.57	0.48	0.29
H₂O	—	0.60	0.54	0.46	—	0.33	0.3	0.26	—	0.25	0.22	0.17
Overcast	2.37	2.81	2.35	1.74	3.42	2.94	2.43	1.76	2.76	2.38	1.96	1.40

Note: These are typical values, not to be taken as absolute.

Table 16 shows calculated values of scene radiance at sensor input for various solar zenith angles and typical Landsat scenery.

These values were calculated with a solar constant of 0.1322 W/cm², two-atmosphere traverse, and an atmospheric transmission of 0.8. These data are shown only as representative examples and should not interpreted as precision values.

1.4.2 Return Beam Vidicon Camera System Specifications*

IDENTIFICATION

Discipline: Resource Observations
Status: Flight Missions

Acronym: RBVC
Instrument Type: Imager
Spacecraft: Landsat -1, -2 & -3
Contractor: RCA Astro Electronics

OBJECTIVES

1. Provide continuous, overlapping multispectral photographic coverage of the earth's surface along the orbital track.

2. Repeated observations of any given area within the minimum time interval possible.

---

*Based on *Handbook of Sensor Technical Characteristics,* NASA Headquarters, Office of Space Technology Applications, Communications Information Division, February 1981.

DESCRIPTION

Summary: The RBVC is a three-camera system spanning the visible spectrum in three bands: 0.475–0.575, 0.580–0.680, and 0.690–0.830 μm. Spectral bands are obtained through use of filters in acquisition optics. An electronically triggered, variable-speed, focal-plane shutter allows picture taking over a wide range of scene brightness and provides uniform exposure of the vidicon. This sensor, a 2-in. return beam vidicon, combines the vidicon and orthicon tube. The video output is derived from the return scanning beam. A photoconductive surface charges the target surface in proportion to the light received. Then as the electron scanning beam traverses the target, the charge modulates this beam, which is then amplified by an electron multiplier. The video output of the system may be fed directly to the modulator of the spacecraft communication system. The cameras are pointed at nadir

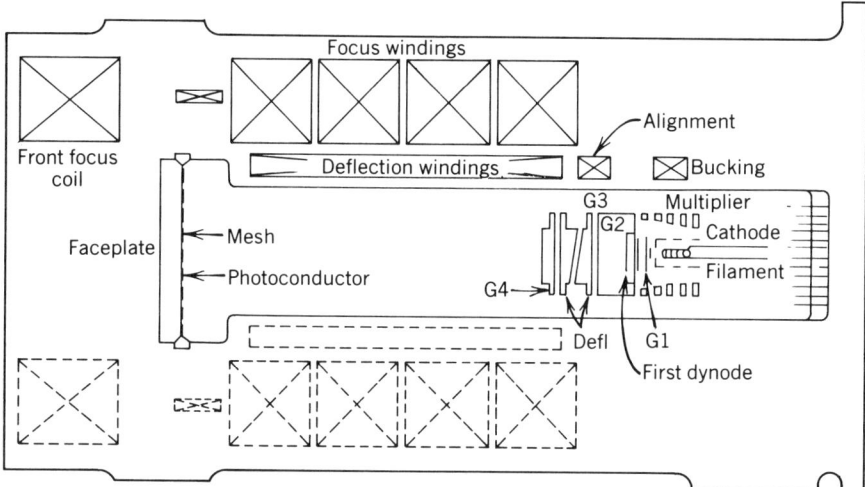

FIGURE 54.   RBV camera system schematic diagram.

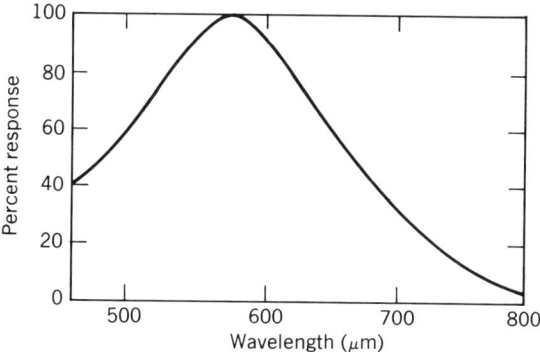

FIGURE 55.   Spectral sensitivity of the RBV camera.

and a new scene is imaged on the photoconductor surfaces every 25 seconds. The resolution capability of the system is 4500 TV lines. Equipped with a 130-mm focal length, $f/2.8$ lens, each frame will cover an area of 100×100 nm at a resolution of about 150 ft per TV line from 496-nm altitude. The sensor is capable of resolving 90 line pairs/mm.

DATA

Data Products: Spectral images

Video data from the RBV were transmitted (2265.5 MHz) in both real-time and tape-recorder modes.

Data can be purchased as image products or as computer-compatible tapes.

Data Archives Location: Earth Resources Data Center, Department of the Interior, Sioux Falls, S. Dak.

Period of Operation: July 23, 1972–January 6, 1978.

BIBLIOGRAPHY

National Space Science Data Center, Fact Sheets on U.S. and weather and earth observation satellites.

SYSTEM CHARACTERISTICS

Weight: 130 lb

Volume: 3.0 ft^3

Average power: 130 W

Peak power: 145 W

Mean time before failure: 12 months

Components: three RBV cameras, recorder, transmitter

Data recovery: delayed telemetry

Frequency of observation: on command

MEASUREMENTS

Spectral range: 0.475–0.83 μm

Field of view: 11.5° × 11.5°

Spectral bands: 0.475–0.575, 0.580–0.680, 0.690–0.830

Ground swath: 185 km × 185 km

POTENTIAL DERIVED PARAMETERS

1. Radiation from the surface of the earth in the visible spectrum
2. High-resolution multispectral photography

1.4.3   Landsat-3 Return Beam Vidicon Response Artifacts*

Background

The return beam vidicon (RBV) sensing systems employed aboard Landsats 1, 2, and 3 have been similar in that they have utilized vidicon tube cameras. These are not mirror-sweep scanning devices such as the Multispectral Scanner (MSS) sensors that have also been carried aboard the Landsat satellites. The vidicons operate more like common television cameras, using an electron gun to read images from a photoconductive faceplate.

In the case of Landsats 1 and 2, the RBV system consisted of three such vidicons, which collected remote sensing data in three distinct spectral bands. Landsat 3, however, utilizes just two vidicon cameras, both of which sense data in a single broad band. The Landsat 3 RBV system additionally has a unique configuration. As arranged, the two cameras can be shuttered alternately, twice each, in the same time it takes for one MSS scene to be acquired. This shuttering sequence results in four RBV "subscenes" for every MSS scene acquired, similar to the four quadrants of a square.

Each subscene represents a ground area of approximately 98 × 98 km. The subscenes are designated A, B, C, and D, for the northwest, northeast, southwest, and southeast quarters of the full scene, respectively. RBV data products are normally ordered, reproduced, and sold on a subscene basis and are in general referred to in this way.

Each exposure from the RBV camera system presents an image which is 98 km on a side. When these analog video data are subsequently converted to digital form, the pixel (picture element) that results is 19 m on a side with an effective resolution element of 30 m. This pixel size is substantially smaller than that obtainable in MSS images (the MSS has an effective resolution element of 73.4 m), and when RBV images are compared to equivalent MSS images, better resolution in the RBV data is clearly evident. It is for this reason that the RBV system can be a valuable tool for remote sensing of earth resources.

Until recently, RBV imagery was processed directly from wide-band videotape data onto 70-mm film. This changed in September 1980 when digital production of RBV data at the NASA Goddard Space Flight Center (GSFC) began. The wideband videotape data are now subjected to analog-to-digital preprocessing and cor-

---

*Based on *Landsat 3 Return Beam Vidicon Response Artifacts*, EROS Data Center, U.S. Geological Survey, Sioux Falls, S. Dak., August 1981.

rected both radiometrically and geometrically to produce high-density digital tapes (HDTs). The HDT data are subsequently transmitted via satellite (Domsat) to the EROS Data Center (EDC), where they are used to generate 241-mm photographic images at a scale of 1:500,000. Computer-compatible tapes of the data are also generated as digital products.

Of the RBV data acquired since September 1, 1980, approximately 2800 subscenes per month have been processed at EDC.

Statement of Problem

Analysis of large volumes of Landsat 3 RBV digital data which have been converted to photographic form has led to the firm identification of several visible artifacts in the imagery. These artifacts have been identified, categorized, and traced directly to specific sensor response characteristics. Furthermore, analysis has determined that none is easily removed. All cases remain under active study for possible image enhancements at some point in the future.

The purpose of this discussion is to describe these artifacts based on the best information available at this time.

To date, seven generic categories of sensor response artifacts have been identified:

1. Shading and stairsteps
2. Corners out of focus
3. Missing reseaus
4. Reseau distortion and data distortion
5. Black vertical line
6. Grain effect
7. Faceplate contamination

An additional category under study, but not yet determined to be caused by sensor response, is a geometric artifact which appears to be peculiar to RBV imagery. These phenomena are only the most common found to date, and it is realized that the list may be by no means comprehensive.

The artifacts can occur singly or in combination. In some cases, their impact on data usability is serious, resulting in total loss of image content. In other cases, only minor defects exist and the overall quality of the images is considered good. Between these two extremes, the case most often found is that a partial reduction in overall image quality occurs. This can create problems for the photo interpreter, such as when two or more visible defects combine to make the discernment of image detail more difficult; or it can result in actual loss of detail in specific areas, such as when significant ground features are suppressed or obscured due to a single type of artifact. In the latter instance especially, scenes that are radiometrically "flat" to begin with (desert scenes, water, cloud-covered areas) exhibit the worst effects.

A large fraction of the RBV data processed and recorded on HDTs to date suffer from this reduction in overall quality. When subsequently converted to film or digital products, therefore, most of the data are downgraded during quality assessment. Because the average background scene radiance varies as it does, the defects in the imagery appear to be random in terms of both magnitude and location in the image area.

The situation clearly presents a problem both for the user and for EDC quality assurance personnel. Users have a right to be concerned about the potential for poor image quality rating for data which may vary unpredictably and whose ultimate usability to some customers is impossible to determine.

Impact on Users

From the serious user's point of view, an image that is only partially affected by artifacts is one that still contains unaffected, usable data. It may be impossible to predict where the artifacts will occur in the image spatially, but many users are willing to take this risk. These same users are also aware that the mere occurrence of a given defect does not necessarily mean that that part of the image automatically becomes "unusable." They may find, for example, that they can work around certain defects once they learn what those defects do to the image. The scientific value of an image for a specific application is therefore something that only the user can determine.

For these reasons, RBV data affected by sensor response artifacts are being made readily available. It is realized that the aesthetic appearance of an image is of little importance to the scientist. In addition, in spite of the ostensible deficiencies, a good amount of usable RBV data is being produced. The philosophy has been taken that as long as users are aware of the problem, its impact will not be as great.

Casual users, of course, may see things differently. Because the majority of data acquired to date includes areas having the flat field radiance response mentioned previously, very few RBV images have been processed that would be suitable for display purposes. Any user ordering RBV imagery should be aware that, although useful scientifically, RBV data may not always be aesthetically pleasing to the eye.

Quality Assurance Procedures

From a quality assurance point of view, the RBV data now being received fit none of the preestablished criteria that are used to judge the quality of other Landsat data. Normally, all Landsat data are subjected to a stringent inspection procedure which is designed to identify the number, type, and extent of visual defects in an image. The image quality rating is established accordingly based on this qualitative assessment. This system works well because the deficiencies are known (having been recognized over the years) and their nature is understood. The impact of each on the usability of final photographic products can be judged with high accuracy.

The RBV data are different. The sensor-caused artifacts are typical of those experienced by other vidicon users; however, their impact on Landsat data users has not yet been completely established due to a lack of experience in dealing with these data by the user community as a whole. The perceived product deficiencies are thus very difficult to categorize. A decision was therefore made not to downgrade RBV image quality because of sensor-caused defects. Instead, the normal qualitative assessment procedure has been replaced by a procedure in which relative image quality is rated based upon the perceived utility of the data. A rating of 8 is given to denote a "good" image; a 5 equals "fair"; a 2 equals "poor"; and a 0 equals "unusable." About 60% of the RBV data received and processed by EDC since September 1, 1980 have been rated good or fair.

The remainder of this discussion is devoted to brief descriptions of the RBV image defects. None of them are easily correctable. Questions on any aspect of the information presented may be directed to:

User Services Section
U.S. Geological Survey
EROS Data Center
Sioux Falls, S. Dak.   57198

Shading and Stairsteps

Prelaunch testing of the RBV cameras involved placing each in front of a light source and measuring camera response as the light level was varied. This test revealed that the responses were neither uniform from camera to camera nor uniform across either camera faceplate. The lack of uniformity across the individual camera faceplates is called *shading*.

The prelaunch tests were performed in order to collect enough information to minimize these differences during ground processing after launch. Although minimization has been accomplished, it has been impossible to remove these effects totally. The result is a residual, spatially varying, gray-level-dependent shading in most RBV images. Shading is unique to each subscene, and it affects some images more than others.

Related to shading is an effect known as *stairsteps*. Some RBV data contain localized areas where the image data have fallen into certain anomalously regular patterns. These patterns are characterized by edges which resemble a staircase viewed from the side. Such patterns can be a series of concentric polygons. Other patterns can occur, but the stairstep characteristic will always be present.

Stairsteps are induced by ground processing and occur as a result of imperfect deshading when the raw input data lie outside established high- and low-radiance limits. Both high- and low-radiance stairsteps therefore occur. Both types are related to camera response characteristics similar to those described above for shading.

Corners out of Focus

For some RBV data, it is obvious that the corners of the image are out of focus when compared to the more central parts of the image. This is a vidicon tube characteristic and can be seen in the raw data. While the optics system could be at fault, the cause is much more likely to be thermal perturbations or a phenomenon known as *beam-landing error*. The effect is aggravated by imperfect deshading and the presence of high-frequency noise (causing spatial distortions) when either of these occurs.

Missing Reseau Marks

The reseau marks that have been inscribed on the faceplate of each camera can appear to be lost in an image when they occur among data that lie at either radiance extreme. For low-radiance data, the reseaus merge into the background and are obscured. For high-radiance data, the reseaus can be washed out or lost due to a vidicon characteristic called *blooming*. The latter effect is probably a result of the gain setting on the RBV sensor and cannot be corrected by ground processing.

Reseau Distortion and Data Distortion

Distortion in reseau marks and image data can occur when the raw data contain time-code-induced anomalies. Some reseau marks are badly curved, and the image data near them, as well as elsewhere, are similarly distorted. Minor data losses in the time code signal are thought to cause this horizontal shifting of data. Although this effect is not restricted to any area in an image, it is usually most pronounced in the corners of the image.

Black Vertical Line

Image data acquired by camera 2 (which acquires subscenes B and D) contain a thin black line along the left vertical edge of the image. This is evident when the left edge of the image contains high-radiance data. This artifact occurs as a result of both hardware and software problems in the ground processing system. Its impact on image quality is strictly cosmetic, and it is mentioned here more for completeness than for any other reason.

Grain Effect

Production of a large volume of RBV data has resulted in the identification of specific images that contain a *graining* or a fabric-type appearance. The grain effect has been found to exist in almost all of the data processed to date. Probably attributable to some form of noise, the grain effect is currently under analysis at GSFC, where attempts are being made to properly characterize it.

Faceplate Contamination

Two categories of faceplate contamination exist. These have been designated *hot spots* and *tears*.

A hot spot correlates to a flaw in the photoconductive layer of the vidicon, appearing as a blurred white spot. Hot spots are repeatable from image to image, and their prominence is dependent upon the gray level of the image data written over them. Visual detection of hot spots is possible only when the average background scene radiance is low enough to permit them to stand out.

The second type of faceplate contamination cannot be seen clearly unless the proper combination of sun angle and scene radiance exists. When these conditions exist, one can detect bright white pinhole-size spots that are each accompanied by a black "shadow" of unknown origin. These are termed *tears*. Tears can be attributed to two causes. In some cases, they are a manifestation of minor flaws in the photoconductive layer of the vidicon. In other cases they represent localized areas of condensate on the camera faceplate. They always occur in the same place, and their detection is dependent upon the gray level of the image data written over them. They appear to be permanent.

Both types of faceplate contamination, hot spots and tears, are uncorrectable by ground processing.

Geometric Anomaly

A small fraction of RBV data has been found to contain a local geometric distortion. Image data in this category can be recognized by the presence of parallel lines restricted to the left one-third of the image. These lines always occur near the top and on the left-hand side of the subscene. They do not extend into any other one-third segment of the image data, and they are abruptly terminated at the boundary between the left one-third of the data and the center one-third.

This artifact is introduced by ground processing and should be capable of being removed by the same means. A study is under way at GSFC to determine how this might best be done.

1.5 FUTURE SENSORS

Over the next two decades NASA expects to launch a variety of earth resources missions. These satellites will carry advanced sensors, which in some cases are still in development. This section describes some of these future sensors. The descriptions are taken from the *NASA Space Systems Technology Model* published by the Office of Aeronautics and Space Technology of NASA Headquarters, Washington, D.C., 1985. Included are VAS, AVHRR, SIR-B, MLA, STIMS, OCI, SIS, Imaging Spectrometer, Multifrequency Lidar Facility, SPOT, and EOS. Many, such as N-ROSS, ERS-1, MOS-1, RADARSAT, and EOSAT, are not described since an exhaustive coverage would be too voluminous. The intent is to convey the growth directions and time scales, that is, to provide a sense of the increasing vigor of this segment of the community.

VISSR Atmospheric Sounder (VAS)

Status: Approved   **Program Phase:** Operational
                                                       Carrier: GOES G, H

PHYSICAL AND GENERAL PROPERTIES

Physical size (m): scanner, 1.5×0.65; electronics, $0.3 \times 0.2 \times 0.3$
Mass (kg): scanner, 69; electronics, 10
Power requirement (W_{elec}): 40
Data rate (kbps): 28,000

PRINCIPAL PURPOSE AND FEATURES

The VISSR Atmospheric Sounder (VAS) is an extension of the original VISSR (Visible Infrared Spin Scan Radiometer) imaging capability and includes additional thermal bands for the determination of atmospheric temperature at various altitudes by spectral selection in the CO_2 absorption bands. Water content is also determined at several altitudes in the H_2O absorption bands. In addition, cloud and earth surface temperatures are measured. The added capabilities of VAS provide the additional atmospheric measurements which enable extensive research in severe weather research. Parameters including spectral band selection, spatial resolution, dwell times, and geographic location are incorporated into the sounding or multispectral imaging modes to provide measurements which best meet the needs of the research scientist. There are three distinct operational modes: the VISSR mode, the dwell-sounding mode, and the multispectral imaging mode. The VISSR operating mode is the same as the regular VISSR system. Both day and night cloud cover imagery will be possible as with VISSR. This is the NOAA operational mode. The dwell-sounding mode is the mode for the VAS experiment, and the multispectral imaging mode can provide normal VISSR imaging plus data in any two selected spectral bands.

PRINCIPAL PHYSICAL PROPERTY MEASURED, RANGE, ACCURACY, AND RESOLUTION

In the VISSR mode, the FOV is 6.9 km with a visible channel resolution of 0.9 km. The dwell-sounding mode covers the range 680 cm^{-1} (14.7 μm) through 2535 cm^{-1} (3.9 μm), and in the spectral range of 703 cm^{-1} (14.2 μm) through 1490 cm^{-1} (6.7 μm), either the 6.9-km or the 13.8-km resolution detectors can be selected.

Advanced Very High Resolution Radiometer (AVHRR):

Status: Approved

Program Phase: Operational
Carrier: NOAA F-J

PHYSICAL AND GENERAL PROPERTIES

Physical size (m): $0.3 \times 0.4 \times 0.8$
Mass (kg): 30
Power requirement (W_{elec}): 29
Data rate (kbps): 750

PRINCIPAL PURPOSE AND FEATURES

To provide global daytime and nighttime sea surface temperature, and information on ice, snow, and cloud formations. This multispectral radiometer operates in the scanning mode and measures emitted and reflected radiation in four spectral intervals. The AVHRR produces an image in each band. The satellite motion is used to provide scanning normal to the rotating mirror's crosstrack scanning. Radiation is reflected off the mirror through an afocal 8-in. Cassegrain telescope and filtered into visible and IR components by dichroic splitters. The visible channels are then filtered by interference filters. The IR detectors are radiatively cooled to 105°K.

PRINCIPAL PHYSICAL PROPERTY MEASURED, RANGE, ACCURACY, AND RESOLUTION

The AVHRR can provide sea surface temperature with an accuracy of about 0.6°C if no clouds contaminate the field of view and if atmospheric effects are carefully removed. Horizontal temperature gradients of about 0.2°C can also be resolved.

ASSOCIATED RANGES AND/OR RESOLUTION

Spectral range: channel 1, 0.55–0.9 μm (visible)
                  channel 2, 0.725–1.3 μm (near IR)
                  channel 3, 10.5–11.5 μm (IR window)
                  channel 4, 3.55–3.93 μm (IR window)
Spatial resolution: 1.1 km

BIBLIOGRAPHY

Tanner, S., *Handbook of sensor technical characteristics,* NASA Reference Publication 1087, July 1982.

Shuttle Imaging Radar–B (SIR-B)

Status: Planned   **Program Phase:** Operational
                                                Carrier: Shuttle Radar Laboratory
                                                            (formerly OSTA-3),
                                                            OSTA-5, OSTA-7

PHYSICAL AND GENERAL PROPERTIES

Physical size (m):   sensor, $1 \times 1 \times 1$; antenna, 10.7×2.16
Mass (kg):   556
Power requirement (W_{elec}):   1000 peak
Data rate (kbps):   46,000

PRINCIPAL PURPOSE AND FEATURES

Imaging radar for earth surface feature applications experiments. The SIR-B antenna can be mechanically tilted while the shuttle's payload bay is facing the earth. This will enable researchers to obtain radar imagery of a specific area at multiple angles of incidence during successive shuttle orbits.

PRINCIPAL PHYSICAL PROPERTY MEASURED, RANGE, ACCURACY, AND RESOLUTION

Principal property measured: radar backscatter coefficient (reflectivity)
Range resolution: 5.8–17 m
Azimuth resolution: 25 m

ASSOCIATED RANGES AND/OR RESOLUTION

Frequency: 1.28 GHz
Wavelength: 23.5 cm
Polarization: HH
Orbital altitude: 225 km
Orbital inclination: 57°
Radar swath (look-angle-dependent): 30–60 km, 15–60°
Number of Looks: 4
Coverage: >40,000,000 km^2

COMMENTS

Optical data collection: 8 h
Digital data collection: >25 h
Real-time data transmission to ground: via Tracking and Data Relay Satellite System

SIR-B is the follow-on to SIR-A, which was flown on Office of Space Technology Applications (OSTA)-1. SIR-B will be flown with the LFC, MAPS, and FILE instruments on STS-17/OSTA-3. Two more flights in the SIR program are planned (OSTA-5 and -7)

BIBLIOGRAPHY

Settle, M. and J. Taranik, Use of the space shuttle for remote sensing research: recent results and future prospects, *Science,* 3 December 1982, pp. 993–995.

Multispectral Linear Array (MLA) Experiment:

Status: Planned

Program Phase: Preliminary Design
Carrier: Shuttle

PHYSICAL AND GENERAL PROPERTIES

Physical size (m): 1.7 × 1.0 × 0.8
Mass (kg): 315
Power requirement (W_{elec}): 310
Data rate (kbps): 120,000 compressed to 50,000, TDRSS-linked

PRINCIPAL PURPOSE AND FEATURES

The MLA will collect data to support observational research in the physical science of the earth's surface (geology, agriculture, biomass, moisture), remote sensing science (pattern recognition, image analysis, improved data precision), and instrument and system technology (multispectral array instrumentation, advanced optics, and data-processing methods).

Data will be obtained over selected portions of the Sahel and rain forest regions of Africa for biomass investigation in addition to off-nadir measurements with the pointable system. The MLA will provide earth observations at higher spectral and spatial resolution than the Thematic Mapper for land cover determination, biomass class delineation, and quantification of bidirectional and atmospheric effects.

PRINCIPAL PHYSICAL PROPERTY MEASURED, RANGE, ACCURACY, AND RESOLUTION

Spectral coverage: 0.45–1.65 μm
Number of bands: 6 (4 VIS*-SWIR, 2 SWIR)
Spatial resolution VNIR*/SWIR* (285 km altitude): 15/30 m (VIS-SWIR/SWIR)

ASSOCIATED RANGES AND/OR RESOLUTION

Cross-track FOV: 6° (30-km ground swath)
Operating altitude: 285–350 km
Pointing capability: Multiposition mirror (cross-track by rolling shuttle)
Calibration: end to end
SWIR detector cooling: solid cryogen (argon)
Modulation transfer function: >0.3
S/N: >100

COMMENTS

Detailed shuttle instrument definition studies have been completed. Development efforts for a multispectral visible focal plane and Schottky barrier monolithic SWIR arrays are in progress. Deliverables from these contracts will be used for the MLA shuttle instrument. A

*VNIR, very near infrared; SWIR, shortwave infrared; VIS, visible.

preliminary design for the instrument has also been developed, and plans for its implementation as an in-house GSFC build have been formulated.

BIBLIOGRAPHY

Vane, D., *Mission/orbit considerations for earth resources research aboard the space shuttle*, NASA/JPL, September 1982.

Shuttle Thermal Infrared Multispectral Scanner (STIMS)

Status: Planned   **Program Phase:**   Concept Formulation
                                                          Carrier: Shuttle

PHYSICAL AND GENERAL PROPERTIES

Physical size (m): 1.5×1.0
Mass (kg):   900
Power requirement (W_{elec}):   300
Data rate (kbps):   15–30

PRINCIPAL PURPOSE AND FEATURES

To find emission features characteristic of silicate rock by measuring spectral emissivity. Will use a 6×500 element array and 0.5-m-diameter telescope. There will be six bands, with $0.1°$ in each band.

PRINCIPAL PHYSICAL PROPERTY MEASURED, RANGE, ACCURACY, AND RESOLUTION

Not specified.

ASSOCIATED RANGES AND/OR RESOLUTION

Instantaneous field of view: 30 m
Ground resolution: 100 m
Swath width: 100 km

Ocean Color Imager (OCI)

Status: Planned   **Program Phase:**   Concept Formulation
                                                          Carrier: NOAA I or J

PHYSICAL AND GENERAL PROPERTIES

Physical size (m): $0.4 \times 1.1 \times 0.6$
Mass (kg):   50
Power requirement (W_{elec}):   62
Data rate (kbps):   618

PRINCIPAL PURPOSE AND FEATURES

The OCI incorporates features from CZCS (Nimbus 7) and proposed NOSS CZCS II. It
is designed to measure total spectral radiance and permit accurate atmospheric turbidity
correction. Upwelled spectral data is used to calculate total chlorophyll pigment concentrations, diffuse attenuation coefficients, and estimate suspended sediment concentration.
Associated data-handling capability will enable global coverage.

PRINCIPAL PHYSICAL PROPERTY MEASURED, RANGE, ACCURACY, AND RESOLUTION

Total spectral radiance: +0.5%, 10-bit digitization, S/N 300 chlorophyll pigments
+30 to −30% (0.01–1.5 μg/liter), +50 to −50% (1–50 μg/liter)
Diffuse attenuation coefficient: +15 to −15%
Aerosol optical depth: +15 to −15%; no clouds, low suspended sediment concentration
Resolution: 1.13 km (local area coverage), 4.5 km (global area coverage)
IFOV: 1.3 mrad

ASSOCIATED RANGES AND/OR RESOLUTION

8 channels: 0.44, 0.49, 0.52, 0.56, 0.59, 0.67, 0.765, 0.87 μm
Bandwidth: 0.02 μm visible, 0.05 μm IR

COMMENTS

Two-year mission, designed to improve accuracy of global and local oceanic primary pro-
ductivity estimates and to augment fisheries science and ocean dynamics studies. To be
flown with AVHRR for sea surface temperature.

BIBLIOGRAPHY

The marine resources experiment program (MAREX), report of the Ocean Color Science
Working Group, prepared for NASA by OAO Corporation, March 1983.

Shuttle Imaging Spectrometer (SIS):

Status: Planned

Program Phase: Concept Definition
Carrier: Shuttle

PHYSICAL AND GENERAL PROPERTIES

Physical size (m): $1.3 \times 0.7 \times 0.6$
Mass (kg): 180–230
Power requirement (W_{elec}): 150 (electronics), 200 (cooling)
Data rate (kbps): 98,400

PRINCIPAL PURPOSE AND FEATURES

To validate the technologies integral to an advanced class of land remote sensing imaging spectrometer systems; to collect high spectral and spatial resolution coverage of the earth's surface; to investigate the utility of high spectral resolution applied to a variety of earth resource problems. The instrument has 128 spectral bands in the range 0.4–2.5 μm. Its sampling rate is 10 nm, 20 nm in the range 1.0–2.5 μm.

PRINCIPAL PHYSICAL PROPERTY MEASURED, RANGE, ACCURACY, AND RESOLUTION

The SIS system utilizes the imaging spectrometer instrument design concept. It provides spectral resolution of 20 nm, spatial resolution in the range of 20–60 m, and swath widths up to 60 km. The instrument uses pushbroom imaging coupled with a spectrometer which disperses each imaging line over the area array focal plane, and area array detectors for both the visual and short infrared wavelengths.

ASSOCIATED RANGES AND/OR RESOLUTION

IFOV: 30 m
Swath: 11.52 km
Noise equivalent reflectance: 0.5%
Coverage with steerable mirror: >250 km
Spectral sample interval:
  VIS/near IR: 10 nm
  SWIR: 20 nm
Spectral range: 0.4–2.5 μm

COMMENTS

The critical technologies leading to the SIS-A are being developed as part of the imaging spectrometer development program. The principal development is an IR area array 64 elements square, which can be butted, or packed, on two sides to form large arrays. Other critical technologies which are being pursued are coolers, optics, and on-board data processing. Feasible approaches to the cooler and optical system have been identified, and current efforts are directed toward understanding all the implications of these approaches.

BIBLIOGRAPHY

Vane, D., *Mission/orbit considerations for earth resources research aboard the space shuttle*, NASA/JPL, September 1982.

Imaging spectrometer fiscal year 1982 progress report, JPL Document 152, 1 October 1982.

Imaging Spectrometer:

Status: Candidate   **Program Phase:**   Concept Formulation
                                                            Carrier: System Z

PRINCIPAL PURPOSE AND FEATURES

To fill in detailed coverage where there are sharp transition zones in geology or biology of the surface, build up a global data set on surface geology, and nest within the coarser continuous coverage an ability to sample at finer scales in more detail.

PRINCIPAL PHYSICAL PROPERTY MEASURED, RANGE, ACCURACY, AND RESOLUTION

Spatial resolution: 30 m

ASSOCIATED RANGES AND/OR RESOLUTION

Spectral range: 0.4–2.5 μm
Spectral resolution: 10 nm

COMMENTS

Capable of looking in all directions within 45° of nadir. This instrument will be a follow-on to the Shuttle Imaging Spectrometer.

Multifrequency Lidar Facility:

Status: Candidate   **Program Phase:**   Concept Formulation
                                                            Carrier: System Z

PHYSICAL AND GENERAL PROPERTIES

Physical size (m):   Not Specified
Mass (kg):   2300
Power requirement (W_{elec}):   3000–10,000
Data rate (kbps):   2000

PRINCIPAL PURPOSE AND FEATURES

To provide principal measurements of atmospheric water vapor for the hydrologic cycle and other concerns and perform altimetry of ice and land surfaces and clouds.

PRINCIPAL PHYSICAL PROPERTY MEASURED, RANGE, ACCURACY, AND RESOLUTION

To measure cloud tops (100-m resolution) and thermodynamic phase; aerosol profiles; moisture profiles (1-km resolution); trace species profiles; wind field profiles (cloud tracking to profiles +2 to −2 m/sec, 1 km); and temperatures/pressure (+1 to −1°C, +1 to −1 mbar, 1–2 km). Will include a 1.2-m diameter, 3-m-long telescope.

SPOT*

The SPOT (Système Probatoire d'Observation de la Terre) system, conceived and designed by the French Centre National d'Études Spatiales (CNES), is being built by French industry in association with European partners Belgium and Sweden. It consists essentially of an earth observation satellite and ground stations for data reception. The satellite consists of two parts: the bus (a standard multipurpose platform) and a payload. The payload is mounted on one of the side panels of the bus. It includes the earth observation instruments and the mission telemetry package. The payload of the first SPOT satellite will consist of two identical high-resolution visible (HRV) imaging instruments and a package comprising two magnetic tape recorders and a telemetry transmitter. The complete satellite, which will weigh approximately 1750 kg at the start of its life, will operate in a circular sun-synchronous near-polar orbit (inclination, 98.7°) at an altitude of 832 km. SPOT will be launched by the Ariane launcher. Characteristic dimensions are 2 × 2 × 3.5 m for the satellite body and 15.6 m for the overall length of the deployed solar panel.

The HRV Instrument

The HRV instrument is designed to operate in either of two modes, in the visible and infrared portions of the spectrum:

*Based on information provided by SPOT Image Corporation, 1150 17th Street, NW, Washington, D.C.

A panchromatic (black-and-white) mode for observation over a broad spectrum

A multispectral (color) mode for observation in three narrower spectral bands

The instrument's sampling mesh corresponds to a ground element that is 10 m × 10 m in the first case and 20 m × 20 m in the second, for nadir viewing. This basic design decision was dictated by the small-scale subdivision of much of the agricultural land in many parts of the world and was required for cartographic applications.

Light from the scene being viewed enters the HRV instrument via a plane mirror that is steerable by ground control. The viewing axis can thus be oriented as required in the plane perpendicular to the orbit. This off-nadir viewing capability covers a range of ±27° relative to the vertical (in 45 steps of 0.6° each). This allows the instrument to image any point within a strip extending 475 km to either side of the satellite ground track. Because of the earth's curvature, the maximum viewing angle at the ground in 33° from the vertical.

For nadir viewing the two HRV instruments can be pointed to cover adjacent fields. In this configuration the total swath width is 117 km and the two fields overlap by 3 km. Since the distance between adjacent ground tracks at the equator is approximately 108 km, complete earth coverage can be obtained with this fixed setting of the instrument fields. The pattern of successive ground tracks is repeated exactly at 26-day intervals.

By appropriately selecting the orientation of the pointing mirror it is possible to observe any region of interest within a 950-km-wide strip centered on the satellite ground track. The width of the swath actually observed varies between 60 km for nadir viewing and 80 km for extreme off-nadir viewing.

The HRV instrument has a 4.13° field of view. In the multispectral mode each line of the image has 3000 pixels. The three spectral bands are 0.50–0.59, 0.61–0.68 and 0.79–0.89 μm. The pixel coding format is 3 × 8 bits. In the panchromatic mode the spectral band is 0.51–0.73 μm and each line has 6000 pixels. The pixel coding format is 6 bits. Transmission employs DPCM (Digital Pulse Code Modulation) a data compression scheme that does not degrade the radiometric accuracy of the image data (256 gray levels). The detectors are CCD (charge coupled device) type. Each array consists of 6000 elements arrayed linearly. This forms a pushbroom scanner since it images a complete line of the ground scene in the cross track direction in one look without any mechanical scanning.

An image-receiving station can receive satellite telemetry when the satellite nadir is within 2600 km. The bit rate is 50 Mbps for the two HRV instruments, using a bandwidth of 100 MHz within the 8–8.4 GHz band. The basic unit for segmenting the image data stream at the ground receiving and preprocessing stations will be the scene which corresponds to the totality of image data for an area 60 km in length (along-track) and 60–80 km in width (cross-track). The basic image data must be processed in a number of ways to make it more directly usable. The correction processes applied include:

Radiometric corrections taking into account the calibration factors for the detectors and the optical and telemetry systems

TABLE 17. EOS Instruments

Instrument	Measurement	Spatial Resolution	Coverage
1. Automated Data Collection and Location System (ADCLS)	Data and command relay and location of remotely sited measurement devices	Location to 1 km for buoys, to 1 m for ice sheet packages	Global, twice daily
SISP—Surface Imaging and Sounding Package			
2. Moderate Resolution Imaging Spectrometer (MODIS)	Surface and cloud imaging in the visible and infrared 0.4–2.2 nm, 3–5 μm, 8–14 μm resolution varying from 10 nm to 0.5 μm	1 km $\times$ 1 km pixels (4 km $\times$ 4 km open ocean)	Global, every 2 days during daytime plus IR nightime
3. High Resolution Imaging Spectrometer (HIRIS)	Surface imaging 0.4–2.2 nm, 10–20 nm spectral resolution	30 m $\times$ 30 m pixels	Pointable to specific targets, 50 km swath width
4. High Resolution Multifrequency Microwave Radiometer (HMMR)	1–94 GHz passive microwave images in several bands	1 km at 36.5 GHz	Global, every 2 days
5. Lidar Atmospheric Sounder and Altimeter (LASA)	Visible and near infrared laser backscattering to measure atmospheric water vapor, surface topography, atmospheric scattering properties	Vertical resolution of 1 km, surface topography to 3-m vertical resolution every 3 km over land	Global, daily atmospheric sounding; continental topography total in 5 years
SAM—Sensing with Active Microwaves			
6. Synthetic Aperture Radar (SAR)	L, C, and X band radar images of land, ocean, and ice surfaces at multiple incidence angles	30 m $\times$ 30 m pixels	200-km swath width daily coverage in regions of shifting sea ice

TABLE 17. EOS Instruments—*Continued*

Instrument	Measurement	Spatial Resolution	Coverage
7. Radar Altimeter	Surface topography of oceans and ice, significant wave height	10 cm in elevation over oceans	Global with precisely repeating ground tracks every 10 days
8. Scatterometer	Sea surface wind stress to 1 m/sec, 10° in direction Ku band radar	One sample at least every 50 km	Global, every 2 days
APACM — Atmospheric Physical and Chemical Monitor			
9. Doppler Lidar	Tropospheric winds to 1 m/sec Doppler shift in laser backscatter	1 km vertical, 2° longitude, 2° latitude	Global, twice daily, surface to 100 mb
10. Upper Atmosphere Wind Interferometers	Upper atmospheric winds to 5 m/sec, Doppler shift in O_2 thermal emissions	3 km vertical, 2° longitude, 2° latitude	Global, daily
11. Tropospheric Composition Monitors	Trace chemical constituents of the troposphere	Varies from total column density to 1 km vertical, from 1° to 0.1° horizontal	Global, daily, surface to 100 mb
12. Upper Atmosphere Composition Monitors	Trace chemical composition passive emission detectors at wavelengths from UV to microwave	3 km vertical 2° longitude, 2° latitude	Tropopause to 120 km global daily day and night coverage
13. Energy and Particle Monitors	Solar emissions from 150 to 400 nm, 1-nm spectral resolution; earth radiation budget Total solar irradiance Particles and fields environment	Total solar output	Roughly continuous sampling, at least twice daily for solar observations

116

Geometric corrections to take account of the viewing angle, earth rotation, and so on.

CNES and the Institut Geographique National are jointly planning to set up a Space Image Rectification Center to archive raw data received by the Toulouse station and to carry out additional standard image data-processing operations.

Earth Observing System

The Earth Observing System (EOS) is a planned NASA program which will support U.S. multidisciplinary earth science studies employing a variety of remote sensing techniques. It will do so in the 1990s as a prime mission using the space station polar platform. Its primary goal is the generation of long-term earth science data sets of measurements in the areas of agriculture, forestry, geology, hydrology, oceanography, snow and ice, troposphere and upper atmosphere chemistry, radiation, and dynamics pertaining to global studies of the earth as a system, emphasizing the interactions and couplings of the atmosphere–ocean–land–cryosphere system. These results will provide the information needed to improve our understanding of the global hydrological cycle, global biochemical cycles, and global climate processes. (See Table 17.)

Chapter **2**

Processing and Analysis Techniques

Eight sections, each covering some aspect of processing and analysis of remotely sensed digital imagery, constitute Chapter 2 of this book. All are somewhat off the beaten track. The reader is assumed to already be familiar with gray scale adjustment, registration, Bayesian and maximum likelihood classification, edge detection, Fourier and Hadamard transforms, template matching, and all of the many image processing topics normally covered in introductory texts on digital image processing. For that reader Chapter 2 is intended to provide insights on how some persistent problems can be attacked in novel ways. Included are algorithms on texture extraction, converting SAR data into imagery, multispectral classification using spatial context information, handling boundary pixels (a problem aggrevated by misregistered channels), assessing the accuracy of map products, and so forth.

In seeking a satisfactory compromise between breadth and depth, some topics are covered in less detail than would be preferred by practitioners of those topics. Rather, the intent was to be expository and thought-provoking for purposes of cross-fertilization.

2.1 DIGITAL PROCESSING OF SYNTHETIC APERTURE RADAR DATA

The conversion of Synthetic Aperture Radar (SAR) data into digital imagery is a computationally intense process, one that many experienced in the algorithms for exploiting remotely sensed imagery are unacquainted with. The algorithm described in this section is one of many that have been developed at various sites around the world. This particular algorithm was developed at NASA Ames Research Center, Mountain View, California. Others are in use at the Jet Propulsion Laboratory, Pasadena, California, the Canadian Center for Remote Sensing, Ottawa, and recently at the NASA Goddard Space Flight Center, Greenbelt, Maryland, using their 16,000+ channel Massively Parallel Processor.

The described algorithm is not represented as the best of breed. Some argue that optical processing, the method used to make images from much of the early SAR data, has advantages over any form of digital processing. Rather, the described algorithm exhibits the essential elements of any digital algorithm and conveys the rationale for all such algorithms being so computationally intensive. As computers get faster and SAR data becomes more commonplace, the algorithms can be expected to continue to evolve.

Synthetic Aperture Radar Concepts*

Consider a radar on board an aircraft flying in a line at an altitude H meters above the ground. The radar antenna is oriented broadside to the aircraft and transmits down and to the side of the aircraft at right angles to the flightpath to illuminate objects on the ground. The distance between an object on the ground and the line of flight as measured in the plane containing the object and perpendicular to the line of flight is called the slant range and will be denoted by R. The direction parallel to the line of flight is called the azimuth direction. The SAR imaging geometry is illustrated in Figure 56.

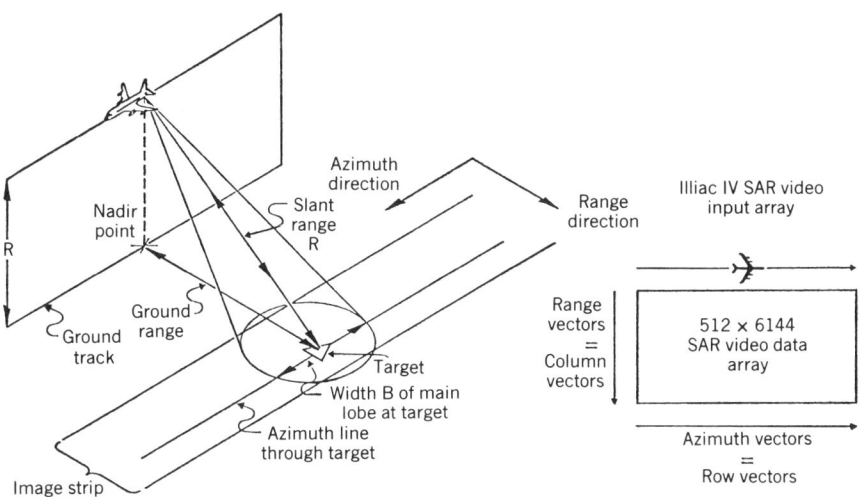

FIGURE 56. SAR imaging geometry and diagram of input data array.

*The next two sections are based on Charles N. Johnson, *Digital Processing of Synthetic Aperture Radar Data on the Illiac IV,* NASA Ames Research Center, June 1978.

The radar echoes received from radar pulses transmitted at regular intervals along the flightpath can be processed to obtain a radar image of a strip on the ground that is parallel to the ground track (nadir) of the aircraft. The image correlates the radar brightness of illuminated objects with their slant range and azimuth positions.

The resolution in the slant range direction is a function of the time–bandwidth product (TBP) of the transmitted pulses. High resolution in the slant range direction can be achieved by high-powered pulses of short duration or by pulses of longer duration and moderate power that have a linear FM modulation applied to the carrier frequency. Radars of the latter type are known as "chirped" radars.

The return signal of a chirped radar has to be processed by a matched filter or some other equivalent process in order to compress the signal and obtain a correlation of power with slant range that is equivalent in resolution to an unmodulated pulse of much shorter duration and higher power. All of the radars considered here are the chirped variety.

The resolution along the ground in the slant range direction is given by

$$p_g = p_r/\sqrt{1 - (H/R)^2}$$

where p_r is the resolution in slant range and is a constant, R is the slant range, and H is the height of the aircraft above ground.

The formula for p_g shows that the ground range resolution is very poor for objects having only a small angular separation from the nadir but improves rapidly as the angular separation from the nadir increases. As the line of sight in the range direction between the radar and the illuminated object approaches the horizontal, the ground range resolution becomes nearly equal to the slant range resolution.

By displaying the echo return of a pulse as a line image, where the distance along the line is correlated with the slant range, a radar image can be made by placing the line images of successive echoes alongside one another to form a strip. The azimuth resolution for an image formed in this way is generally not very good since it is limited by the width of the radar beam's main lobe. The width B of the main lobe is given approximately by

$$B = 2\lambda R/D$$

Where λ is the wavelength of the carrier, R is the slant range, and D is the width of the antenna in the azimuth direction. From the formula for B we see that a large value for the ratio D/λ is needed in order to achieve high azimuth resolution. Unfortunately, the ratio D/λ for most airborne radar systems is limited to values of about 100. This means that at a slant range of 10 km, the azimuth resolution cannot be expected to be much better than 200 m.

By recording in analog or digital form both the phase and the amplitude histories of the return echoes of an airborne radar, it is possible by processing the phase and amplitude information to obtain an image with a much higher resolution in the azimuth direction. A SAR is an airborne radar where the phase and amplitude information from the return signals are processed to obtain a radar image having a high resolution in both range and azimuth.

A stationary object on the ground that is being imaged by a SAR is illuminated in succession by a large number of individual pulses. This number (for main lobe illumination) is approximately equal to the width of the main lobe of the beam at the object divided by the distance the radar moves between pulses. Each pulse reflected back by the object is received by the antenna at different positions along the line of flight. Because the phase and amplitude histories of the received signals are recorded and stored, the signals reflected back by the object can be processed to simulate the signal that would have been produced if all the signals reflected from the object had been reflected simultaneously and had been received by a long linear array of antennas placed along the line of flight, the length of the array being approximately equal to the width of the beam's main lobe along the object's azimuth line. The long array of antennas is equivalent to a single antenna of the same length. It is from this synthetic antenna that the SAR gets its name.

The azimuth resolution p_a of the image produced by the SAR at slant range R is approximately equal to the width of the main lobe B_s of the synthetic beam at slant range R. Therefore, we find that p_a is approximately equal to

$$2\lambda R/B_s = 2\lambda R/(2\lambda R/D) = D$$

Where D is the length of the real antenna in the azimuth direction. It is an interesting fact that the azimuth resolution of the SAR is essentially independent of both the slant range R and the wavelength λ of the carrier.

The reader is directed to references 1 and 2 for a more detailed exposition of SAR concepts. A nice expository article on airborne SAR can be found in the October 1977 issue of *Scientific American*.[3]

Digital Processing of SAR Data

By SAR video data we mean SAR signal data after demodulation with the carrier. SAR imagery can be produced by recording the SAR video data on film and then processing it by optical means or by digitizing the SAR video data and processing them digitally. We will discuss only the second method and refer the reader to references 3 and 4 for details of optical processing.

We now assume that the SAR video data are in digital form and that each sample is an 8-bit word. We will regard the sequence of words produced by sampling the return signal of a single pulse as a vector. We will call all such vectors range vectors. We will also assume that N of these vectors are processed at a time to obtain an image of a piece of the strip illuminated by the radar. We will refer to the image produced by N consecutive range vectors as a frame. The entire strip image can be obtained by piecing frames together.

If each range vector has M elements, then we can view the digital SAR data as an $M \times N$ array. The row vectors of such an array will be called azimuth vectors. The position of each element in the array is correlated with a specific slant range and azimuth position.

One method for determining the amount of signal received from the position correlated with the array location (i, j) is to (1) compute (up to a scale factor) the expected value of the elements in a subarray containing (i, j) that would have been produced by a point target located at that position and then (2) multiply the elements of the subarray by the values of the corresponding elements for the point target and then sum the products. The probability that a radar point target is located at the position correlated with (i, j) is proportional to the absolute magnitude of the resulting number. The entire array could be processed in this way to obtain the image array. This method is in most cases not a practical way to process SAR data on a digital computer due to the large number of computations required.

We will now describe a procedure for processing SAR data that is equivalent to the method above but which requires much less computation. First, for each range vector, we compute the Fourier transform vector. The components of the transform vector are then multiplied by the corresponding components of the range filter vector. The inverse Fourier transform is then applied to the resulting vector to obtain a new range vector which replaces the old one. This process is called range correlation and is equivalent to the convolution of the original range vector with the complex conjugate of the vector corresponding to the chirped transmitted pulse after demodulation with the carrier. The Fourier transform was used in place of doing the convolution because by using a fast Fourier transform (FFT) algorithm to perform the transforms, less computation is required.

Each azimuth vector in the new array has associated with its row index a specific slant range, as before, but now the data in the row are correlated with the same slant range as well. This means that the data for a point target at a slant range corresponding to the row index i will be contained within row i.

The second phase of the process is to correlate the azimuth data with azimuth position. Each azimuth vector is now processed in the same way as the row vectors. The matched azimuth filter vector depends on slant range. Normally the coefficients of the azimuth filter vector change slowly with respect to change in the slant range, so that the azimuth filter vector needs to be recomputed only once every k vectors.

A digital radar map can be obtained from the array of correlated azimuth vectors by replacing each element by its amplitude squared. The array may be further processed to correct for known geometric distortions and to normalize the power by computing the average power for elements of a row and then dividing each of the elements in the row by the average. A visual image can be produced from the array on a digital image recorder.

SAR Program

During 1977, the NASA Ames Institute for Advanced Computation (IAC) in cooperation with the Jet Propulsion Laboratory and Lockheed Missiles and Space Company developed an experimental SAR data-processing program that has run successfully a number of times. The SAR processing program was developed for the purpose of gaining experience with digital techniques for processing SAR data that might be applicable to the NASA Seasat program. Another reason for developing the program was to be able to experiment with various algorithms for azimuth correlation of SAR data.

The program was designed to be highly modular so that various parts could be easily changed. The input to the program is a 512 × 6144 array of SAR video data. Each column of the array is a range vector of length 512.

Each data element in the array is an 8-bit word. The program processes the data using Fourier techniques as discussed. There are three main parts to the program:

Part I. First, the input data are converted from an 8-bit representation to a 64-bit complex number representation with 32-bit real and imaginary parts. The imaginary part is initially set to zero. Secondly, all the range vectors are correlated using a matched filter.

Part II. The 512 × 6144 array is broken into blocks of size 64 × 64 and the data in each block are transposed.

Part III. The azimuth vectors are correlated in blocks of 16. The azimuth filter vector is recomputed every 8 azimuth vectors. After an azimuth vector is correlated, the squared amplitude of each of its components is computed and converted to an 8-bit representation. The output is a 512 × 6144 array of 8-bit words.

The Illiac IV run time for this program is 55 seconds. An additional 6 minutes is required for data movement from the external system to and from I4DM,* so the total time is 7 minutes.

For an airborne radar operating at a frequency of approximately 1300 MHz (L band), a 512 × 6144 array of SAR video data corresponds to an image area of approximately 1.8 km × 2.4 km. Thus, a large amount of digital data and a lot of computation are required to produce a high-resolution image covering a modest area. It is for this reason that most SAR data have until now been processed optically.

REFERENCES

1. Cutrona, L. J., Synthetic aperture radar, in *Radar handbook,* ed. M. I. Skolnik, McGraw-Hill, New York, 1970.

2. Moore, R. K., Microwave remote sensors, in *Manual of remote sensing,* Ed. R. L. Janza, American Society of Photogrametry, Falls Church, Va., 1975.

3. Jensen, H. et al., Side-looking airborne radar, *Scientific American,* October 1977, pp. 84–95.

4. Harger, R. O., *Synthetic aperture radar systems theory and design,* Academic Press, New York, 1970.

*I4DM is an acronym for Illiac IV Disk Memory.

2.2 COASTAL ZONE COLOR SCANNER ALGORITHM COMPONENTS

This section describes government processing of raw Coastal Zone Color Scanner (CZCS) data, including manual steps for screening out scenes that are totally cloud-covered or land only. The processing is done by the NASA Goddard Space Flight Center and includes segmenting into standard length scenes, geographically locating pixels, preparing header information, and performing a variety of calibration steps.

The calibration efforts have generally been disappointing. An internal calibration lamp and blackbody provide data to the telemetry stream, and calibration/correction processes are applied as described here. But serious work that requires accurate absolute values rather than relative values is today done in conjunction with sea surface and subsurface sampling.

There are two reasons for including this material here. The first is to inform the user about how the data he acquires differs from raw observations. If processes have been applied which interfere with his investigations, he may have to consider a deprocessing step. The second, more general reason for this section is to provide an appreciation of what elements constitute preprocessing, how time-consuming it is, and how accurate it can be expected to be.

The government's Coastal Zone Color Scanner program processes CZCS data at two levels, as diagrammed in figures 57 and 58. The operations performed in the boxes of these flowcharts are described in correspondingly identified paragraph.

Level I Processing*

I.1. *Raw data* consisting of computer compatible digital counts (8-bit words) from all channels, plus housekeeping and calibration data (cal lamp counts, voltage reference staircase, tilt, gain, timing data, etc.).

I.2. *M1:* A man-interactive screening of the data on a color CRT display, the purpose of which is to eliminate segments of the data which either are totally cloud covered and/or show only land areas. Selection would be made by the operator using either a light pen or cursor to delineate image segments to be processed or eliminated. The operator might also block image segments into standard-length scene files for convenient process scheduling.

I.3. *Image location data* including spacecraft ephemeris (subsatellite point) and three-axis pointing angles to be used in locating pixels geographically across each scan. It would be convenient to include solar zenith and azimuth angles in the image location data base, but these angles can also be calculated internally to I.4.

I.4. *Pixel location:* Using image location data (I.3) provided externally, the geographic position (latitude & longitude), viewing zenith and azimuth

*NASA, Goddard Space Flight Center, Greenbelt, Maryland.

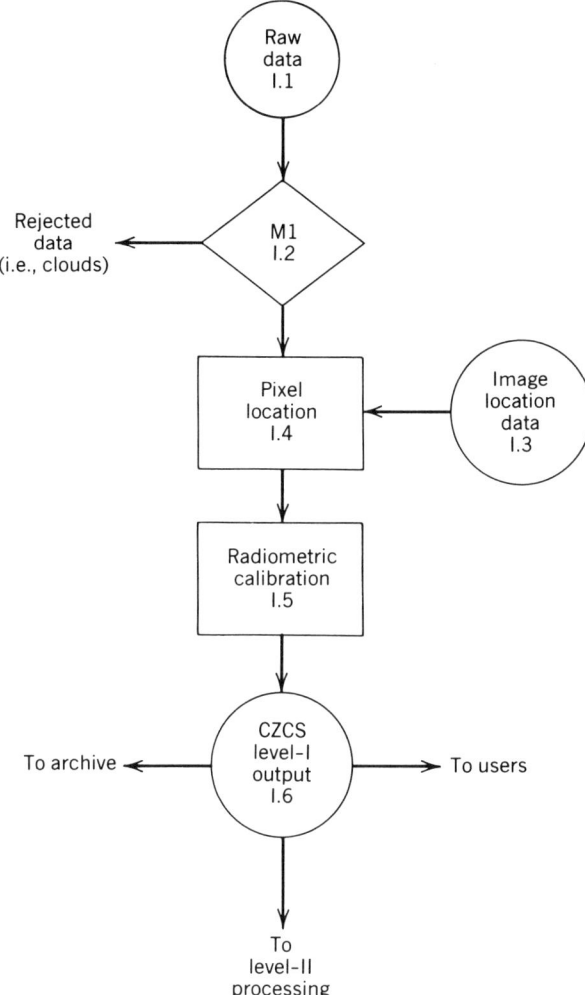

FIGURE 57. CZCS, level I.

angles, and solar zenith and azimuth angles will be calculated at enough pixels across each scan to permit interpolation to a specified precision at each pixel in the scan. The location and geometric data will be inserted in a header record for each scan line.

I.5. *Radiometric calibration:* The voltage staircase and internal visible calibration source (test lamp) will be tested against a priori (current) expected values, and if it is found to differ, new calibration coefficients will be determined, using least squares, to convert counts to radiance in the visible and near infrared channels. A similar procedure using the CZCS internal calibration blackbody source will be used to convert counts to equivalent blackbody temperature in the thermal infrared channel(s).

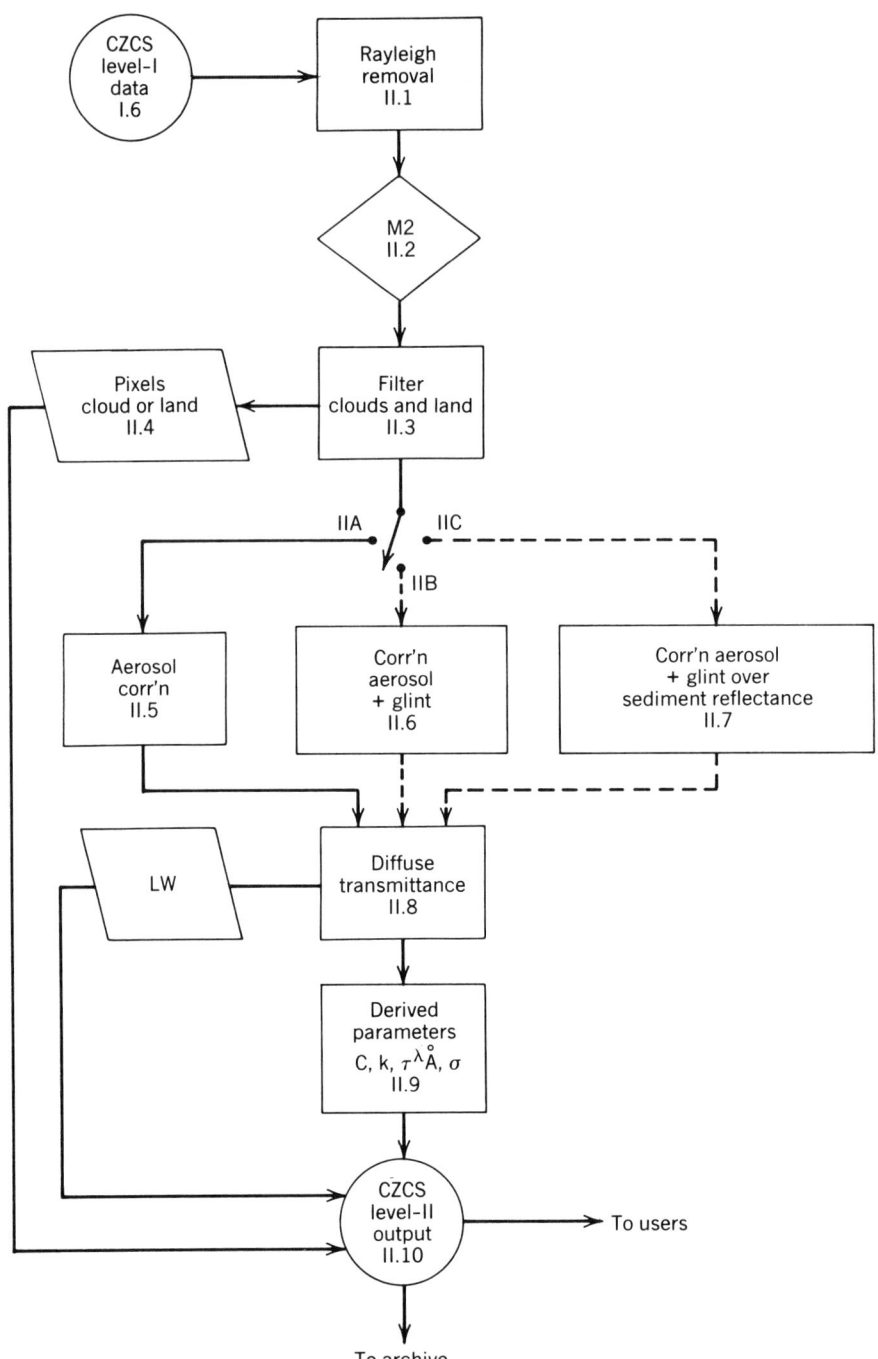

FIGURE 58. CZCS, level II.

I.6. *CZCS level I output* will consist of a header record (location and geometric parameters plus housekeeping data), calibrated aperture radiance for the visible and near IR channels, calibrated equivalent blackbody temperature for the thermal IR channel(s), and raw counts for a version of one near IR channel to be utilized as a cloud/land filter in level II and III processing.

Level II Processing

II.1. *Rayleigh removal:* A single scattering estimate of Rayleigh backscattered radiance will be computed for enough points in each image to allow interpolation to all pixels with a precision to be specified. Typically, points every degree of latitude and every 5° of scan should suffice for this purpose. Rayleigh radiance estimates are then subtracted from aperture radiance (level I) in each visible and near IR channel at each pixel. The cloud/land mask and thermal infrared data are passed through this step untouched.

II.2. *M2:* A second man-interactive data screening to ascertain which correction pathway will be followed for each image segment. The operator will use a light pen or cursor and automated programs to:
   i. Select pixels (10 to 20) from which data will be used in the sun glint algorithm (II.6).
   ii. Block out areas of high in-water reflectance in near IR channels due to suspended sediments.
   iii. Determine or verify the threshold (water radiance level) to be used in the cloud/land mask (II.3).

II.3. *Cloud/land filter:* Raw counts in the low-gain near IR channel will be thresholded (II.2) to zero out any pixels which are open water. All nonzero pixels in this channel are thus automatically classified as cloud or land. For pixels thus flagged, the Rayleigh-removed radiances are passed through directly to the level II output (II.10).

II.4. *Cloud/land pixels:* An arbitrary scaling and false color combination of the Rayleigh-removed radiance can be easily made to display land in earth tones (browns) and clouds white. This provides a "natural" masking in any subsequent level III processing by users, should they desire it. Automated spectral cloud-vs.-land discrimination schemes can also be formulated using the Rayleigh removed radiances, but these will not be implemented in the National Oceanic Satellite System Picture Processing Facility (NOSS PPF) operational data-processing algorithms.

II.5. *Aerosol correction:* This branch is used to correct open-water pixels in image segments for which sun glint has been determined *not* to be a significant factor (in II.2) and where high suspended sediment reflectance in the near IR is not a factor. On a pixel-by-pixel basis, this algorithm computes estimates of upwelled radiance just beneath the sea surface as

$$L_w^\lambda = t^\lambda[L^\lambda - L_R^\lambda - r^{\lambda,\lambda_0}(L^{\lambda_0} - L_R^{\lambda_0})]$$

where t^λ is combined diffuse transmittance of the atmosphere and radiance transmittance through the sea surface (applied in II.8, below) at wavelength λ, L_R^λ is Rayleigh radiance (applied in II.1 above), λ_0 is a near-IR wavelength where the ocean is assumed to be black (so that $L^{\lambda_0} - L_R^{\lambda_0}$ is an estimate of radiance contributions due to the presence of atmospheric aerosols), and r^{λ,λ_0} is a scaling factor defined in terms of ratios of specified values for incident solar fluxes, aerosol optical thicknesses and phase functions, and ozone transmittance at the two wavelengths. The use of "typical" open-water upwelled radiances, in conjunction with pixels selected in M2 (II.2, above), is a possible candidate method for refining r^{λ,λ_0} in each image.

II.6. *Correction for aerosol and sun glint:* Where sun and scan geometry indicate that sun glint may contaminate an image segment, the algorithm II.5 above must be modified to account for glint contributions and aerosol radiance simultaneously. A candidate method, which involves a three-iteration solution of an exponential equation for aerosol optical thickness $\tau_A^{\lambda_0}$ and the standard deviation of wave slopes σ, is under study using Nimbus-7 CZCS data.

II.7. *Correction for aerosol and sun glint in presence of sediment reflectance:* In regions where suspended sediments make the ocean nonblack at λ_0 algorithms II.5 and II.6 won't work. Such areas can be detected as part of the screening in M2 (II.2, above), but a method for correcting the data in such regions has not yet been developed or validated.

II.8. *Diffuse transmittance:* This is the point where t^λ in II.5, II.6, and II.7 is actually applied to the data to obtain L_w^λ, which values are passed to the level II output and the derived parameter module.

II.9. *Derived parameters:* Chlorophyll-*a* (+ phaeopigments-*a*) C and diffuse attenuation coefficients k are estimated as functions of ratios of L_w^λ. In Nimbus-7 versions, these equations are of the form, for example,

$$C = a(L_w^{\lambda_1}/L_w^{\lambda_2})^b$$

If the II.6 branch path is used, $\tau_A^{\lambda_0}$ and σ are computed for each pixel by the algorithm. If not, the σ is not determined and $\tau_A^{\lambda_0}$ is estimated from $L^{\lambda_0} - L_R^{\lambda_0}$.

II.10. *Level II output:* For each scan line there are a header record containing location and housekeeping data and then for each pixel a data vector containing

$$\text{Output} = \begin{cases} L_w^{\lambda_1}, \ldots, L_w^{\lambda_6}, f^{\lambda_7}, T_s, C, k, \tau_A^{\lambda_7} & \text{for water pixels} \\ (L^{\lambda_1} - L_R^{\lambda_1}), \ldots, (L^{\lambda_6} - L_R^{\lambda_6}), f^{\lambda_7} & \text{(5 blank words) for} \\ & \text{cloud/land pixels} \end{cases}$$

(For computational time estimates, see Table 18.)

TABLE 18. CZCS Computational Time Estimates (360/91 min/orbit)

Algorithm Step[a]	Man Interaction $t_s + (t_p)$[b]	360/91 CPU Time Maximum/Average[c]	Estimated Core Size (kbytes)
I.2	5 + (10)	—	—
I.4	—	28/17	400
I.5	—	67.5/40.5	400
II.1	—	27.5/16.5	400
II.2	5 + (25)	—	—
II.3 + II.5 @ 60% = 26[d]	—	—	—
II.6 @ 30% = 37.5	—	76.5/46	350
II.7 @ 10% = 12.5	—	—	—
II.8 + II.9	—	41/24.5	350
Subtotals:	10 + (35)	240.5/144.5	
Cumulative throughput time per orbit:		250.5/154.5	

[a]No time loads have been assigned for input/output (I/O) operations, on the assumption that asynchronous I/O will be used to avoid impact on throughput.

[b]Man-interactive step times are expressed as $t_s + (t_p)$, where t_s is time in sequence with CPU time (and therefore affects throughput), whereas (t_p) is time for operations which can be performed in parallel with CPU processing of previous image segments from the same orbit (and therefore does not impact data throughput).

[c]*Maximum* CPU estimates are based on a full 25 minutes of CZCS data per orbit, whereas *average* CPU estimates assume an average rejection of image segments due to excessive cloud cover at a rate of 40%.

[d]Computational percentage distribution assumes that 10% of image segments from each orbit are contaminated with excessive sediment reflectance, 30% are sun-glint-contaminated, and 60% require correction only for aerosol radiance.

2.3 TEXTURE EXTRACTION: MAX-MIN AND SPATIAL DEPENDENCY*

Two texture extraction algorithms are described. The MAX-MIN algorithm was developed at Purdue and has been employed more by the military image-processing community than by the civilian sector. The spatial dependency or cooccurrence algorithm, on the other hand, seems to be the texture algorithm of choice in the earth resources image analysis domain. Both have their uses.

It has been shown that classification on the basis of textural signatures alone is not, in general, as successful as classification on the basis of spectral signatures alone; classification based on both types of information is generally better than using either alone.

These two algorithms are selected from a large literature on texture analysis. These were chosen in part due to their popularity in one subdiscipline or another, but also because they are both amenable to parallel implementation, a character-

*Based on a report, R. M. Brown and M. J. Hannah, *Texture Extraction on the ILLIAC IV*, NASA/Ames Research Center, 15 May 1979.

istic of growing importance. Some other texture analysis literature is included in the references.

Introduction

Texture is an innate characteristic of all scenes. It can be described with words like *smooth, coarse, linear,* and *random,* and most people will have a good idea of the meaning. Texture contains important information about scenes and, except for color, may have more information content than any other characteristic. Texture is a two-dimensional statistical measure on an image, and a good texture extraction algorithm will be able to separate texture information that has one orientation from that with another.

Another property of texture is that is characterizes the image in aggregate. Because it does have a statistical nature, it cannot measure the image at the pixel level, but must combine a number of pixels into a window and then characterize the texture of the window. The size, shape, and location of the window is of importance in determining the texture. The window size is of particular importance for, if the window is too small, the statistics gathered have little significance, while if the window is too large, the process may combine adjacent, but disparate, textural regions.

The Texture Algorithms

Texture is one of many properties of a scene which can be used to segment the scene into different objects, to identify the objects pictured, and to identify identical points in separate images in the scene. Because of the computational intensity involved in the extraction of texture information, texture has not, in the past, been a substantial element in many pattern recognition or scene classification applications. However, this is changing.

Mitchell, Myers, and Boyne[1] have proposed and developed a new technique for image texture analysis, which they call MAX-MIN. The algorithm used is computationally simple and can be implemented in hardware for real-time analysis. Their algorithm was based on an intuitive determination that important texture information is contained in the relative frequency of local extremes in the intensity.

Mitchell et al. characterize the texture of the image at a given point by counting the number of extrema in a window centered at that point. Through the use of a thresholding parameter, they have introduced a hysteresis-like smoothing to ignore reversals of amplitude smaller than the threshold.

Haralick, Shanmugam, and Dinstein[2] have presented a different procedure for extracting textural properties from windows of image data. This procedure consists of first computing spatial dependence probability distribution matrices from the gray levels in the image, then calculating a set of statistical textural features from these matrices. The result is information on properties such as the homogeneity, gray level linear structure, contrast, number and nature of boundaries present, and complexity of the image.

Calculation of the spatial dependence (SPADEP) matrices is a simple but data-intensive computational task. Given a scan direction and a distance to neighbor, the output for each window of data is a matrix containing frequency of adjacency counts between pixel values, that is, each matrix element $M(I,J)$ is the count of the number of times that a pixel of gray level I was adjacent to a pixel of gray level J in the given scan direction at the given neighbor distance.

Input Images

In order to describe the implemented algorithms further, it is first useful to describe the data. Figure 59 illustrates typical image dimensions. Figure 60 shows a window of size WS surrounding a match point (X_m, Y_m). As shown in the expanded picture of the window, each window is square and symmetrically located around the match point in question. This means that the window size (WS) is limited to the odd integers between 3 and 127. In the production runs, a window size of 55 was used for both textures.

Scan Directions

One of the features of these texture processes is their ability to extract the texture information along four separate directions. These directions, shown in Figure 61, are denoted by the angle which the scan direction makes with the positive x axis, measured counterclockwise. Note that in all the directions, the windows of data being examined by the algorithms are identical and that therefore the scans along the diagonal directions have differing lengths. Also notice that, while every pixel in the window is examined in the 0° and 90° angle scans, this is not true in the 45° and 135° scans. Because of this, caution must be used when comparing data taken along diagonal scans to those taken along axial scans.

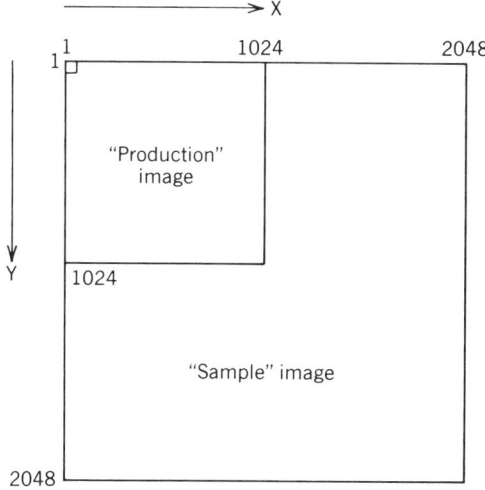

FIGURE 59. Input images.

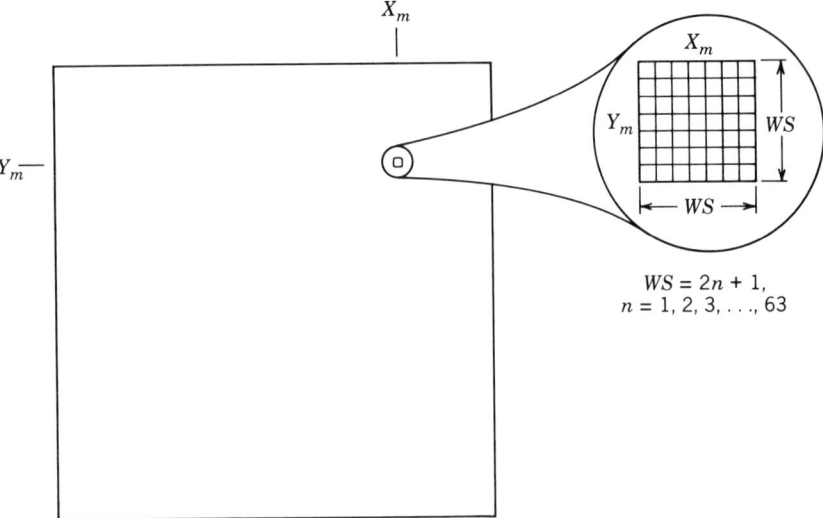

FIGURE 60. Window of size *WS*.

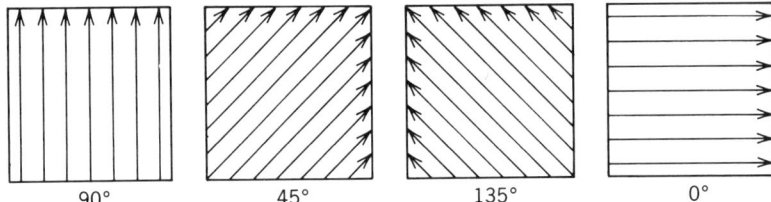

FIGURE 61. Scan directions.

Data Conversion

I4MXMN and I4SPDP* were designed to be tolerant of recognized erroneous data in the images, in that such data are ignored in the texture-extraction process. The key, of course, is the detection of the erroneous data. With these algorithms this is done quite simply. Incoming images are recorded at 6 bits/pixel, so the maximum value that a pixel can have is 63. However, computers are most effective when dealing with data having 8, 32, or 64 bits/pixel. For convenience in coding, and for compatibility with existing software, a standard pixel size of 8 bits was chosen, giving a possible range of 0 to 255 for the pixel value. To protect the texture extraction processes against possible garbling of the data, the data conversion function shown in Figure 62 was used. Note that this function leaves all legal values of the pixel intensity (0–63) unchanged, while converting all other values to 255. This value (8 bits of all ones) is recognized by the MAX-MIN and SPADEP algorithms as being invalid.

*I4MXMN is the name of the computer program that performs MAX-MIN algorithms; I4SPDP is the name of the computer program that performs spatial dependency algorithms.

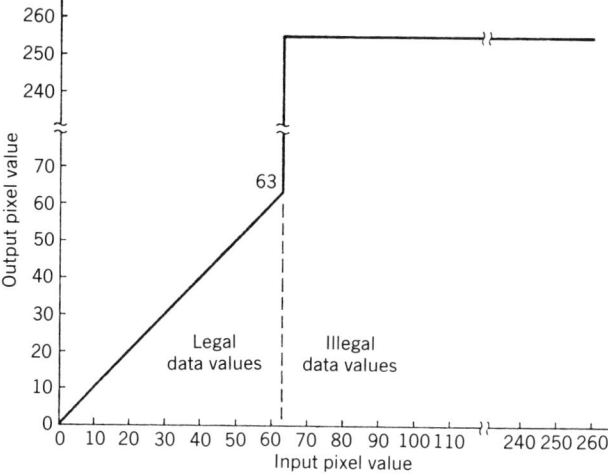

FIGURE 62. Separation of legal from illegal data.

Algorithm Descriptions

In this section we present a more detailed look at the algorithms.

The MAX-MIN Algorithm

As mentioned above, the MAX-MIN process proposed by Mitchell, Myers, and Boyne attempts to characterize the texture of the image at a given point by counting the number of extrema located in a window centered about that point. To do this, a number of parallel scans are made through the window, and the number of extrema counted (according to a given criterion) are summed together for all scan lines to give an overall measure of the texture at that point.

The variables involved in the processes are:

1. The coordinates of the point in question
2. The size of the window area to be examined
3. The direction of the scans made through the image
4. The threshold value that determines whether an extremum is large enough to be counted.

A logical formulation of the MAX-MIN algorithm is shown further on. This formulation is a computer-language-like description of the process shown in Figure 63. In this figure, the pixel values along the scan are shown in the heavy lines, while the lighter lines that bound the pixel values denote the upper and lower bounds of the hysteresis limits that provide a criterion for the MAX-MIN counting process. The arrows at the top indicate instances where the MAX-MIN threshold has reversed direction.

The hysteresis limits are set as follows. The upper limit is initially given the value of the starting pixel. The lower limit is simply equal to the upper limit minus the threshold. Now, going along the scan, the upper limit will be increased

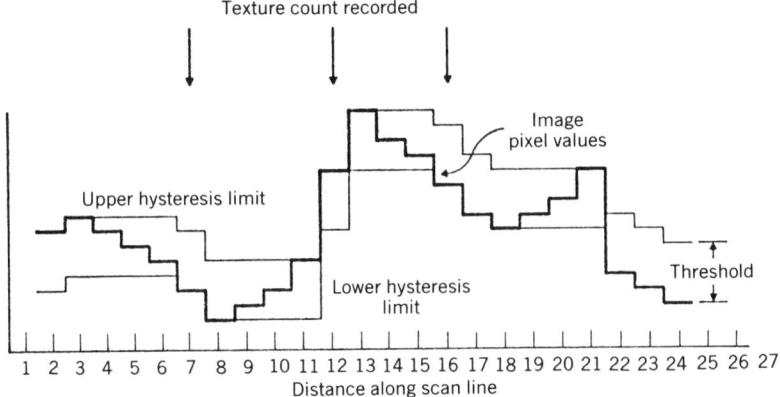

FIGURE 63. Operation of the MAX-MIN algorithm.

(pulling the lower limit up with it) as required to contain the pixel information. When the pixel value drops below the lower limit, the lower limit will be reduced (pulling the upper limit down with it) as required to contain the pixel information. At each reversal of the direction of "drift" of the interval, one threshold crossing count will be registered. The process continues counting alternate decreases and increases of the thresholds until the end of the scan line is reached. Note that a count is not recorded if the pixel value merely equals the limit; a pixel must exceed the limit to register a count.

In the example shown, the threshold has a value of 4. In the MAX-MIN implementation, however, the threshold can be any value between 1 and 63. On any given run of MAX-MIN, up to 16 measurements (corresponding to different thresholds) can be computed for each point in each direction. Up to four directions can be specified from the set of 0°, 45°, 90°, and 135°. The program will process up to 4096 match points per run, where each point forms the center of a data window of the given window size, which can range from 3 to 127 pixels square.

The SPADEP Algorithm

As previously mentioned, the algorithm described in Haralick, Shanmugam, and Dinstein consists of two parts — computing the spatial dependence probability distribution matrices and calculating a set of statistical textural features from these matrices.

The variables involved in the I4SPDP process are:

1. The coordinates of the point in question
2. The size of the window area to be examined
3. The direction of the scans made through the image
4. The distance to the neighboring pixels to be examined

A logical formulation of the I4SPDP algorithm is contained below in the form of a computer-language-like description of the process. The algorithm is essentially a histogramming technique, which counts in the matrix element (A, B) the number of times that a pixel of value A is adjacent to a pixel of value B in the given direction and at the given neighbor distance.

The algorithm described in Haralick et al. calls for information to be extracted in both orientations along each direction vector specified. However, the information produced in the positive direction along any vector is the transpose of that produced in the negative direction. Consider the 0° direction vector. For every A with a B to its right [which increments $M(A, B)$], there will be a B with an A to its left [which will increment $M(B, A)$]. Recognizing the symmetry, we have simplified the task. Instead of calculating the adjacencies for both directions of a vector, we calculate only the matrix for one of the directions and then add this matrix to its transpose to produce the desired result.

SPADEP can process up to 4096 match points in a given run, with information extracted in up to four directions. Each point forms the center of a data window of the given window size, which can range from 3 to 127 pixels square. The matrices can be extracted at neighbor distances of up to 8 pixels. The input data can be up to 8 bits (256 gray levels); the resulting matrices can be up to 64 × 64 in size, corresponding to 6-bit output data.

MAX-MIN Example

Figure 64 shows the pixel data for a 15 × 15 window. Using the threshold value 10, the MAX-MIN texture was extracted for the 45° and 90° scan directions. Both the manual and program-driven texture processes determined that there were 18 threshold crossing counts in each direction. The dots in Figure 64 indicate the locations where the hysteresis-filtered extrema were manually detected when scanning in the 90° direction.

```
26   32.  35.  35.  36.  39.  39.  40.  39.  44   45   42   41   41.  41
38.  41   45   45   46   47   46   45   45   49   50   47   46   44   43
45   47   48   48   47   48   49   50   50   52   51   50   50   49   47
48   49   49   49   49   51   52   51   50   50   51   52   51   52   51
49   50   49   49   51   53   52   50   49   49   50   50   49   50   48
49   50   50   51   51   53   50   49   51   49   50   47   45   46   41
49   47   49   50   52   52   49   49   49   49   46   44   41.  38.  34
49   48   48   48   50   49   48   45   45   45   41   38   35   30   30
46   44   42   41   41   45   41   38   33   36   40   35   27.  27.  30
40.  36.  36   34   35   40   33   30   27   29   33   37   37   30   35
28   25   28   30   29   39   38   30   30   31   34   40   39   33   39
25   27.  25   28   32   39   43   35   33   35   36   42   42   36   39
30.  32   29   29   34   42   43   35   34   35   37   39   38   33   33
39   37   35   32   33   42   42   33   33   31   36   40   39   35   34
42   39   33   32   32   39   38   33   35   37   41   41   42   39   38
```

Pixel values about point 1474,1767 (RECO1)

FIGURE 64. Pixel values about point 1474,1767 (RECO1).

MAX-MIN Code Details

The variables used in the description of the algorithm are as follows:

Integers	Description
i	Current index along scan line, $1 \leq i \leq S(j)$
j	Current index of scan line, $1 \leq j \leq J$
J	Number of scan lines of length greater than 1
$S(j)$	Length of the jth scan line
n	Index of the current match point, $1 \leq n \leq NP$
NP	Number of match points to be examined
Z	Number of scan directions to be examined
z	Index of the current scan direction, $1 \leq z \leq Z$
NT	Number of threshold crossings to be examined, $1 \leq NT \leq 16$
t	Index of the current threshold crossing, $1 \leq t \leq NT$
$V(j,i)$	Initial pixel value, $0 \leq V(j,i) \leq 255$
$L(j,i)$	Checked pixel value: $= V(j,i)$ if $0 \leq V(j,i) \leq 63$ $= 255$ if $V(j,i)$ undefined or >63
$C(z,j,t)$	Number of threshold crossings detected on the jth scan line
$M(z,t)$	Output matrix whose components are the sum over j from 1 to J of $C(z,j,t)$
$D(j,i)$	Direction of hysteresis: $= 1$ if $L(j,i)$ last passed through the top of the hysteresis range $= 0$ if $L(j,i)$ last passed through the bottom of the hysteresis range

Real Number	Description
$T(t)$	Value of the threshold parameter, $1 \leq T(t) \leq 63.0$
$A(j,i)$	Value of the top of the threshold range
$B(j,i)$	Value of the bottom of the threshold range

The texture extraction algorithm is as follows:

```
For n ← 1, step 1 until NP do begin "loop on n"
    Locate the window associated with the match point
    For z ← 1, step 1 until Z do begin "loop on z"
        Calculate J, S(j)
        For t ← 1, step 1 until NT do begin "loop on t"
            For j ← 1, step 1 until J do begin "loop on j"
                C(z, j, t) ← 0
                M(z, t) ← 0
```

$D(j, 1) \leftarrow 1$
If $L(j, 1) = 255$, then $L(j, 1) \leftarrow 0$
$A(j, 1) \leftarrow L(j, 1)$
  $B(j, 1) \leftarrow L(j, 1) - T(t)$
For $i \leftarrow 2$, step 1 until $S(j)$ do begin "loop on i"
    If $L(j, i) = 255$, then
    (the data point is bad)
    begin $A(j, i) \leftarrow A(j, i - 1)$
          $B(j, i) \leftarrow B(j, i - 1)$
          $D(j, i) \leftarrow D(j, i - 1)$
    end
    If $255 > L(j, i) > A(j, i - 1)$, then
    (the point is above the upper threshold)
    begin $A(j, i) \leftarrow L(j, i)$
          $B(j, i) \leftarrow L(j, i) - T(t)$
          $D(j, i) \leftarrow 1$
          $C(z, j, t) \leftarrow C(z, j, t) + 1 - D(j, i - 1)$
    end
    If $0 \leq L(j, i) < B(j, i - 1)$, then
    (the point is below the lower threshold)
    begin $A(j, i) \leftarrow L(j, i) + T(t)$
          $B(j, i) \leftarrow L(j, i)$
          $C(z, j, t) \leftarrow C(z, j, t) + D(j, i - 1)$
    end
    If $A(j, i - 1) \geq L(j, i) \geq B(j, i - 1)$ then
    (the point is within the threshold limits)
    begin $A(j, i) \leftarrow A(j, i - 1)$
          $B(j, i) \leftarrow B(j, i - 1)$
          $D(j, i) \leftarrow D(j, i - 1)$
    end
  End "loop on i"
 End "loop on j"
$M(z, t) \leftarrow$ Sum over j from 1 to J of $C(z, j, t)$
 End "loop on t"
End "loop on z"
Output coordinates of match point plus components of $M(z, t)$
End "loop on n"

SPADEP Code Details

The variables used in the description of the algorithm are as follows:

Integer Arrays	Description
DATA(1:127, 1:127)	Window of pixel data
DI, DJ(1:4)	Vectors to neighbors, one for each scan direction; vectors are of length NDIST

SPADEP(1:4, 0.63, 0:63)   Counters for spatial dependency measure; can do
                                 4 directions, up to 6-bit data

Integer	Description
BADCT	Counter for invalid points in window
CENT	Pixel value of current data point
I, J, K	Loop indices for row, column, and direction
II, JJ	Position of current neighbor
NABOR	Pixel value of current neighbor
NDIR	Number of scan directions
NDIST	Distance to neighbors, from 1 to 8 pixels
WINSIZ	Window diameter

The texture extraction algorithm is as follows:
FOR each image block DO BEGIN "BLOCK-LOOP"
   Initialize for block, get data, preprocess as needed;
   FOR each match point in this block DO BEGIN "MATCH-LOOP"
      DATA(1:WINSIZ, 1:WINSIZ) ← window of data centered at match point;
      BADCT ← 0; (Initialize counter for invalid points in window)
      SPADEP(*, *, *,) ← 0; (and spatial dependence matrices)
      FOR I ← 1 STEP 1 UNTIL WINSIZ DO BEGIN "I-LOOP"
        FOR J ← 1 STEP 1 UNTIL WINSIZ DO BEGIN "J-LOOP"
      (For each data point in the window, get point's value)
          CENT ← DATA(I, J);
          IF CENT = 255, THEN BADCT ← BADCT + 1, ELSE BEGIN
          (If point is invalid, count it and skip to next point,
            otherwise process the data)
            FOR K ← 1 STEP 1 UNTIL NDIR DO BEGIN "K-LOOP"
              (For each direction around this point)
              II ← I + DI(K);
              JJ ← J + DJ(K);
              NABOR ← DATA(II, JJ);
              IF 1 ≤ II ≤ WINSIZ AND
                1 ≤ JJ ≤ WINSIZ AND
              NABOR NEQ 255 THEN
              (If the neighboring point is in the window and
              it is a valid point, count the adjacency)
              SPADEP(K, CENT, NABOR) ← SPADEP
                          (K, CENT, NABOR) + 1;
          END "K-LOOP";
        END;
      END "J-LOOP";
    END "I-LOOP";

Form the true spatial dependence matrices by adding each matrix
  to its transpose;
Put together matrix headers for this match point's data;
Move data to its place in the output area;
  END "MATCH-LOOP";
END "BLOCK-LOOP":
Move output data set to PDP-10;

REFERENCES

1. Mitchell, O. R., C. R. Myers, and W. Boyne, A MAX-MIN measure for image texture analysis, *IEEE Trans. Comp.* C-26, 1977: 408–414.
2. Haralick, R. M., K. Shanmugam, and I. Dinstein, Textural features for image classification, *IEEE Trans. Systems, Man, Cyb.* SMC-3(6), 1973:

BIBLIOGRAPHY

Carton, E. J., J. S. Weszka, and A. Rosenfeld, *Some Basic Texture Analysis Techniques,* Technical Report TR-288, University of Maryland Computer Science Center, College Park, Md., 1976.

Conners, R., M. Trevedi, and C. Harlow, Segmentation of a high resolution urban scene using texture operators, *Computer Vision, Graphics, and Image Processing* 25(3), 1984:

Hsu, S. Y., The Mahalanobis classifier with the generalized inverse approach for automated analysis of imagery texture data, *Computer Graphics and Image Processing* 9, 1979: 117.

Shen, H. C., and A. K. C. Wong, Generalized texture representation and metric, *Computer Vision, Graphics, and Image Processing* 23(2), 1983:

Thompson, W. B., *The role of texture in computerized scene analysis,* U.S.C. IPI Report 550 University of Southern California, 1974.

2.4 SELECTING BAND COMBINATIONS WITH THEMATIC MAPPER DATA*

This section addresses the important but rarely discussed topic of how to enhance digital satellite imagery to optimize visual interpretability.

Background

Since the human eye employs three primary colors and the Thematic Mapper returns seven bands of data, one obvious problem that arises in making color-

*Based on a note by Charles Sheffield, Presented at IEEE AIPR WORKSHOP, 1983. Reproduced with permission from *Photogrammetric Engineering and Remote Sensing*. Copyright 1985 by the American Society of Photogrammetry.

composite images is the choice of bands. The choice is nontrivial, since three bands can be selected from 7 in 35 ways. Also, any band can be assigned any color. This gives a total of 210 different possible color presentations of TM three-band images. In this section we present a way of reducing that 210 to a single choice, decided uniquely by the statistics of a scene or subscene and taking full account of any correlations that exist between different bands.

We should remark here that one well-known and widely used approach to this problem of choice is through the use of principal component images. However, such methods offer a new problem as great as the one that they solve. For although the first three principal components contain in a statistical sense as much information as can be presented using three colors, the resulting scene is completely data-dependent. It is thus difficult for an interpreter to apply any previous experience of color–surface relationships to the analysis of a principal components image.

Definition of the Method

Consider the 7×7 variance–covariance matrix M for the scene or subscene, ignoring for the moment the fact that the thermal band is of inherently lower resolution than the rest. Any triplet of bands will be represented within this 7×7 matrix by a 3×3 submatrix.

Considering now the three-dimensional subspace spanned by any particular band triplet, the associated variance–covariance matrix defines an ellipsoid within the subspace. Further, the sum of the squared principal axes of this ellipsoid represents the total variance accounted for by these three bands (see Figure 65). One could plausibly (but as we shall see, wrongly) argue that the best three bands are those that have the largest sum of squared principal axes and hence that account for the largest total variance. This is, after all, exactly the argument applied in employing principal component images. Since the trace of a matrix is invariant under rotational transformations, and since the sum of squared principal axes is equal to that trace, the band triplet that accounts for the greatest possible variance can be

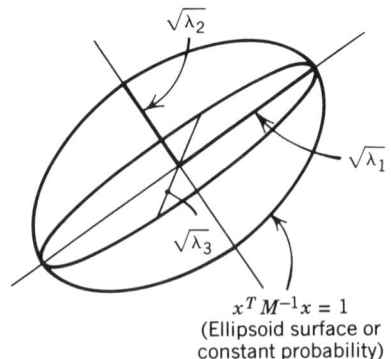

$$x^T M^{-1} x = 1$$
(Ellipsoid surface or
constant probability)

FIGURE 65. The variance–covariance ellipsoid.

found from the original variance–covariance matrix simply by selecting the three bands with the largest diagonal elements. There is no need to examine all 35 band combinations.

To see what is wrong with this approach, consider an extreme case where there happens to be perfect correlation between a pair of bands. For convenience, suppose that those bands are 1 and 2, and suppose that the variance of band 1 (and therefore of 2) is larger than that of any other band. The 7×7 matrix M then has the form

$$\begin{bmatrix} a & a & \vdots & \cdot & \cdot & \cdot & \cdot & \cdot \\ a & a & \vdots & \cdot & \cdot & \cdot & \cdot & \cdot \\ \hline \cdot & \cdot & & b & & & & \\ \cdot & \cdot & & & c & & & \\ \cdot & \cdot & & & & \ddots & & \\ \cdot & \cdot & & & & & \ddots & \\ \cdot & \cdot & & & & & & \ddots \end{bmatrix}$$

where $a > b, c, \ldots$.

The rotation matrix that will diagonalize the upper left 2×2 submatrix then has the form

$$\left(\begin{array}{cc|c} 1/\sqrt{2} & 1/\sqrt{2} & \\ -1/\sqrt{2} & 1/\sqrt{2} & \quad 0 \\ \hline & 0 & I \end{array} \right)$$

and thus *after* rotation the upper left 2×2 submatrix will have the form

$$\begin{pmatrix} 2a & 0 \\ 0 & 0 \end{pmatrix}$$

As expected, one eigenvalue is zero; but the other is the sum of the variances from the original bands 1 and 2. Since a is assumed to be large, both bands 1 and 2 will be included in the triplet that accounts for maximum variance—despite the fact that if either one of them is used, adding the other contributes no new information.

The problem lies in the use of *total variance* as the measure for the information content of the band triplets. This is equivalent to use of the sum of squares of ellipsoid principal axes, and there is no penalty associated with a very small principal axis provided that it occurs in association with a large axis (see Figs. 66 and 67), as was the case for the above example.

We propose the use of a new measure for the information content of the triplet, one that avoids the undesirable property demonstrated above. We will select the

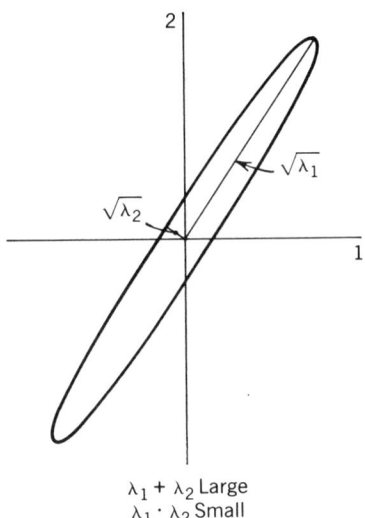

$\lambda_1 + \lambda_2$ Large
$\lambda_1 \cdot \lambda_2$ Small

FIGURE 66. High correlation, bands 1 and 2.

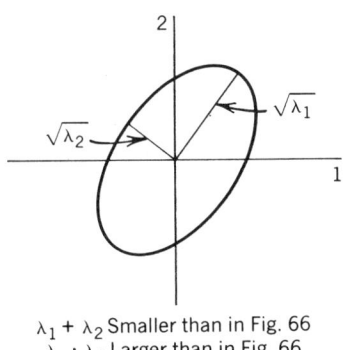

$\lambda_1 + \lambda_2$ Smaller than in Fig. 66
$\lambda_1 \cdot \lambda_2$ Larger than in Fig. 66

FIGURE 67. Low correlation, bands 1 and 2, but lower individual variances.

ellipsoid of *maximum volume*. This discourages selection of pairs of bands with high correlation, since in such cases one eigenvalue will be close to zero and the corresponding ellipsoid volume will be small.

Since the ellipsoid volume is simply $\frac{4}{3}$. πabc, where a, b, and c are the principal axes of the ellipsoid, the volume of the ellipsoid associated with a particular band triplet is a constant multiple of the square root of the product of the eigenvalues for the 3×3 variance–covariance matrix of that triplet. However, under rotational transformation the product of the eigenvalues is equal to the determinant of the original 3×3 submatrix. Thus we can select the band triplet that provides the ellipsoid of maximum volume simply by computing and ranking in order the determinants of each 3×3 principal submatrix of the original matrix M. The band triplets associated with these determinants will then be ranked in order of decreasing overall information content. Given the original matrix M, the total computation to achieve this ranking is trivial. It requires a few hundred multiplications, followed by a sort of a list of 35 items. A BASIC program to perform this is given at the end of this section.

This procedure gives the best triplet, but the assignment of colors is still to be made. Now we can make use of the actual variances (the diagonal elements of M). Since the eye is most sensitive to green, next to red, and least to blue, we will assign green to the band triplet member of maximum variance (i.e., most variation within the image), red to the triplet member of second largest variance, and blue to the triplet member of smallest variance. The definition of bands for production of a color image is now complete.

Examples and Comments

The procedure has been applied to a number of scenes of very different ground cover, including Washington, D.C., Death Valley, and Cement, Oklahoma. The results for Washington and for Death Valley are given in Tables 19–22, together with the associated variance–covariance matrices. The following comments apply to all scenes studied to date.

1. The band combination $1, 4, 5$ (in the order blue, red, green) is usually, but not always, the selected triplet. In cases where it does not rank first, it ranks second or third.

2. The natural color combination $1, 2, 3$ and the standard false color combination $2, 3, 4$ both place far down in the rankings. In the case of Washington, the natural color combination is 29th (lower than anything except some thermal band combinations, which are low for another reason to be discussed shortly); the $2, 3, 4$ combination was ranked in 16th place. For Death Valley, the $1, 2, 3$ natural color combination ranked 32nd, and the $2, 3, 4$ combination just above it, at 31st. This is presumably a consequence of the very high correlations between the first four bands.

3. Triplets that rank high always include either band 5 or band 6. (*Note:* the bands here are ordered by increasing wavelength, so the thermal band is

TABLE 19. *Variance–Covariance Matrix for the Washington, D.C. Scene*

53.32	27.41	35.74	5.86	36.04	33.56	7.77
27.41	17.01	21.35	11.36	29.35	21.29	4.13
35.74	21.35	31.66	20.01	46.56	31.03	6.69
5.86	11.36	20.01	131.71	131.64	38.14	8.26
36.04	29.35	46.56	131.64	210.83	86.25	19.1
33.56	21.29	31.03	38.14	86.25	50.01	11.51
7.77	4.13	6.69	8.26	19.1	11.51	9.8

Note: Selection of best three bands based on ellipsoid volume. This is the Washington scene with reduced variance on the thermal channel. Variance–covariance matrix; thermal band is band 7.

TABLE 20. Ranked Results for the Washington, D.C. Scene

Rank	Determinant	Combination
1	433858	145
2	205811	345
3	138551	146
4	124784	245
5	101638	456
6	71723	156
7	62960	346
8	49759	135
9	39992	134
10	39609	246
11	36060	356
12	22847	125
13	21953	256
14	16732	124
15	11646	235
16	9709	234
17	7967	136
18	5094	457
19	4752	157
20	3634	126
21	3606	147
22	2294	467
23	2194	357
24	1945	347
25	1616	236
26	1386	567
27	1348	257
28	1130	247
29	727	123
30	688	167
31	276	367
32	215	137
33	175	267
34	84	127
35	43	237

Note: Selection of best three bands based on ellipsoid volume. This is the Washington scene with reduced variance on the thermal channel.

band 7.) This emphasizes the great importance of these new bands on general information-bearing grounds.

4. The triplet selected is not always or even usually the triplet with the greatest individual variances, though large variances are naturally preferred somewhat in the selection process.

TABLE 21. Variance–Covariance Matrix for the Death Valley Scene

251.64	146.31	198.55	176.41	246.36	144.63	5.3
146.31	90.4	125.4	112.95	178.63	105.16	10.33
198.55	125.4	181.12	163.27	276.44	162.93	22.5
176.41	112.95	163.27	159.7	262.74	152.99	14.79
246.36	178.63	276.44	262.74	627.47	366.9	75.38
144.63	105.16	162.93	152.99	366.9	223.38	48.73
5.3	10.33	22.5	14.79	75.38	48.73	69.89

Note: Selection of best three bands based on ellipsoid volume. This is the Death Valley scene with reduced thermal variance. Variance–covariance matrix; thermal band is band 7.

Other Considerations and Comments

1. The statistical analysis performed here used *P* tapes in which the original histograms had already been modified by the gains and offsets. It would be preferable to work with data that have had no gains or offsets applied, that is, with *A* tapes prior to any radiometric correction. If band selection of this type becomes common, it would be nice to have *A* tapes generally available from the EROS Data Center.

2. The thermal band is of lower resolution than the rest; thus it would not be appropriate to give it the same weight in the selection process. How should one therefore deweight it? One argument runs as follows: The maximum information that a scene can contain is given by the number of pixels, since in the ultimate case there would be no correlation between pixels, and each would carry independent information about some feature of the surface. In such a case, the amount of information that the thermal band can contribute is only 1/16th that of the other bands, because there are 16 times fewer pixels in that band. Therefore one should deweight the thermal channel by a factor of 16. Such deweighting was performed in the experiments reported here. However, we should also note that this made no difference at all to the preferred band triplets, since even without deweighting we found no case where a triplet involving the thermal channel was in the top five.

3. It is obvious when one looks at images created from the triplet 1, 4, 5 that for some applications this combination will be much inferior to others, such as natural color and standard false color. This restates the old truth that one man's noise is another man's signal. However, the preferred triplets have another advantage: They provide images of unusual clarity, with far less residual striping than is seen in, for example, the natural color images.

4. Although combinations such as 1, 4, 5 produce images that are at first sight unfamiliar and unusual, the assigned colors are not scene-dependent. Thus in contrast to the scene-dependent colors of principal component or ratio images, the interpreter quickly learns to associate colors with particular ground condition. We therefore believe that there are definite advantages to seeking

TABLE 22. Ranked Results for the Death Valley Scene

Rank	Determinant	Combination
1	1462581	145
2	859695	156
3	684248	135
4	601687	146
5	432952	345
6	346425	157
7	328331	356
8	319827	245
9	275534	456
10	263989	136
11	219239	256
12	204146	125
13	167450	346
14	137060	357
15	127643	246
16	121117	167
17	107494	457
18	103781	235
19	89506	257
20	76827	126
21	75913	134
22	49163	367
23	40621	467
24	39230	236
25	37614	147
26	31621	267
27	21579	137
28	21322	124
29	20256	567
30	9168	347
31	8118	234
32	7895	123
33	7197	247
34	5037	127
35	2407	237

Note: Selection of best three bands based on ellipsoid volume. This is the Death Valley scene with reduced thermal variance.

color composites from the original bands rather than through band ratios or band combinations.

The Best-Band Program

```
20  PRINT "SELECTION OF BEST THREE BANDS BASED ON ELLIPSOID
    VOLUME"
30  PRINT "DEATH VALLEY WITH REDUCED THERMAL VARIANCE"
40  DIM R(36), Q(36)
```

```
50   DIM U(36), V(36)
60   DIM M(8, 8)
```

70 REMARK: M is the variance-covariance matrix for the scene or subscene.
80 REMARK: The arrays R, Q, U and V are storage arrays used in the program.
90 REMARK: Note that the program assumes that band 7 is the thermal data, and band 6 is the 2.2-micrometer data.
100 REMARK: The instructions 190 to 230 (except for 220, which sets a count) reduce the variance of the thermal channel to allow for the lower spatial resolution of the thermal channel pixel.

```
190    FOR I = 1 TO 6
200  M(I, 7) = M(I, 7) / 4
210    NEXT
220  C = 1
230  M(7, 7) = M(7, 7) / 16
240    PRINT "RANK          DETERMINANT          COMBINATION"
250    FOR I = 1 TO 5
260    FOR J = I + 1 TO 6
270    FOR K = J + 1 TO 7
280  D1 = M(I, I) * (M(J, J) * M(K, K) − M(J, K) − 2)
290  D2 = M(I, J) * (M(J, K) * M(I, K) − M(I, J) * M(K, K))
300  D3 = M(I, K) * (M(I, J) * M(J, K) − M(I, K) * M(J, J))
310  DT = D1 + D2 + D3
```

315 REMARK: The next instruction makes the determinant an integer; this is not necessary, it is done for convenience of output only.

```
320  DT = INT (DT)
330  N = 100 * I + 10 * J + K
340  R(C) = DT:Q(C) = N
350  C = C + 1
360    NEXT
370    NEXT
380    NEXT
```

385 REMARK: The next piece of code sorts the determinant into descending order.

```
390    FOR I = 1 TO 35
400  N = 0
410    FOR J = 1 TO 35
420    IF R(I) | R(J) THEN 440
430  N = N + 1
440    NEXT
450  U(N) = R(I):V(N) = Q(I)
460    NEXT
470    FOR I = 1 TO 35
480    PRINT I, U(I), V(I)
490    NEXT
500    PR# 0
510    END
```

2.5 HADAMARD TRANSFORM IMAGE MATCHING*

Image matching is, of course, a very timely topic. It is an important subdiscipline in what has today come to be called multisensor data fusion, the synergistic employment of information obtained from diverse sensors, at various times, and from different perspectives to resolve a remote sensing exploitation issue.

This section describes a novel method for image matching. Many practitioners have used Hadamard transforms, but usually Hadamard transforms are used for data compression.

A task that is becoming increasingly investigated in the field of digital image processing is that of image matching. Here either one wishes to establish whether any two given images are, by and large, identical, or one is interested in determining whether a given subimage forms a portion of a larger image. The latter problem seems to have commanded somewhat more attention, 1–3 and it is investigated here, too. In either case, a rapid image-matching technique is required. One of the main reasons for the new interest in digital image matching stems from the fact that earth satellite imagery, used for temporal change detection, is so abundant. The objectives of matching procedures are (a) to establish whether two images taken at different times cover a similar area and then (b) to highlight the differences, if any, between them. The work to be described is more directly related to the first objective.

It should be noted that while the establishment of match criteria as such is of importance, the fact remains that digital image matching operations can be extremely time-consuming, especially if the images are large. Thus in reference 1 the objective has been to establish a threshold for matching such that if during the point-by-point comparison (the points being chosen at random) the cumulative error exceeds this threshold, a mismatch is deemed to have occurred. Unlike conventional correlation techniques, in which all image points on both images are investigated, the rationale here is to spend less time in determining whether or not a match exists. A similar attitude is taken in reference 3, where points within the images to be examined first are those whose associated gray level probabilities are minimal. Here, too, some cutoff threshold beyond which no further investigation is to be performed is required. Both approaches comprise two subproblems: that of selecting the image points to be considered in the first instance and that of establishing the match–mismatch threshold. The work described here falls into the domain of the first subproblem.

Because it is costly in terms of computer operations to determine whether or not a match exists between images and because images can be considered to consist of slowly varying gray level patterns, it would appear useful initially to reconstruct the images in question so that they consist of small uniform regions, thus giving information on global rather than on a local level. One way of accomplishing this is to break up the images into small squares in which the gray level is constant.

*Based on E. S. Deutsch and Z. F. Wan, *On the use of the Hadamard transform in digital image matching,* Technical Report 216, Computer Science Center, University of Maryland, January 1973.

This will have the effect of reducing the difference errors due to localized detail. Thus by employing such preprocessing for matching, a coarse match can be established first, whereas finer matching can be resorted to subsequently. This approach reduces the number of operations required to produce a match.

Low-pass filtering in the Hadamard domain is particularly useful for our approach, since the original images can be reconstructed into the required form much more cheaply than by the usual averaging method. The use of the Hadamard transform also affords a convenient flexibility in obtaining the reconstructed images, as will be seen.

We have departed somewhat from the traditional use of the Hadamard transform. Unlike most applications, in which the fidelity of the reconstructed image is important, the aim here is to deliberately degrade the images so that they become void of local detail. Matching on a global scale facilitates the use of sampling, which leads to the reduction in the operations required.

In some matching applications, especially where satellite imagery is concerned, one or both of the images may contain noise. The noise characteristics of each of the images might not always be the same; this is particularly true of the latter of the two types of noise considered here, namely random noise and herringbone noise. It has been shown empirically that low levels of random noise can be considered as detail and that by using this (filtering) approach its effect on matching performance is reduced.

It was mentioned above that the task of change detection is first to align the two images (one image of which is termed the reference image) so that they correspond in some sense. The process of extracting the dissimilarities between the preselected and now matched image portions can then follow. Note that the images do not have to be of identical size; however, it will be assumed that they are both square. The notation used below follows, essentially, that used traditionally.

Specifically, let S be the reference image of dimensions $L \times L$. Square images are used throughout for simplicity of the presentation: the argument is, however, true for a rectangular image of any size. Elements of S are denoted as $s(i,j)$ with

$$0 \leq s(i,j) \leq g$$
$$1 \leq i, j \leq L - 1$$

where g is the gray level value. For the type of imagery used here,

$$0 \leq g \leq 63$$

Let W be a square window image of dimension $M \times M$, $M < L$. Elements of W are referred to as $w(l,m)$, with

$$0 \leq w(l,m) \leq g$$
$$1 \leq l, m \leq M$$

The aim of any matching technique is to locate (if at all possible) within S the $M \times M$ subarea which matches with W. There are a total of $(L - M + 1)^2$ possible different positions within a confined area of S at which W can be wholly placed; at some such position a match might occur. See Figure 68A.

Let $S_{i,j}^M$ refer to the wholly contained $M \times M$ subarea (corresponding to the size of W) within S. Let the top-and-leftmost coordinate of $S_{i,j}^M$ be (i, j). Then, any element within $S_{i,j}^M$ is given by

$$S_{i,j}^M(l, m) = s(i + l - 1, j + m - 1)$$

such that

$$1 \le l, m \le M$$

$$1 \le i, j \le L - M + 1$$

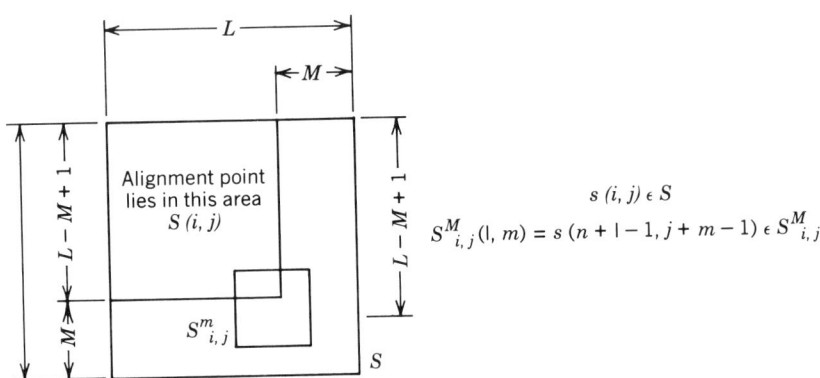

$$s(i, j) \epsilon S$$
$$S_{i,j}^M(l, m) = s(n + l - 1, j + m - 1) \epsilon S_{i,j}^M$$

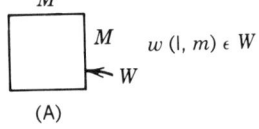

$$w(l, m) \epsilon W$$

(A)

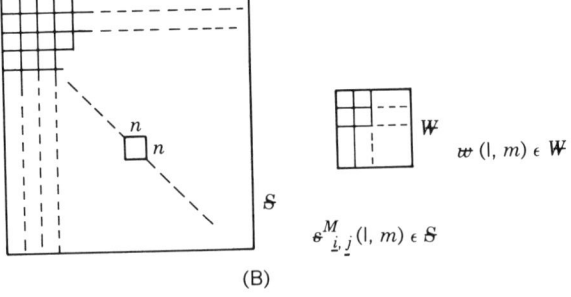

$$w(l, m) \epsilon W$$

$$s_{i,j}^M(l, m) \epsilon S$$

(B)

FIGURE 68. Nomenclature.

Using the point (i,j) as a reference point for alignment purposes on the subimage, the set of all (i,j) pairs contains all potential alignment points, subject, of course, to the last constraint given above. The techniques described below are all aimed at finding the alignment point, denoted by (i^*,j^*); that is, they attempt to find the picture point (i,j) in S which will satisfy the equation

$$(i,j) = (i^*,j^*) \quad \text{such that} \quad S^M_{i^*,j^*} = W$$

Since it is undesirable to examine all (i,j) positions within S, a transformation is applied to both S and W, yielding the new images $\mathscr{S}$ and $\mathscr{W}$. $\mathscr{S}$ and $\mathscr{W}$ are obtained by low-pass-filtering the two-dimensional Hadamard transform[7] of S and W, respectively. $\mathscr{S}$ and $\mathscr{W}$ now consist of $(L/n)^2$ and $(M/n)^2$ $n \times n$ subsquares respectively, the gray level in each subsquare being the same. See Figure 68B. It is hence necessary to examine only one point within each $n \times n$ subsquare; these points (i,j) form the new set of potential match points, the coarse match point derived therefrom being denoted by (i^*,j^*). Correspondingly, elements of $S^m_{i,j}$ are now given by

$$\mathscr{S}^M_{i,j}(l,m) = s(i + l - 1, j + m - 1)$$

$$1 \le l, m \le M$$

where

$$s(i + Kn, j + K'n) = s(i + Kn + 1, j + K'n + 1)$$

$$\vdots$$

$$= s(i + n(K + 1) - 1, j + n(K' + 1) - 1)$$

where $K, K' = 0, 1, 2, \ldots, L/n - 1$. Elements of $\mathscr{W}$ are given by $w(l,m)$ such that

$$w(Kn + 1, K'n + 1) = (Kn + 1, K'n + 2)$$

$$\vdots$$

$$= w(n(K + 1), n(K' + 1))$$

where $K, K' = 0, 1, 2, \ldots, M/n - 1$.

Thus using $\mathscr{S}$ and $\mathscr{W}$ instead of S and W in the matching procedure requires that only a subset of the set of all potential alignment points (i,j) has to be examined. This may result in the true alignment point $(i,j) = (i^*,j^*)$ being missed altogether; however, the approach is justifiable if a match point (i^*,j^*) situated very close to the true match point using $S \times W$ if found. Thus (i,j) will indicate the rough location of the match position of W in S provided the radial distance between the coarsely located match point and the true match points, r, given by

$$r^2 = (i^* - i^*)^2 + (j^* - j^*)^2$$

is minimal, so that if

$$S_{i^*,j^*}^M = W$$

then

$$S_{i^*,j^*}^M = W$$

Given (i^*,j^*), (i^*,j^*) can then be found exactly using S and W.

Conclusions

The use of reconstructed images derived via filtering of the Hagamard transform in matching applications has been investigated. It is concluded that there are considerable advantages in this approach in that:

1. On the average, fewer computer operations are required to establish a match.
2. The error in the location of the coordinates of the match position are greatly reduced.

Using such imagery can also reduce transmission costs for matching purposes, since only the gray levels for each subsquare within the imagery need be transmitted.

REFERENCES

1. Barnea, D. I., and H. F. Silverman, A class of algorithms for fast digital image registration *Trans. IEEE Computers*, C-21(2), 1972: 179–186.
2. Pope, E. J., J. L. Sedwick, and D. S. McCormack, *Image correlation and sampling study*, Report No. MDC-A1740 (NAS 5-21662), McDonnel Aircraft Corporation, Palo Alto, Calif.
3. Nagel, R., and A. Rosenfeld, Ordered search techniques in template matching, *Proc. IEEE* (letters) 60, 1972: 242–244.
4. Goldstein, A. J., L. D. Harmon, and A. B. Lesk, Identifications of human faces, *Proc. IEEE* 59, 1971: 748–760.
5. Andrews, H. C., *Computer techniques in image processing*, Academic Press, New York, 1970, pp. 136–137.
6. Palgen, J. J. O., I. Tamches, and E. S. Deutsch, *Weakly stationary noise filtering of satellite acquired imagery*, Technical Report No. 3 (NAS 5-21617), Allied Research Associates, Inc., Friendship Airport, Baltimore, Md. December 1971.
7. Hord, R. M., *Digital image processing of remotely sensed data*, Academic Press, New York, 1982, pp. 122–123.

2.6 HANDLING BOUNDARY PIXELS*

The USDA/NASA/NOAA Agristars research program has determined that over 65% of the small grains and corn pixels in a single Landsat MSS scene are boundary pixels, having portions of more than one field mixed together. When multiple scenes from different dates are registered (even if there is no misregistration), the fraction of boundary pixels increases. Hence, under some circumstances, processing an image means handling boundary pixels. Traditionally, this problem has challenged analysts; this section reports one of the approaches to a solution. Other approaches are in the literature; write Agristars, Johnson Space Center, SK–Documentation Manager, Houston, Texas 77058.

Introduction

Multispectral classification of Landsat imagery has been employed for some time to label pixels by land use category, say vegetated vs. nonvegetated. Since the ground area of a Landsat pixel may be partly vegetated and partly nonvegetated — for example, when the pixel falls on an agricultural field boundary — some pixels cannot accurately be labeled as falling into either category. The boundary pixels are generally a biased portion of the scene, and therefore they are potential error sources in proportion estimation. In this analysis the boundary pixels are divided into three classes. The relative size of each class is tabulated, and several methods are proposed for estimating the proportion of small grains in the set of boundary pixels.

Terms such as *border, edge, mixed,* and *misregistered* have been used to describe various classes of boundary pixels. All these terms, as well as the term *boundary pixel,* have ambiguous and confusing definitions.

In this section, the boundary pixels as observed in the spatial clustering algorithms described in reference 1 are considered. To define the boundary pixels, each acquisition to be processed is stratified into vegetated and nonvegetated subsets. After this stratification is complete, interior and boundary pixels of each subset are found. The interior pixels of these subsets are the pixels which have the property that all four horizontal and vertical neighbors belong to the same subset as the pixel in question. On a single acquisition, the boundary pixels are simply the noninterior pixels. The complete set of boundary pixels considered in this section are the pixels which fail to be interior pixels on any of the acquisitions used.

Types of Boundary Pixels

The data used in this study consisted of eight Landsat segments taken from the northern and southern U.S. Great Plains during crop years 1977 and 1978. The segments and acquisitions used are listed in Table 23.

---

*Based on T. B. Dennis, *Investigation of boundary pixel handling procedures,* Report No. JSC-16838, SR-LO-04013, NASA Johnson Space Center, Houston, Tex., December 1980.

TABLE 23. Segment Locations and Acquisitions

Segment	Crop Year	County, State	APU[a]	Acquisition Julian Dates
1005	1977 or phase III	Cheyenne, Colorado	10	76254 76326 77159 77177
1059	1977 or phase III	Ochiltree, Texas	9	76307 76325 77121 77157
1520	1977 or phase III	Big Stone, Minnesota	19	77120 77156 77174
1803	1977 or phase III	Shannon, South Dakota	17	76255 77123 77159 77178
1003	1978 or transition year	Adams, Colorado	10	77268 77304 78138 78227
1047	1978 or transition year	Stanton, Kansas	9	77266 78081 78117 78135
1154	1978 or transition year	Jones, South Dakota	17	78190 78226
1380	1978 or transition year	Redwood, Minnesota	15	78115 78169 78204 78241

[a]APU, acquisition parcel units.

The boundary pixels in this data set were then divided into three categories as follows:

1. The perimeter boundary pixels are boundary pixels that are easily associated with a vegetated or nonvegetated field and lie in the perimeter of that field. These can be easily identified in an automated procedure because they are simply the boundary pixels which are neighboring on an interior pixel.

2. The linear boundary pixels are boundary pixels that are not readily identified with any field, but tend to lie in a line falling between two fields. To determine which boundary pixels are linear in nature, the perimeter

boundary pixels are first removed from the general class of boundaries, and then the maximal connected subsets of the remaining boundary pixels are found. (In reference 1, it is stated that a set of pixels is connected if for each pair of points p and q in the set there exists a sequence $\{X_i\}_{i=0}^n$ of points in the set such that $X_0 = p$, $X_n = q$, and X_i and X_{i-1} are either horizontal or vertical neighbors for $i > 0$.) The linear boundary pixels are those that belong to connected subsets of the nonperimeter boundaries which have no interior pixels.

3. The remaining boundary pixels then belong to connected subsets of the non-perimeter boundaries which do have interiors. These are called interior boundaries herein, and they can be interpreted as being those boundary pixels which lie in contiguous regions large enough to represent fields.

To begin this study, each acquisition of each segment in Table 23 was treated independently. The boundary pixels in each acquisition were classified into the three categories described above, and the number of pixels in each category was recorded.

Table 24 gives the segment numbers, the acquisition numbers, the total percentage of each scene determined to be in boundary pixels, and the relative proportion of each type of boundary pixel. A cursory analysis of these data suggests that, for single acquisitions, the boundary pixels are quite manageable.

On the average, 26.8% of each scene consists of boundary pixels between vegetated and nonvegetated areas, but 78.3% of these boundary pixels are perimeter, and they are readily identified with a particular field. In fact, the boundary pixels can be made completely manageable on single acquisitions by using the boundary pixels to slightly alter the vegetation and nonvegetation stratification of each image.

This is done by the subroutine REGCL as described in reference 1, but the basic idea is that the nonperimeter boundary pixels of the vegetated subset are reassigned to the nonvegetated subset. This restratification causes a slight difference in the set of boundary pixels that are obtained. In the new set of boundary pixels, all of the nonperimeter boundary pixels are in the nonvegetated subset. These nonperimeter boundary pixels are then reassigned to the set of vegetated pixels, and this creates a vegetation and nonvegetation stratification in which the boundary pixels become almost entirely perimeter-type boundaries. Table 25 shows that, with this altered stratification, the average scene contains only 19.1% boundary pixels and that, on the average, 99.8% of these are perimeter boundary pixels and can therefore be assigned to fields.

Unfortunately, this does not solve the boundary pixel problem but only produces a method of assigning boundary pixels to groups of pure pixels on a single acquisition. The problem that remains is the difficulty introduced by misregistration of boundary pixels from acquisition to acquisition. When a boundary detected in each of two acquisitions is not registered, the set of multitemporal boundary pixels must consist of the boundary pixels on each acquisition. The boundary then grows in thickness with each additional acquisition.

TABLE 24. Boundary Pixels between Vegetated and Nonvegetated Areas of Single Acquisitions

Segment	Acquisition	Percent of Scene in Boundary	Percent of Boundary That Is Perimeter	Percent of Boundary That Is Linear	Percent of Boundary That Is Interior
1005	76254	24.2	79.7	11.3	9.0
	76326	24.9	79.9	13.8	6.3
	77159	28.0	80.6	12.4	7.0
	77177	29.0	78.4	11.1	10.5
1059	76307	24.0	76.0	13.9	10.1
	76325	22.8	76.0	15.2	8.8
	77121	26.4	73.0	13.9	13.1
	77157	25.2	78.6	12.0	9.4
1520	77120	35.1	75.8	14.7	9.5
	77156	35.8	80.2	12.2	7.6
	77174	37.4	76.8	15.4	7.8
1803	76255	27.1	73.0	14.2	12.8
	77123	16.6	78.1	12.9	9.0
	77159	19.7	75.4	13.4	11.2
	77178	22.9	75.7	16.2	8.1
1003	77268	24.6	77.4	12.7	9.9
	77304	25.0	75.9	12.8	11.3
	78138	24.5	82.5	11.0	6.5
	78227	16.7	83.1	11.3	5.6
1047	77266	24.3	83.0	10.7	6.3
	78081	20.5	88.4	8.6	3.0
	78117	22.0	86.6	9.5	3.9
	78135	25.2	86.8	9.5	3.7
1154	78190	24.7	81.3	10.7	8.0
	78226	26.7	79.4	15.0	5.6
1380	78115	38.6	60.5	15.4	24.1
	78169	30.1	79.0	14.0	7.0
	78204	36.2	72.9	13.5	13.6
	78241	38.0	76.4	13.8	9.8

As a result, the proportion of linear and interior boundaries increases as new acquisitions are considered. Table 26 shows this increase in both total and non-perimeter boundary pixels when two, three, and four acquisitions are processed. Note the decrease in the percentage of boundary pixels in the perimeter class. For processing four acquisitions, this shows that the set of boundary pixels has increased to an average of 48.8% of the scene, while the average proportion of perimeter boundaries has dropped to 66.4%. This leaves approximately 16% of each scene in the categories of linear and interior boundary pixels.

The perimeter boundary pixels can readily be assigned to fields of multitemporally pure pixels, but a considerable bias can exist in the proportions of a given

TABLE 25. Boundary Pixels between Vegetated and Nonvegetated Areas of Single Acquisitions after Restratification

Segment	Acquisition	Percent of Scene in Boundary	Percent of Boundary That Is Perimeter	Percent of Boundary That Is Linear	Percent of Boundary That Is Interior
1005	76254	18.3	99.9	0.1	0.0
	76326	17.6	99.8	0.2	0.0
	77159	20.5	99.9	0.1	0.0
	77177	20.7	99.7	0.3	0.0
1059	76307	16.6	99.9	0.1	0.0
	76325	15.4	99.8	0.2	0.0
	77121	17.1	99.8	0.2	0.0
	77157	18.1	99.9	0.1	0.0
1520	77120	24.1	99.8	0.2	0.0
	77156	26.0	99.9	0.1	0.0
	77174	26.0	99.9	0.1	0.0
1803	76255	18.3	99.8	0.2	0.0
	77123	11.9	99.9	0.1	0.0
	77159	13.2	99.9	0.1	0.0
	77178	15.3	99.8	0.2	0.0
1003	77268	17.3	99.9	0.1	0.0
	77304	17.2	99.9	0.1	0.0
	78138	18.7	99.9	0.1	0.0
	78227	12.7	100.0	0.0	0.0
1047	77266	19.0	99.9	0.1	0.0
	78081	17.2	99.9	0.1	0.0
	78117	17.7	99.8	0.2	0.0
	78135	20.6	99.9	0.1	0.0
1154	78190	18.4	99.9	0.1	0.0
	78226	19.6	99.8	0.2	0.0
1380	78115	24.6	99.8	0.2	0.0
	78169	20.8	99.8	0.2	0.0
	78204	24.6	99.7	0.3	0.0
	78241	26.6	99.7	0.3	0.0

crop type within the remaining boundary pixels when compared to the proportions within the full segment. Thus, the pure pixels themselves can be a biased subset of the scene.

Estimating Proportions

In this study, the bias in the small grains proportions among the set of boundary pixels is considered. The perimeter boundary pixels need not be considered because they are readily assigned to fields, and they can be used in estimating the small grains proportions within the pure areas. The boundary pixels under consid-

TABLE 26. Accumulation of Multitemporal Boundaries

Segment	Two acquisitions	Three acquisitions	Four acquisitions
	Percent of Scene That Is Boundary by Number of Acquisitions		
1005	31.7	43.9	53.2
1059	25.8	37.2	46.8
1520	42.4	52.9	—
1803	27.6	34.6	42.0
1003	29.1	40.9	46.3
1047	30.8	37.8	43.9
1380	38.7	51.0	60.8
	Percent of Boundary That Is Perimeter by Number of Acquisitions		
1005	88.9	75.2	65.9
1059	81.2	62.8	68.8
1520	83.3	67.0	—
1803	90.2	80.1	73.1
1003	84.6	76.5	69.1
1047	85.0	71.8	63.5
1380	82.0	69.2	58.2
	Percent of Boundary That Is Linear by Number of Acquisitions		
1005	8.2	10.4	8.7
1059	13.3	9.1	7.3
1520	10.3	9.9	—
1803	7.3	10.2	9.0
1003	12.7	12.4	9.2
1047	13.1	12.2	9.9
1380	10.3	0.7	5.3
	Percent of Boundary That Is Interior by Number of Acquisitions		
1005	2.9	14.4	26.4
1059	5.5	28.1	23.9
1520	6.4	23.1	—
1803	2.5	9.7	17.9
1003	2.7	11.1	21.7
1047	1.9	16.0	26.6
1380	7.7	20.1	36.5

eration then are the linear and interior types. As was noted, these nonperimeter boundary pixels are the result of misregistration of the perimeter boundaries obtained on individual acquisitions.

In some cases, these boundaries can become very thick. It seems reasonable that, for field-size estimation purposes, some decisions could be made concerning field membership of these pixels, even though spectrally they are not samples of any one field. Here the assumption is made that if one of these nonperimeter boundary pixels' neighbors is adjacent to only one field, then that field has been undersampled because of registration problems (since the pixel must be in the perimeter of that field on one of the acquisitions). Therefore, the nonperimeter boundary pixels which neighbor on only one field can be logically assigned to that field for proportion estimation purposes.

In practice, this procedure reduces the size of the unassigned boundary pixels to an average of around 8% of the scene. The ways of estimating the small grains proportions within these remaining boundary pixels will now be considered, assuming that no problem exists in obtaining proportions for the part of the image assigned to fields.

Four methods of estimating the proportion of small grains in the remaining boundary area are considered. Each method requires that the fields obtained in the clustering procedure be labeled. The four methods are:

1. Total the populations of those fields labeled small grains and divide this by the number of pixels assigned to fields. This assumes that there is no bias included by the field extraction.

2. Count the number of perimeter boundary pixels belonging to fields labeled small grains which are adjacent to the remaining boundary pixels and divide this by the total number of perimeter boundary pixels which are adjacent to the remaining boundary pixels. This is an attempt to model the contribution of field-size differences to the bias in the boundary sample. Figure 69 gives a hypothetical example of the kind of bias this attempts to remove. The pixels marked with an × are used to estimate the proportions in the shaded areas.

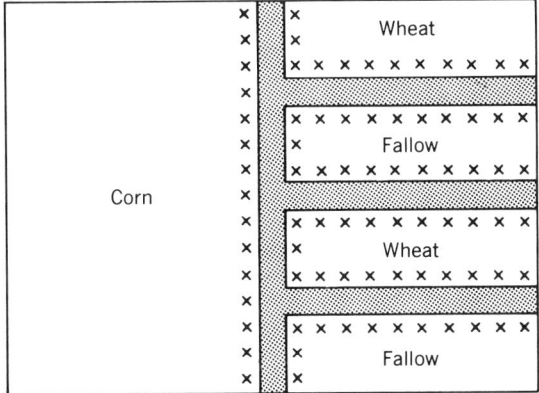

FIGURE 69. Modeling field size contribution to boundary pixel bias.

3. Divide the number of fields labeled small grains by the number of fields. This treats the boundary pixels as if they were fields having the same probability of being small grains that had been observed in the usual fields.

4. The final method is a combination of methods 2 and 3, which is to stratify the remaining boundary pixels into linear and interior types. Method 2 is applied to the linear boundary pixels, and method 3 is applied to the interior boundary pixels, producing a stratified estimate.

These methods were applied to the boundary pixels which remained after clustering the data shown in Table 23 using the spatial methods discussed here and in reference 1. Each field was assigned the label corresponding to the predominant ground truth crop code of the pixels within the field. Table 27 shows the ground truth proportion of small grains among the boundary pixels for each segment, and the estimates obtained from the four methods discussed above with the bias and root mean square error (RMSE) observed using each method. The stratified method, method 4, is clearly superior to the first three methods in terms of overall bias and rmse.

After using these methods to estimate the small grains proportions among the boundary pixels and using method 1 to estimate the small grains proportions among the pixels assigned to fields, a stratified estimate of the small grains proportions in each scene can be obtained. The final estimates obtained using these methods to estimate proportions among the boundary pixels are presented in Table 28, along with the bias and RMSE associated with each method. These results imply that, when aggregated across segments, any reasonable method will work equally well when given the size of the boundary class used here.

TABLE 27. Performance of Estimation Methods Applied to Boundary Pixels

Segment	Ground Truth Proportion of Small Grains in the Boundary	Method 1 Estimate of Boundary	Method 2 Estimate of Boundary	Method 3 Estimate of Boundary	Method 4 Estimate of Boundary
1005	43.82	36.07	40.99	42.07	41.78
1059	42.28	46.98	49.00	43.55	44.83
1520	27.92	29.10	29.17	25.69	26.73
1803	3.91	3.86	3.91	2.49	2.97
1003	14.29	20.60	16.15	15.94	16.00
1047	39.21	36.23	42.84	30.74	34.02
1154	30.59	19.74	27.22	37.16	33.19
1380	10.42	11.71	12.89	11.98	12.10
Bias		−1.08	+1.22	−0.35	−0.10
RMSE		5.60	3.33	4.06	2.56

TABLE 28. Performance of Estimation Methods Applied to the Segment
Proportion Estimation

Segment	Ground Truth Proportion of Small Grains in the Segment	Method 1 Segment Estimate	Method 2 Segment Estimate	Method 3 Segment Estimate	Method 4 Segment Estimate
1005	37.25	36.07	36.49	36.59	36.57
1059	45.76	46.98	47.12	46.73	46.83
1520	30.02	29.10	29.12	28.79	28.88
1803	3.20	3.86	3.88	3.78	3.81
1003	21.09	20.60	20.27	20.29	20.29
1047	35.40	36.23	36.72	35.82	36.06
1154	20.70	19.74	20.20	20.82	20.57
1380	9.83	11.17	11.37	11.27	11.28
Bias		+0.06	+0.24	+0.10	+0.13
RMSE		0.99	1.05	0.88	0.90

Conclusions and Recommendations

The problems encountered with boundary pixels seem to be related more to registration problems than to a mixture of classes within pixels. This has a significant bearing on the accuracy that can be obtained from pixel labeling procedures. Two problems should be noted here.

First, the registration error that occurs between boundary pixels also causes pure pixels to shift within a field to the same extent. Therefore, pure pixels may serve as samples of different areas within the same field from acquisition to acquisition. Sampling a good stand of wheat on one acquisition and an adjacent small waterhole on the next could degrade the spectral values for such a pixel. Any pixel labeling procedure would be difficult in this situation without the benefit of information about the entire field from which this pure pixel was taken.

If this possibility seems too remote to worry about, consider the following crude approximation of the size of the problem. Suppose two acquisitions of the same segment are given. Let A denote the percentage of the full scene which appears as a boundary pixel in the first acquisition but not the second. Let B denote the percentage of the full scene which appears as a boundary in both acquisitions, and let C denote the portion of the full scene which appears as a boundary in the second acquisition but not in the first.

Table 25 shows that the average size of the boundary class in one acquisition is 19.1%, and Table 26 shows an average of 32.3% combined boundaries using two acquisitions. This yields the following approximately true equations:

$$A + B = 19.1$$

$$B + C = 19.1$$

$$A + B + C = 32.3$$

Solving this system of equations yields $B = 5.9$. Thus, $5.9/32.3 = 18.3\%$ of the combined boundary detected from two acquisitions is detected on both acquisitions. The nonperimeter boundary pixels that are observed using two acquisitions must be a subset of this 18.3%, since only perimeter-type boundary pixels occur within single acquisitions.

The nonperimeter boundary accumulated using two acquisitions averages 15.0% of the total boundary given in Table 26. This says that $15.0/18.3$ or about 82% of the pixels observed on both acquisitions shifted positions and created nonperimeter boundaries. This indicates that the probability is about 80% that any pixel does not represent the same area of land from one acquisition to the next.

A second problem that occurs in pixel-labeling procedures is that the ground truth that is used to estimate accuracy has the same registration problems. If it is assumed that the ground truth takes the form of another acquisition and that the ground truth small grains on that acquisition are the only vegetated pixels on that acquisition, then this leads to the assumption that 19% of the scene consists of pixels which lie in a perimeter of areas of ground truth small grains or other and that 80% of those shift one or more pixels when compared to the imagery used in labeling. If it is assumed that 50% of those that shift change classes and 50% remain in the same class, this leads to a conservative estimate that 7% of the ground truth is labeled inappropriately. Therefore, some sort of quasi-field labeling seems to be a logical alternative. Bryant (reference 2) reinforces this conclusion.

The problem of sampling bias included by field-finding algorithms does not appear to be significant in this sudy. Perhaps this is because the algorithms used here reduced the boundary class considerably before trying to estimate this bias, or perhaps there is no significant boundary bias in this particular crop type. If in the future there does appear to be a significant bias involved in such a subsetting operation, the methods presented appear to be a promising solution.

REFERENCES

1. Dennis, T. B., Spatial/color sequence proportion estimation techniques. LEMSCO-15641, NASA/JSC, Houston. (To be published.)
2. Bryant, J., *The easy remote sensing problem,* NASA-9-14689-9S, August 1979.

2.7 CLASSIFICATION USING SPATIAL–SPECTRAL INFORMATION*

Context has been recognized for a long time as a source of interpretation evidence. Early attempts at using information from neighboring pixels to improve the correct classification probability of any arbitrary pixel were effectively results-smoothing;

*Based on C. B. Chittineni, *Utilization of spectral–spatial information in the classification of image data,* Lockheed Electronics and Management Services Company, Report No. JSC-16335, NASA Johnson Space Center, Houston, Tex., June 1980.

a one-pixel cornfield in the middle of a lake was reclassified as water, for example. Later context analysis grew into field classifiers and relaxation labeling.

Introduction

Classification of multichannel imagery data is typically done by applying a decision rule to each resolution element or picture element (pixel) and classifying it based on spectral information. This procedure ignores spatial information. Most of the imagery data contain much spatial information which can be used to improve computer-assisted classification.

The use of contextual information in pattern classification has attracted the attention of many researchers, mainly in the area of character recognition.[1,2] Generally, one of two basic approaches has been used: the table lookup method or the Markov approach. The table lookup method is based on the assumption that every word to be recognized is selected from a known finite table. A word is classified by comparing it with every word of the same length in the table and finding the best match.

The Markov approach is based on the assumption that the true category of a character is related in a probabilistic manner to the true categories of a small number of surrounding characters. Its use requires the estimation of the probability of occurrence of all possible pairs, triplets of characters, and so forth from the sample text. Abend[3] derived optimal procedures when a Markov dependence exists between the states of nature of neighboring characters, and Raviv[4] gives the results of applying such procedures for the recognition of English text.

The use of contextual information in speech recognition is considered by Alter.[5] Chow,[6] using a nearest-neighbor-dependence method, obtained the structure and parameters of a recognition network for patterns represented by binary matrices.

Several researchers attempted to use spatial information in the classification of imagery data. Kettig and Landgrebe[7] developed a technique called extraction and classification of homogeneous objects (ECHO), which segments a scene into homogeneous objects and uses sample classification to assign each object as a whole rather than by its individual pixels. Haralick et al.[8] used textural features based on gray tone spatial dependence matrices to characterize a local scene texture and experimentally showed them to be useful for classification purposes. Swain[9] developed a cascade model for classifying a pattern based on multiple observations in a time-varying environment. Welch and Salter[10] presented a method for the contextual classification of imagery data. Chittineni[11] discusses the use of context with linear classifiers. Toussaint[12] gives a brief review of the use of context in pattern recognition and presents an extensive list of references on the subject.

All of the approaches proposed in the literature either use arbitrarily selected transition probabilities or estimate them from a sample and treat them as global. For imagery data such as those obtained in remote sensing, the transition probabilities very often not only vary from one image to the other but also vary from one local neighborhood to the other in the same image. It is difficult to obtain global estimates of transition probabilities because of the varying nature of imagery and the nonavailability of true classes of pixels of images.

It is the purpose of this section to develop methods for locally estimating transition probabilities and to use these estimates in contextual classification. It is assumed that the classifier is trained on representative data from the image and, for every pixel of the image, the a posteriori probabilities of the classes are estimated from spectral information. Thus, the incorporation of contextual information into classification is treated as a postprocessing operation.

The number of transition probabilities to be estimated increases as the square of the number of classes. Mathematical expressions for contextual classification become complex with the increase in the size of the local neighborhood. Thus, making the estimation of transition probabilities is computationally expensive. In this analysis the transition probabilities are modeled in terms of a single parameter θ, under reasonable assumptions, and methods are developed for the estimation of θ. The estimated θ is then used for the incorporation of spatial information into classification.

Modeling Transition Probabilities

The models for the transition probabilities of the classes of the neighboring pixels, in terms of a single parameter θ, are developed under reasonable assumptions. Let i and j be the neighboring pixels as shown in Figure 70.

Let X_i and X_j be the pattern vectors and ω_i and ω_j be the labels (classes) of pixels i and j, respectively. Let ω_i and ω_j take values of $1, 2, \ldots, M$, where M is the number of classes.

A linear model describing the dependency between the neighboring pixels in terms of a single parameter θ for different r and s is given in equation (1).

$$P(\omega_i = r \mid \omega_j = s) = (1 - \theta)P(\omega_i = r)$$
$$P(\omega_i = r \mid \omega_j = r) = (1 - \theta)P(\omega_i = r) + \theta \qquad (1)$$
$$0 \leqslant \theta \leqslant 1$$

For $\theta = 0$, equation (1) becomes

$$P(\omega_i = r \mid \omega_j = s) = P(\omega_i = r)$$
$$P(\omega_i = r \mid \omega_j = r) = P(\omega_i = r) \qquad (2)$$

Equation (2) is the case where the labels of neighboring pixels are independent. For $\theta = 1$, equation (1) becomes

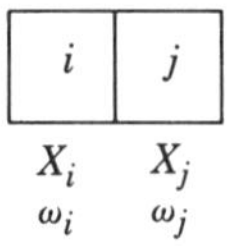

$$
\begin{array}{|c|c|}
\hline
i & j \\
\hline
\end{array}
$$

$$X_i \quad X_j$$
$$\omega_i \quad \omega_j$$

FIGURE 70. Neighboring pixels i and j.

$$P(\omega_i = r \,|\, \omega_j = s) = 0$$
$$P(\omega_i = r \,|\, \omega_j = r) = 1 \tag{3}$$

Equation (3) is the case where the labels of the neighboring pixels are completely dependent. Notice that the linear transition probabilities model of equation (1) is a linear interpolation in terms of a single parameter θ between the extremes of equations (2) and (3). It can be easily shown that the model of equation (1) satisfies the postulates of probabilities. That is,

$$0 \leqslant P(\omega_i = r \,|\, \omega_j = s) \leqslant 1$$
$$\sum_{r=1}^{M} P(\omega_i = r \,|\, \omega_j = s) = 1 \tag{4}$$

Using a quadratic interpolation between the extremes of equations (2) and (3), a quadratic model describing the dependencies between the labels of neighboring pixels can be written in terms of a single parameter θ as

$$P(\omega_i = r \,|\, \omega_j = s) = (1 - \theta)^2 P(\omega_i = r)$$
$$P(\omega_i = r \,|\, \omega_j = r) = (1 - \theta)^2 P(\omega_i = r) + \theta(2 - \theta) \tag{5}$$
$$0 \leqslant \theta \leqslant 1$$

The model of equation (5) also satisfies the postulates of probabilities.

However, it is to be noted that the dependencies between the neighboring pixels can be modeled through some other parameter. For example, by replacing θ with $\alpha/(1 + \alpha)$, the dependencies are described in terms of α, $0 \leqslant \alpha \leqslant \infty$; by replacing θ with $e^{-\beta}/(1 + e^{-\beta})$, the dependencies are described in terms of β, $-\infty \leqslant \beta \leqslant \infty$.

The transition probabilities between the classes of the neighboring pixels i and j also can be modeled to satisfy the following characteristics of dependencies, resulting in a nonlinear model. Some of the general characteristics of dependencies between neighboring pixels i and j can be written as follows:

a. If the label $\omega_i = r$ of pixel i frequently occurs concurrently with the label $\omega_j = s$ of pixel j, then

$$P(\omega_i = r \,|\, \omega_j = s) > P(\omega_i = r) \tag{6}$$

and, if they always occur concurrently, then

$$P(\omega_i = r \,|\, \omega_j = s) = 1 \tag{7}$$

b. If the label $\omega_i = r$ of pixel i rarely occurs concurrently with the label $\omega_j = s$ of pixel j, then

$$P(\omega_i = r \,|\, \omega_j = s) < P(\omega_i = r) \tag{8}$$

and, if they never occur concurrently, then

$$P(\omega_i = r \,|\, \omega_j = s) = 0 \tag{9}$$

c. If the label $\omega_i = r$ of pixel i occurs independently of the label $\omega_j = s$ of pixel j, then

$$P(\omega_i = r \,|\, \omega_j = s) = P(\omega_i = r) \tag{10}$$

A model satisfying characteristics a, b, and c can be written in terms of a single parameter θ for different r and s as

$$P(\omega_i = r \,|\, \omega_j = s) = \frac{(1 - \theta)P(\omega_i = r)}{(1 - \theta) + \theta P(\omega_j = s)}$$

$$P(\omega_i = r \,|\, \omega_j = r) = \frac{P(\omega_i = r)}{(1 - \theta) + \theta P(\omega_j = r)} \tag{11}$$

$$-\infty \leqslant \theta \leqslant 1$$

It can be easily shown that the transition probabilities described by equation (11) satisfy the postulates of probabilities. Also, notice that requirements a, b, and c on

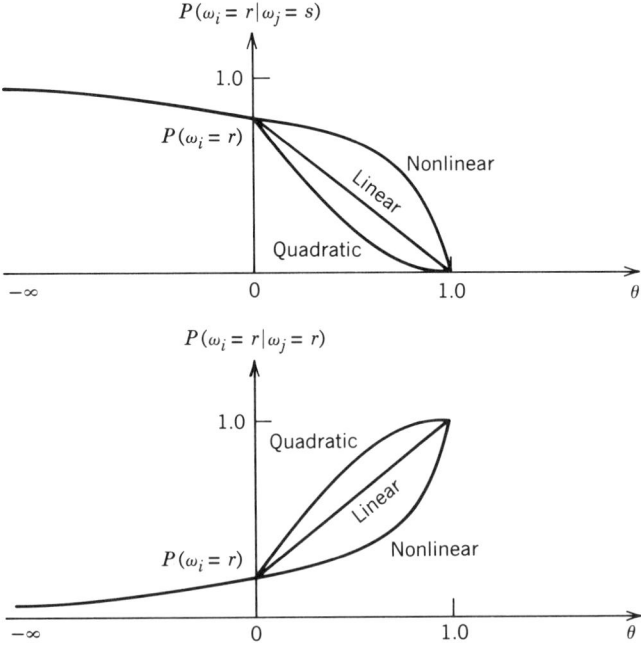

FIGURE 71. Illustrations of spatial dependency models.

the transition probabilities correspond to the cases where $\theta < 0$, $\theta > 0$, and $\theta = 0$, respectively. The model of equation (11) is referred to in this paper as the nonlinear transition probabilities model. Figure 71 illustrates the linear, quadratic, and nonlinear transition probabilities models.

In the remainder of this discussion only the linear model of equation (1) and the nonlinear model of equation (11) are considered.

Local Neighborhood Estimation of Transition Probabilities

In this subsection, techniques are developed for the estimation of transition probabilities in the local neighborhood of the pixel under consideration for use in its contextual classification. The criterion used for their estimation is the likelihood function. That is, the transition probabilities are estimated as those that maximize the likelihood function of observed spectral vectors if their spatial relationships are as given in the local neighborhood.

A General Expression for the Likelihood Function

An expression for the likelihood function of N patterns in a general local neighborhood is developed in the following. Let $X_1, X_2, \ldots, X_N$ be the patterns in the general neighborhood. The likelihood function of these patterns can be written as

$$
\begin{aligned}
L' &= p(X_1, X_2, \ldots, X_N) \\
&= \sum_{i_1=1}^{M} \sum_{i_2=1}^{M} \cdots \sum_{i_N=1}^{M} p(X_1, \omega_1 = i_1; X_2, \omega_2 = i_2; \ldots; X_N, \omega_N = i_N) \\
&= \sum_{i_1=1}^{M} \sum_{i_2=1}^{M} \cdots \sum_{i_N=1}^{M} p(X_1, X_2, \ldots, X_N \mid \omega_1 = i_1, \omega_2 = i_2, \ldots, \omega_N = i_N) \\
&\quad \times P(\omega_1 = i_1, \omega_2 = i_2, \ldots, \omega_N = i_N)
\end{aligned}
\tag{12}
$$

where M is the number of classes. In the following, it is assumed that (a) the probability density function of a pattern, given its label, is independent of other patterns and their labels; and (b) the labels of the patterns are independent of the labels of their nonneighbors. By repeatedly using assumption (a), the following is obtained. Consider

$$
\begin{aligned}
&p(X_1, X_2, \ldots, X_N \mid \omega_1 = i_1, \omega_2 = i_2, \ldots, \omega_N = i_N) \\
&\quad = p(X_1 \mid X_2, \ldots, X_N; \omega_1 = i_1, \ldots, \omega_N = i_N) \\
&\qquad \times p(X_2, \ldots, X_N \mid \omega_1 = i_1, \ldots, \omega_N = i_N) = i_N) \\
&\quad = p(X_1 \mid \omega_1 = i_1) p(X_2 \mid X_3, \ldots, X_N; \omega_1 = i_1, \ldots, \omega_N = i_N) \\
&\qquad \times p(X_3, \ldots, X_N \mid \omega_1 = i_1, \ldots, \omega_N = i_N) \\
&\quad = p(X_1 \mid \omega_1 = i_1) p(X_2 \mid \omega_2 = i_2) p(X_3, \ldots, X_N \mid \omega_1 = i_1, \ldots, \omega_N = i_N) \\
&\quad = \prod_{j=1}^{N} p(X_j \mid \omega_j = i_j)
\end{aligned}
\tag{13}
$$

Using equation (13) in equation (12) results in

$$L' = \sum_{i_1=1}^{M} \sum_{i_2=1}^{M} \cdots \sum_{i_N=1}^{M} \left[\prod_{j=1}^{N} p(X_j \,|\, \omega_j = i_j) \right] P(\omega_1 = i_1, \ldots, \omega_N = i_N) \qquad (14)$$

Since $[\prod_{j=1}^{N} p(X_j)]$ is independent of the transition probabilities, dividing both sides of equation (14) by it yields the criterion L to be used for estimating the transition probabilities. That is,

$$L = \sum_{i_1=1}^{M} \sum_{i_2=1}^{M} \cdots \sum_{i_N=1}^{M} \left[\prod_{j=1}^{N} \frac{p(\omega_j = i_j \,|\, X_j)}{P(\omega_j = i_j)} \right] P(\omega_1 = i_1, \omega_2 = i_2, \ldots, \omega_N = i_N) \qquad (15)$$

Now $P(\omega_1 = i_1, \ldots, \omega_N = i_N)$ depends on the particular local neighborhood and will be considered in detail in the following.

Spatially Uniform Context—Four Neighbors

The pixel under consideration, pixel 0, and its four neighbors in a two-dimensional local neighborhood are shown in figure 72.

By repeatedly using assumption (b), equation (16) is obtained. From equations (15) and (16), an expression for L for the local neighborhood of figure 72 is obtained, as shown in equation (17).

$$
\begin{aligned}
P(\omega_0 = i_0, \omega_1 &= i_1, \ldots, \omega_4 = i_4) \\
&= P(\omega_0 = i_0)P(\omega_1 = i_1, \ldots, \omega_4 = i_4 \,|\, \omega_0 = i_0) \\
&= P(\omega_0 = i_0)P(\omega_1 = i_1 \,|\, \omega_0 = i_0, \omega_2 = i_2, \ldots, \omega_4 = i_4) \\
&\quad \times P(\omega_2 = i_2, \ldots, \omega_4 = i_4 \,|\, \omega_0 = i_0) \\
&= P(\omega_0 = i_0)P(\omega_1 = i_1 \,|\, \omega_0 = i_0)P(\omega_2 = i_2 \,|\, \omega_0 = i_0, \omega_3 = i_3, \omega_4 = i_4) \\
&\quad \times P(\omega_3 = i_3, \omega_4 = i_4 \,|\, \omega_0 = i_0) \\
&= P(\omega_0 = i_0) \prod_{j=1}^{4} P(\omega_j = i_j \,|\, \omega_0 = i_0)
\end{aligned}
\qquad (16)
$$

$$
\begin{aligned}
L &= \sum_{i_0=1}^{M} \sum_{i_1=1}^{M} \cdots \sum_{i_4=1}^{M} \left[\prod_{j=0}^{4} \frac{p(\omega_j = i_j \,|\, X_j)}{P(\omega_j = i_j)} \right] \left[P(\omega_0 = i_0) \prod_{l=1}^{4} P(\omega_l = i_l \,|\, \omega_0) \right] \\
&= \sum_{i_0=1}^{M} p(\omega_0 = i_0 \,|\, X_0) \left\{ \prod_{j=1}^{4} \left[\sum_{i_j=1}^{M} \frac{p(\omega_j = i_j \,|\, X_j)}{P(\omega_j = i_j)} P(\omega_j = i_j \,|\, \omega_0 = i_0) \right] \right\}
\end{aligned}
\qquad (17)
$$

An Expression for L with Linear Transition Probabilities Model

Since a priori probabilities are position-independent, when a linear model of equation (1) is used for transition probabilities in equation (17), L becomes

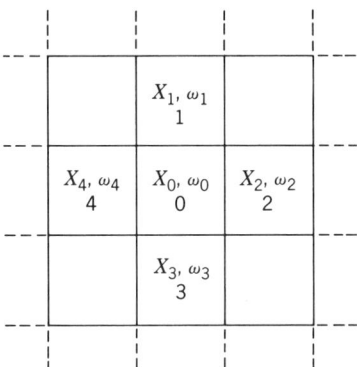

FIGURE 72.   Four neighbors of pixel 0.

$$L = \sum_{i_0=1}^{M} p(\omega_0 = i_0 | X_0) \prod_{j=1}^{4} \left[\sum_{\substack{i_j=1 \\ i_j \neq i_0}}^{M} \frac{p(\omega_j = i_j | X_j)}{P(\omega_j = i_j)} P(\omega_j = i_j | \omega_0 = i_0) \right.$$

$$\left. + \frac{p(\omega_j = i_0 | X_j)}{P(\omega_j = i_0)} P(\omega_j = i_0 | \omega_0 = i_0) \right]$$

$$= \sum_{i_0=1}^{M} p(\omega = i_0 | X_0) \left\{ \prod_{j=1}^{4} \left[(1 - \theta) \sum_{i_j=1}^{M} p(\omega = i_j | X_j) + \theta \frac{p(\omega = i_0 | X_j)}{P(\omega = i_0)} \right] \right\} \qquad (18)$$

$$= \sum_{i_0=1}^{M} p(\omega = i_0 | X_0) \left\{ \prod_{j=1}^{4} \left[(1 - \theta) + \theta \frac{p(\omega = i_0 | X_j)}{P(\omega = i_0)} \right] \right\}$$

$$= (1 - \theta)^4 + \theta(1 - \theta)^3 A + \theta^2(1 - \theta)^2 B + \theta^3(1 - \theta)C + \theta^4 D$$

where

$$A = \sum_{i_0=1}^{M} \frac{p(\omega = i_0 | X_0)}{P(\omega = i_0)} [p(\omega = i_0 | X_1) + p(\omega = i_0 | X_2)$$

$$+ p(\omega = i_0 | X_3) + p(\omega = i_0 | X_4)]$$

$$B = \sum_{i_0=1}^{M} \frac{p(\omega = i_0 | X_0)}{p^2(\omega = i_0)} [p(\omega = i_0 | X_1)p(\omega = i_0 | X_2) + p(\omega = i_0 | X_1)p(\omega = i_0 | X_3)$$

$$+ p(\omega = i_0 | X_1)p(\omega = i_0 | X_4) + p(\omega = i_0 | X_2)p(\omega = i_0 | X_3)$$

$$+ p(\omega = i_0 | X_2)p(\omega = i_0 | X_4) + p(\omega = i_0 | X_3)p(\omega = i_0 | X_4)]$$

$$C = \sum_{i_0=1}^{M} \frac{p(\omega = i_0 | X_0)}{P^3(\omega = i_0)} [p(\omega = i_0 | X_1)p(\omega = i_0 | X_2)p(\omega = i_0 | X_3)$$

$$+ p(\omega = i_0 | X_1)p(\omega = i_0 | X_2)p(\omega = i_0 | X_4)$$

$$+ p(\omega = i_0 | X_1)p(\omega = i_0 | X_3)p(\omega = i_0 | X_4)$$

$$+ p(\omega = i_0 | X_2)p(\omega = i_0 | X_3)p(\omega = i_0 | X_4)]$$

$$D = \sum_{i_0=1}^{M} \frac{p(\omega = i_0|X_0)}{P^4(\omega = i_0)}[p(\omega = i_0|X_1)p(\omega = i_0|X_2)p(\omega = i_0|X_3)p(\omega = i_0|X_4)]$$

An Expression for L With Nonlinear Transition Probabilities Model

Using the nonlinear transition probabilities model of equation (11) in equation (18) gives the following expression for L:

$$L = \sum_{i_0=1}^{M} p(\omega_0 = i_0|X_0) \prod_{j=1}^{4} \left\{ \frac{1}{(1-\theta) + \theta P(\omega_0 = i_0)} \right.$$

$$\left. \times \left[(1-\theta) \sum_{\substack{i_j=1 \\ i_j \neq i_0}}^{M} p(\omega_j = i_j|X_j) + p(\omega_j = i_0|X_j) \right] \right\}$$

$$= \sum_{i_0=1}^{M} p(\omega = i_0|X_0) \left\{ \prod_{j=1}^{M} \left[\frac{(1-\theta) + \theta p(\omega = i_0|X_j)}{(1-\theta) + \theta P(\omega = i_0)} \right] \right\} \qquad (19)$$

Sequential Neighborhood — General Case

A general N-pixel sequential neighborhood is shown in Figure 73. The transitions for which the transition probabilities are applied in the sequential neighborhood are indicated in Figure 73. An expression is developed here for the likelihood function L of the patterns in a general sequential neighborhood of N pixels. Only the pixels immediately adjacent to each pixel are treated as its neighbors. Consider

$$P(\omega_1 = i_1, \ldots, \omega_N = i_N)$$
$$= P(\omega_2 = i_2)P(\omega_1 = i_1, \omega_3 = i_3, \ldots, \omega_N = i_N | \omega_2 = i_2)$$
$$= P(\omega_2 = i_2)P(\omega_1 = i_1 | \omega_2 = i_2, \omega_3 = i_3, \ldots, \omega_N = i_N) \qquad (20)$$
$$\times P(\omega_3 = i_3, \ldots, \omega_N = i_N | \omega_2 = i_2)$$
$$= P(\omega_2 = i_2)P(\omega_1 = i_1 | \omega_2 = i_2)P(\omega_3 = i_3, \ldots, \omega_N = i_N | \omega_2 = i_2)$$

Assumption (b) is used in obtaining equation (20). The Bayes rule is used to obtain the following:

$$P(\omega_3 = i_3, \omega_4 = i_4, \ldots, \omega_N = i_N | \omega_2 = i_2) = \frac{P(\omega_2 = i_2, \ldots, \omega_N = i_N)}{P(\omega_2 = i_2)} \qquad (21)$$

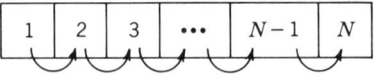

FIGURE 73. A general N-pixel sequential neighborhood.

Proceeding in a manner similar to equation (20), the numerator of equation (21) can be written as

$$
\begin{aligned}
P(\omega_2 &= i_2, \omega_3 = i_3, \ldots, \omega_N = i_N) \\
&= P(\omega_3 = i_3)P(\omega_2 = i_2 \mid \omega_3 = i_3)P(\omega_4 = i_4, \ldots, \omega_N = i_N \mid \omega_3 = i_3)
\end{aligned}
\tag{22}
$$

Continuing in a similar manner obtains the following result:

$$
P(\omega_{N-1} = i_{N-1}, \omega_N = i_N) = P(\omega_{N-1} = i_{N-1})P(\omega_N = i_N \mid \omega_{N-1} = i_{N-1})
\tag{23}
$$

The following is obtained from the Bayes rule:

$$
P(\omega_2 = i_2 \mid \omega_3 = i_3) = \frac{P(\omega_2 = i_2)}{P(\omega_3 = i_3)}P(\omega_3 = i_3 \mid \omega_2 = i_2)
\tag{24}
$$

Equation (25) is obtained from equations (20) through (24):

$$
\begin{aligned}
P(\omega_1 &= i_1, \omega_2 = i_2, \ldots, \omega_N = i_N) \\
&= P(\omega_1 = i_1)P(\omega_2 = i_2 \mid \omega_1 = i_1)P(\omega_3 = i_3 \mid \omega_2 = i_2)\cdots \\
&\quad \times P(\omega_N = i_N \mid \omega_{N-1} = i_{N-1})
\end{aligned}
\tag{25}
$$

Substitution of equation (25) into equation (15) results in an expression for the criterion L for a general sequential neighborhood. That is,

$$
\begin{aligned}
L = \Bigg\{ \sum_{i_1=1}^{M} \sum_{i_2=1}^{M} &\cdots \sum_{i_N=1}^{M} \left[\sum_{j=1}^{N} \frac{p(\omega_j = i_j \mid X_j)}{P(\omega_j = i_j)} \right] \\
&\times [P(\omega_1 = i_1)P(\omega_2 = i_2 \mid \omega_1 = i_1) \\
&\times P(\omega_3 = i_3 \mid \omega_2 = i_2)\cdots P(\omega_N = i_N \mid \omega_{N-1} = i_{N-1})] \Bigg\}
\end{aligned}
\tag{26}
$$

The Likelihood Function L of Patterns in a Sequential Neighborhood with the Linear Transition Probabilities Model

In this subsection, equation (26) is expressed in a polynomial form in terms of θ for a four-pixel sequential neighborhood, using the linear transition probabilities model of equation (1). The four-pixel sequential neighborhood considered is shown in Figure 74.

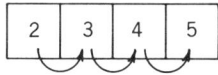

FIGURE 74. A four-pixel sequential neighborhood.

The likelihood function of equation (26) for the neighborhood of Figure 74 becomes

$$
L = \sum_{i_2=1}^{M} p(\omega_2 = i_2 | X_2) \sum_{i_3=1}^{M} \frac{p(\omega_3 = i_3 | X_3)}{P(\omega_3 = i_3)} P(\omega_3 = i_3 | \omega_2 = i_2)
$$

$$
\times \sum_{i_4=1}^{M} \frac{p(\omega_4 = i_4 | X_4)}{P(\omega_4 = i_4)} P(\omega_4 = i_4 | \omega_3 = i_3) \sum_{i_5=1}^{M} \frac{p(\omega_5 = i_5 | X_5)}{P(\omega_5 = i_5)}
$$

$$
\times P(\omega_5 = i_5 | \omega_4 = i_4)
$$

$$
= \sum_{i_2=1}^{M} p(\omega_2 = i_2 | X_2) \left\{ (1-\theta)^3 + \theta(1-\theta)^2 \left[a_{45} + a_{34} + \frac{p(\omega_3 = i_2 | X_3)}{P(\omega_3 = i_2)} \right] \right.
$$

$$
+ \theta^2(1-\theta) \left[a_{345} + a_{45} \frac{p(\omega_3 = i_2 | X_3)}{P(\omega_3 = i_2)} \right. \tag{27}
$$

$$
+ \frac{p(\omega_3 = i_2 | X_3) p(\omega_4 = i_2 | X_4)}{P(\omega_3 = i_2) P(\omega_4 = i_2)} \Bigg]
$$

$$
\left. + \theta^3 \frac{p(\omega_3 = i_2 | X_3) p(\omega_4 = i_2 | X_4) p(\omega_5 = i_2 | X_5)}{P(\omega_3 = i_2) P(\omega_4 = i_2) P(\omega_5 = i_2)} \right\}
$$

$$
= (1-\theta)^3 + \theta(1-\theta)^2 (a_{23} + a_{34} + a_{45})
$$

$$
+ \theta^2(1-\theta)(a_{23}a_{45} + a_{234} + a_{345}) + \theta^3 a_{2345}
$$

Since the a priori probabilities are position independent in the local neighborhood, the different quantities in equation (27) can be shown to be

$$
a_{23} = \sum_{i_2=1}^{M} \frac{p(\omega_2 = i_2 | X_2) p(\omega_3 = i_2 | X_3)}{P(\omega = i_2)}
$$

$$
a_{34} = \sum_{i_3=1}^{M} \frac{p(\omega_3 = i_3 | X_3) p(\omega_4 = i_3 | X_4)}{P(\omega = i_3)}
$$

$$
a_{45} = \sum_{i_4=1}^{M} \frac{p(\omega_4 = i_4 | X_4) p(\omega_5 = i_4 | X_5)}{P(\omega = i_4)}
$$

$$
a_{234} = \sum_{i_2=1}^{M} \frac{p(\omega_2 = i_2 | X_2) p(\omega_3 = i_2 | X_3) p(\omega_4 = i_2 | X_4)}{P^2(\omega = i_2)} \tag{28}
$$

$$
a_{345} = \sum_{i_3=1}^{M} \frac{p(\omega_3 = i_3 | X_3) p(\omega_4 = i_3 | X_4) p(\omega_5 = i_3 | X_5)}{P^2(\omega = i_3)}
$$

$$
a_{2345} = \sum_{i_2=1}^{M} \frac{p(\omega_2 = i_2 | X_2) p(\omega_3 = i_2 | X_3) p(\omega_4 = i_2 | X_4) p(\omega_5 = i_2 | X_5)}{P^3(\omega = i_2)}
$$

Using a linear model of equation (1) for the transition probabilities and the definitions in equation (28) in equation (26), expressions for the likelihood function for different sizes of sequential neighborhoods can be easily written and are listed in Table 29.

TABLE 29. Expressions for the Likelihood Function for Different Sizes of Sequential Neighborhoods with the Linear Transition Probabilities Model

Example	Neighborhood Size (pixels)	Illustration	Expression for the Likelihood Function
1	2	1 2	$L(\theta) = (1 - \theta) + \theta a_{12}$
2	3	1 2 3	$L(\theta) = (1 - \theta)^3 + \theta(1 - \theta)^2 (a_{12} + a_{23}) + \theta^2 a_{123}$
3	4	1 2 3 4	$L(\theta) = (1 - \theta)^3 + \theta(1 - \theta)^2 (a_{12} + a_{23} + a_{34}) + \theta^2(1 - \theta) (a_{12}a_{34} + a_{123} + a_{234}) + \theta^3 a_{1234}$
4	5	1 2 3 4 5	$L(\theta) = (1 - \theta)4 + \theta(1 - \theta)^3 (a_{12} + a_{23} + a_{34} + a_{45}) + \theta^2(1 - \theta)^2$ $\times (a_{123} + a_{234} + a_{345} + a_{12}a_{34} + a_{12}a_{45} + a_{23}a_{45}) + \theta^3(1 - \theta)$ $\times (a_{1234} + a_{2345} + a_{12}a_{345} + a_{45}a_{123}) + \theta^4 a_{12345}$
5	6	1 2 3 4 5 6	$L(\theta) = (1 - \theta)^5 + \theta(1 - \theta)^4(a_{12} + a_{23} + a_{34} + a_{45} + a_{56}) + \theta^2(1 - \theta)^3$ $\times [a_{123} + a_{234} + a_{345} + a_{456} + a_{12}(a_{34} + a_{45} + a_{56}) + a_{23}(a_{45} + a_{56}) + a_{34}a_{56}]$ $+ \theta^3(1 - \theta)^2[a_{1234} + a_{2345} + a_{3456} + a_{123}(a_{45} + a_{56}) + a_{234}a_{56} + a_{12}$ $\times (a_{456}) + a_{23}a_{456} + a_{12}a_{34}a_{56}] + \theta^4 (1 - \theta)$ $\times (a_{12345} + a_{23456} + a_{12}a_{3456} + a_{56}a_{1234} + a_{123}a_{456}) + \theta^5 a_{123456}$

Expressions for the Likelihood Function L of a Sequential Neighborhood with Nonlinear Transition Probabilities Model

Using the nonlinear transition probabilities model of equation (11) in equation (26), expressions for the likelihood function for several sequential neighborhoods can be easily derived. These are illustrated for three-, four-, and five-pixel sequential neighborhoods in the following expressions. In order for the transition probabilities model to hold true, the transitions in the neighborhood must be as indicated in Figures 75, 76, and 77. Define

$$a_5(i_4, \theta) = \frac{(1 - \theta) + \theta p(\omega = i_4 | X_5)}{(1 - \theta) + \theta P(\omega = i_4)}$$

$$a_{45}(i_3, \theta) = \frac{(1 - \theta)\left[\sum_{i_4=1}^{M} p(\omega = i_4 | X_4)a_5(i_4, \theta)\right] + \theta p(\omega = i_3 | X_4)a_5(i_3, \theta)}{(1 - \theta) + \theta P(\omega = i_3)}$$

$$a_{345}(i_2, \theta) = \frac{(1 - \theta)\left[\sum_{i_3=1}^{M} p(\omega = i_3 | X_3)a_{45}(i_3, \theta)\right] + \theta p(\omega = i_2 | X_3)a_{45}(i_2, \theta)}{(1 - \theta) + \theta P(\omega = i_2)}$$

$$a_{2345}(i_1, \theta) = \frac{(1 - \theta)\left[\sum_{i_2=1}^{M} p(\omega = i_2 | X_2)a_{345}(i_2, \theta)\right] + \theta p(\omega = i_1 | X_2)a_{345}(i_1, \theta)}{(1 - \theta) + \theta P(\omega = i_1)}$$

(29)

The likelihood function for the three-pixel sequential neighborhood of Figure 75 is given by

$$L_3(\theta) = \sum_{i_3=1}^{M} p(\omega = i_3 | X_3)a_{45}(i_3, \theta)$$

(30)

FIGURE 75. A three-pixel sequential neighborhood.

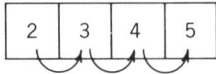

FIGURE 76. A four-pixel sequential neighborhood.

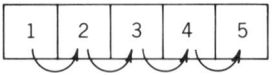

FIGURE 77. A five-pixel sequential neighborhood.

The likelihood function for the four-pixel sequential neighborhood of Figure 76 is given by

$$L_4(\theta) = \sum_{i_2=1}^{M} p(\omega = i_2 | X_2) a_{345}(i_2, \theta) \tag{31}$$

The likelihood function for the five-pixel sequential neighborhood of Figure 77 is given by

$$L_5(\theta) = \sum_{i_1=1}^{M} p(\omega = i_1 | X_1) a_{2345}(i_1, \theta) \tag{32}$$

Computation of θ by the Maximization of Likelihood Function
With both linear and nonlinear transition probabilities models, the likelihood function is a continuous function of the parameter θ. The parameter θ that maximizes the likelihood function with the nonlinear transition probabilities model can be obtained using a one-dimensional bounded search, since the parameter θ is bounded and the likelihood function is nonlinear. With the linear transition probabilities model, the likelihood function is a polynomial in the parameter θ. The flow diagram[13-16] of Figure 78 can be used to find the optimal θ (θ_{opt}) for the linear transition probabilities model in the range $0 \leq \theta \leq 1$, which gives the global maximum for the likelihood function.

Updating A Posteriori Probabilities
Using the transition probabilities models, methods are developed in this section for incorporating contextual information into the classifier decision process.

Updating the A Posteriori Probabilities of a Pixel Using Information from a Single Neighbor
Expressions are developed for updating the a posteriori probabilities of the labels of a pixel using information from its single neighbor. These are used to exploit contextual information from large local neighborhoods. Let the pixel under consideration be X_n and its neighbors be X_{n-1} and X_{n+1}. Figure 79 shows the positions of these pixels.

The assumptions used for updating the a posteriori probabilities are the same as those made earlier — namely, (a) the probability density function of a pattern, given its label, is independent of other patterns and their labels; (b) the labels of the pixels are independent of the labels of their nonneighbors. These assumptions are used in the rest of the section. The information contained in the pattern X_{n-1} regarding the label of the pattern X_n can be written in terms of transition probabilities as

$$p(\omega_n = k | X_{n-1}) = \sum_{i=1}^{M} p(\omega_n = k, \omega_{n-1} = i | X_{n-1})$$

$$= \sum_{i=1}^{M} P(\omega_n = k | \omega_{n-1} = i) p(\omega_{n-1} = i | X_{n-1}) \tag{33}$$

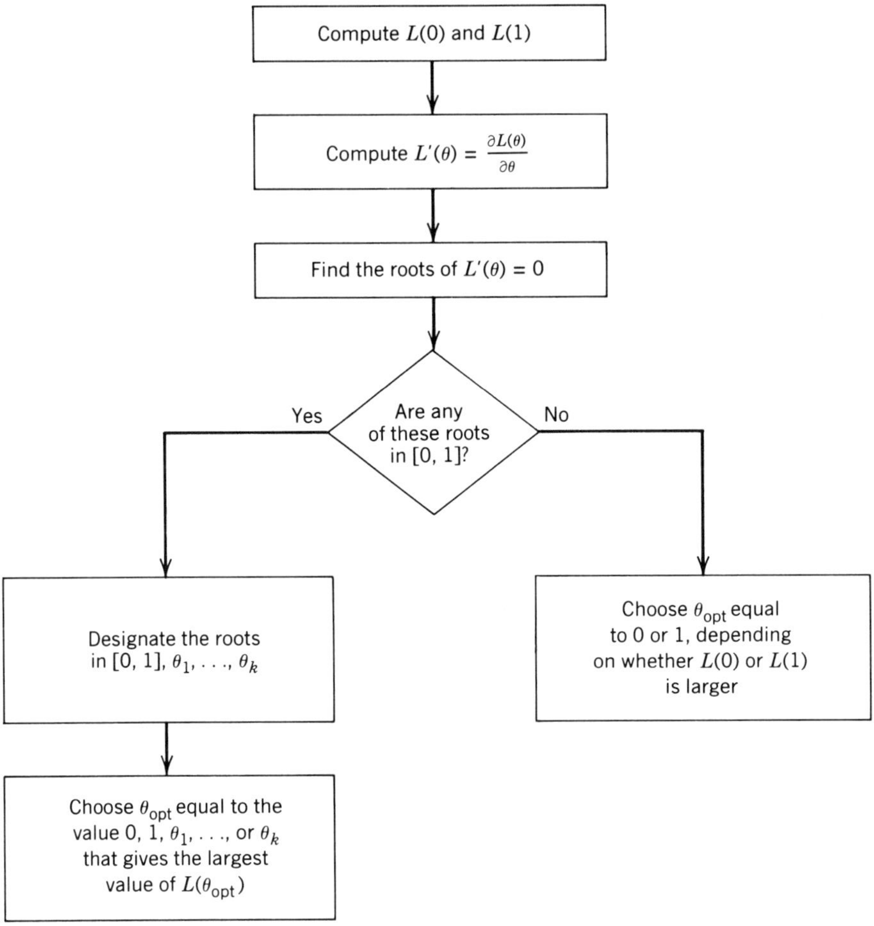

Compute $L(0)$ and $L(1)$

Compute $L'(\theta) = \dfrac{\partial L(\theta)}{\partial \theta}$

Find the roots of $L'(\theta) = 0$

Are any of these roots in [0, 1]?

Yes — Designate the roots in [0, 1], $\theta_1, \ldots, \theta_k$

Choose θ_{opt} equal to the value 0, 1, $\theta_1, \ldots,$ or θ_k that gives the largest value of $L(\theta_{\text{opt}})$

No — Choose θ_{opt} equal to 0 or 1, depending on whether $L(0)$ or $L(1)$ is larger

FIGURE 78. Procedure for finding θ_{opt}.

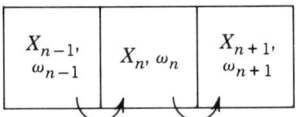

FIGURE 79. Illustration of pixel n under consideration and its neighbors.

Similarly, the following is obtained:

$$p(X_n \mid X_{n-1}) = \sum_{i=1}^{M} p(X_n, \omega_n = i \mid X_{n-1})$$

$$= \sum_{i=1}^{M} p(X_n \mid \omega_n = i) p(\omega_n = i \mid X_{n-1})$$

(34)

Now, the a posteriori probabilities of the labels of the pattern X_n are updated using the information from the patterns X_n and X_{n-1} and their spatial relationship as follows, using the assumptions (a) and (b) above:

$$p(\omega_n = k \mid X_{n-1}, X_n) = \frac{p(X_n \mid \omega_n = k)p(\omega_n = k \mid X_{n-1})}{p(X_n \mid X_{n-1})}$$

$$= \frac{\dfrac{p(\omega_n = k \mid X_n)}{P(\omega_n = k)} \sum\limits_{i=1}^{M} P(\omega_n = k \mid \omega_{n-1} = i)p(\omega_{n-1} = i \mid X_{n-1})}{\sum\limits_{k=1}^{M} \dfrac{p(\omega_n = k \mid X_n)}{P(\omega_n = k)} \sum\limits_{i=1}^{M} P(\omega_n = k \mid \omega_{n-1} = i)p(\omega_{n-1} = i \mid X_{n-1})} \tag{35}$$

Using the linear transition probabilities model of equation (1) in equation (35) yields

$$p(\omega_n = k \mid X_{n-1}, X_n)$$

$$= \frac{(1 - \theta)p(\omega_n = k \mid X_n) + \theta\dfrac{p(\omega_n = k \mid X_n)p(\omega_{n-1} = k \mid X_{n-1})}{P(\omega_n = k)}}{(1 - \theta) + \theta\sum\limits_{k=1}^{M} \dfrac{p(\omega_{n-1} = k \mid X_{n-1})p(\omega_n = k \mid X_n)}{P(\omega_n = k)}} \tag{36}$$

The information in the pattern X_n, in obtaining the label of pattern X_{n+1}, can be written as follows:

$$p(\omega_{n+1} = j \mid X_n) = \sum_{i=1}^{M} p(\omega_{n+1} = j, \omega_n = i \mid X_n)$$

$$= \sum_{i=1}^{M} P(\omega_{n+1} = j \mid \omega_n = i)p(\omega_n = i \mid X_n) \tag{37}$$

Similarly, the following is obtained:

$$p(X_{n+1} \mid X_n) = \sum_{j=1}^{M} p(X_{n+1}, \omega_{n+1} = j \mid X_n)$$

$$= \sum_{j=1}^{M} p(X_{n+1} \mid \omega_{n+1} = j)p(\omega_{n+1} = j \mid X_n) \tag{38}$$

Using the patterns X_n and X_{n+1}, one has

$$p(\omega_n = k \mid X_n, X_{n+1}) = \sum_{j=1}^{M} p(\omega_n = k, \omega_{n+1} = j \mid X_n, X_{n+1})$$

$$= \frac{\sum_{j=1}^{M} p(X_{n+1} \mid \omega_{n+1} = j)P(\omega_{n+1} = j \mid \omega_n = k)p(\omega_n = k \mid X_n)}{p(X_{n+1} \mid X_n)}$$

$$= \frac{p(\omega_n = k|X_n) \sum_{j=1}^{M} P(\omega_{n+1} = j|\omega_n = k)\dfrac{p(\omega_{n+1} = j|X_{n+1})}{P(\omega_{n+1} = j)}}{\sum_{i=1}^{M} p(\omega_n = i|X_n) \sum_{j=1}^{M} P(\omega_{n+1} = j|\omega_n = i)\dfrac{p(\omega_{n+1} = j)|X_{n+1})}{P(\omega_{n+1} = j)}}$$

(39)

Using the linear transition probabilities model of equation (1) in equation (39) yields the following:

$$p(\omega_n = k|X_n, X_{n+1}) = \frac{p(\omega_n = k|X_n)\left[(1 - \theta) + \theta\dfrac{p(\omega_{n+1} = k|X_{n+1})}{P(\omega_{n+1} = k)}\right]}{\left[(1 - \theta) + \theta\sum_{j=1}^{M}\dfrac{p(\omega_{n+1} = j|X_{n+1})}{P(\omega_{n+1} = j)}p(\omega_n = j|X_n)\right]}$$

(40)

Use of Single-Neighbor Updating Equations for Large Local Neighborhoods

This subsection shows how single-neighbor updating equations can be used repeatedly to exploit spatial information in large local neighborhoods.

Spatially Uniform Context—Four Neighbors

Consider the pixel under consideration, pixel 0, and its neighbors in the local neighborhood shown in Figure 72. In this subsection, expressions are developed for obtaining the a posteriori probabilities of the classes of pixel 0, using information from its local neighborhood. Consider equation (41), where $f = p(X_0, X_1, \ldots, X_4)$. Using equations (13), (16), and (17) in equation (41) yields equation (42).

$$p(\omega_0 = i_0|X_0, X_1, \ldots, X_4) = \frac{p(\omega_0 = i_0, X_0, X_1, \ldots, X_4)}{p(X_0, X_1, \ldots, X_4)}$$

$$= \sum_{i_1=1}^{M} \cdots \sum_{i_4=1}^{M} \frac{p(\omega_0 = i_0, X_0, \omega_1 = i_1, X_1, \ldots, \omega_4 = i_4, X_4)}{f}$$

$$= \sum_{i_1=1}^{M} \cdots \sum_{i_4=1}^{M}$$

$$\times \frac{p(X_0, X_1, \ldots, X_4|\omega_0 = i_0, \omega_1 = i_1, \ldots, \omega_4 = i_4)P(\omega_0 = i_0, \omega_1 = i_1, \ldots, \omega_4 = i_4)}{f}$$

(41)

$$p(\omega_0 = i_0|X_0, X_1, \ldots, X_4)$$

$$= \frac{P(\omega_0 = i_0)p(X_0|\omega_0 = i_0)\prod_{j=1}^{4}\left[\sum_{i_j=1}^{M}p(X_j|\omega_j = i_j)P(\omega_j = i_j|\omega_0 = i_0)\right]}{f}$$

$$= \frac{p(\omega_0 = i_0|X_0)\prod_{j=1}^{4}\left\{\sum_{i_j=1}^{M}\left[\dfrac{p(\omega_j = i_j|X_j)}{P(\omega_j = i_j)}P(\omega_j = i_j|\omega_0 = i_0)\right]\right\}}{\sum_{i_0=1}^{M}p(\omega_0 = i_0|X_0)\prod_{j=1}^{4}\left\{\sum_{i_j=1}^{M}\left[\dfrac{p(\omega_j = i_j|X_j)}{P(\omega_j = i_j)}P(\omega_j = i_j|\omega_0 = i_0)\right]\right\}}$$

(42)

From equations (39) and (42), the following is easily understood. Updating the a posteriori probabilities of the classes of pixel 0, using information from its neighbors as shown in Figure 72, is equivalent to using the single-neighbor updating equation (39) repeatedly, taking one neighbor at a time. The sequence in which the neighbors are used is immaterial.

Sequential Neighborhood—General Case

This subsection considers the problem of updating the a posteriori probabilities of the classes of the pixel under consideration, pixel j, in a general sequential neighborhood. The location of pixel j in a general sequential neighborhood of N pixels is shown in Figure 80.

The transitions for which the estimated transition probabilities apply in the whole sequential neighborhood are indicated in Figure 80. Consider equation (43), where $NU(i_j)$ is the numerator of the first expression in equation (43). Using equations (13) and (25) in the numerator of equation (43) results in equation (44).

$$
\begin{aligned}
p(\omega_j = i_j | X_1, \ldots, X_j, \ldots, X_N) &= \frac{p(\omega_j = i_j, X_1, \ldots, X_N)}{p(X_1, \ldots, X_N)} \\
&= \frac{\sum_{i_1=1}^{M} \cdots \sum_{i_{j-1}=1}^{M} \sum_{i_{j+1}=1}^{M} \cdots \sum_{i_N=1}^{M} p(X_1, \ldots, X_N | \omega_1 = i_1, \ldots, \omega_j = i_j, \ldots, \omega_N = i_N)}{} \\
&\quad \times P(\omega_1 = i_1, \ldots, \omega_N = i_N) / p(X_1, \ldots, X_N)
\end{aligned}
$$

$$
= \frac{NU(i_j)}{\sum_{i_j=1}^{M} NU(i_j)}
\tag{43}
$$

$$
\begin{aligned}
NU(i_j) &= \sum_{i_1=1}^{M} \cdots \sum_{i_{j-1}=1}^{M} \sum_{i_{j+1}=1}^{M} \cdots \sum_{i_N=1}^{M} \left[\prod_{l=1}^{N} p(X_l | \omega_l = i_l) \right] \\
&\quad \times \left[P(\omega_1 = i_1) P(\omega_2 = i_2 | \omega_1 = i_1) \cdots P(\omega_N = i_N | \omega_{N-1} = i_{N-1}) \right] \\
&= p(X_j | \omega_j = i_j) \left\{ \sum_{i_1=1}^{M} \cdots \sum_{i_{j-1}=1}^{M} \left[\prod_{l=1}^{j-1} p(X_l | \omega_l = i_l) \right] \right. \\
&\quad \times \left[P(\omega_1 = i_1) P(\omega_2 = i_2 | \omega_1 = i_1) \right. \\
&\quad \left. \cdots P(\omega_j = i_j | \omega_{j-1} = i_{j-1}) \right] \bigg\} \\
&\quad \times \left\{ \sum_{i_{j+1}=1}^{M} \cdots \sum_{i_N=1}^{M} \left[\prod_{l=j+1}^{N} p(X_l | \omega_l = i_l) \right] \right\}
\end{aligned}
\tag{44}
$$

FIGURE 80. The pixel under consideration, pixel j, and its general sequential neighborhood.

$$\times \left[P(\omega_{j+1} = i_{j+1} \mid \omega_j = i_j) \cdots P(\omega_N = i_N \mid \omega_{N-1} = i_{N-1})\right]\Big\}$$

$$= p(X_j \mid \omega_j = i_j) \left[\sum_{i_{j-1}=1}^{M} p(X_{j-1} \mid \omega_{j-1} = i_{j-1})P(\omega_j = i_j \mid \omega_{j-1} = i_{j-1}) \right.$$

$$\left. \cdots \sum_{i_1=1}^{M} p(X_1 \mid \omega_1 = i_1)P(\omega_1 = i_1)P(\omega_2 = i_2 \mid \omega_1 = i_1) \right]$$

$$\times \left[\sum_{i_{j+1}=1}^{M} p(X_{j+1} \mid \omega_{j+1} = i_{j+1})P(\omega_{j+1} = i_{j+1} \mid \omega_j = i_j) \right.$$

$$\left. \cdots \sum_{i_N=1}^{M} p(X_N \mid \omega_N = i_N) \times P(\omega_N = i_N \mid \omega_{N-1} = i_{N-1}) \right]$$

If the numerator and denominator of equation (43) are divided by $[\prod_{l=1}^{N} p(X_l)]$, the numerator of equation (43) can then be written as shown in equation (45).

$$NU(i_j) = \frac{p(\omega_j = i_j \mid X_j)}{P(\omega_j = i_j)} \left[\sum_{i_{j-1}=1}^{M} P(\omega_j = i_j \mid \omega_{j-1} = i_{j-1}) \frac{p(\omega_{j-1} = i_{j-1} \mid X_{j-1})}{p(\omega_{j-1} = i_{j-1})} \right.$$

$$\left. \cdots \sum_{i_1=1}^{M} P(\omega_2 = i_2 \mid \omega_1 = i_1)p(\omega_1 = i_1 \mid X_1) \right] \tag{45}$$

$$\times \left[\sum_{i_{j+1}=1}^{M} P(\omega_{j+1} = i_{j+1} \mid \omega_j = i_j) \frac{p(\omega_{j+1} = i_{j+1} \mid X_{j+1})}{P(\omega_{j+1} = i_{j+1})} \right.$$

$$\left. \cdots \sum_{i_N=1}^{M} P(\omega_N = i_N \mid \omega_{N-1} = i_{N-1}) \frac{p(\omega_N = i_N \mid X_N)}{P(\omega_N = i_N)} \right]$$

The term in the first set of brackets of equation (45) is the contribution from pixels to the left of pixel j (see Fig. 80), the term in the second set of brackets is the contribution from pixels to the right of pixel j, and the first term is the contribution from pixel j to the a posteriori probabilities of the classes of pixel j. These contributions appear in multiplicative form in equation (45).

An examination of equations (35), (39), and (45) reveals that the single-neighbor updating equations (35) and (39) can be used repeatedly to update the a posteriori probabilities of the classes of pixel j, using information from its sequential neighborhood as follows. Equation (39) is used to update the a posteriori probabilities of pixel $(N - 1)$, using the a posteriori probabilities of pixels $(N - 1)$ and N. The updated a posteriori probabilities of pixel $(N - 1)$ and the a posteriori probabilities of pixel $(N - 2)$ are used to update those of pixel $(N - 2)$. Proceeding in a similar manner, the updated a posteriori probabilities of pixel $(j + 1)$ and the a posteriori probabilities of pixel j are used to update those of pixel j. Similarly, equation (35) is used to update the a posteriori probabilities of pixel j, using information from pixels to the left of pixel j. The a posteriori probabilities of pixels 1 and 2 are used to update those of pixel 2. The updated a posteriori probabilities of pixel 2 and the a posteriori probabilities of pixel 3 are used to update the a posteriori probabilities of pixel 3. The process is repeated until the updated a posteriori probabilities of pixel $(j - 1)$ and the previously updated ones of pixel j are used to update those of pixel j.

Application of Sequential Context to Two-Dimensional Neighborhoods

The expressions for the likelihood function and updating equations become complex with the increase in the size of the local neighborhood. Hence, it is proposed to use sequentially the sequential context for two-dimensional local neighborhoods. It is desirable that the updating be independent of the sequence of the sequential neighborhoods in which the updating is done. From equation (45) it is seen that, with the use of sequential neighborhoods (centering on the pixel under consideration), the updating is independent of the sequence of the sequential neighborhoods in which the updating is done. The sequential neighborhoods to be used in updating, then, are the ones centering on the pixel under consideration in four directions: 0°, 45°, 90°, and 135°. A few typical two-dimensional local neighborhoods composed of these sequential neighborhoods are illustrated in Figure 81.

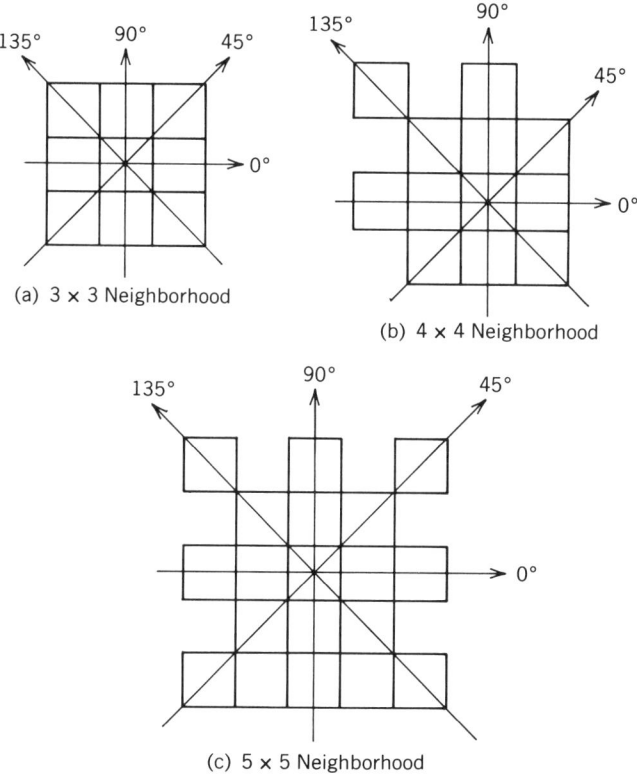

FIGURE 81. Some typical neighborhoods and updating directions.

Experimental Results

In this subsection, some results are obtained by applying the theory developed in the previous sections to the classification of the remotely sensed Landsat MSS data. Several segments* were processed in the following manner. The image was overlaid with a rectangular grid of 209 grid intersections, and the labels of the pixels or dots corresponding to each grid intersection were acquired. Two classes are in the image: Class 1 is wheat, and class 2 is nonwheat, designated "other." A linear classifier is trained on one-half of the labeled data. The remaining one-half of the labeled data is used as a test set. The a posteriori probabilities of the classes of the pixels are estimated by normalizing the discriminant function values of the classes.

Computational Results for a Typical 5 × 5 Neighborhood

The a posteriori probabilities of the classes of the pixels in a typical 5×5 neighborhood from an MSS image of segment 1739 are listed in Table 30. This segment is in Teton County, Montana.

The pixel under consideration is the central pixel of the neighborhood. The a priori probabilities are estimated as an average of the a posteriori probabilities in the neighborhood. Consider the following:

$$p(\omega = i) = \int p(\omega = i, X) \, dX$$

$$= \int p(\omega = i \mid X) p(X) \, dX$$

$$= \frac{E}{p(X)} [p(\omega = i \mid X)]$$

$$= \frac{1}{N} \sum_{j=1}^{N} [p(\omega = i \mid X_j)]$$

(46)

where X_j ($j = 1, 2, \ldots, N$) are the pixels in the local neighborhood. The a poste-

TABLE 30. The A Posteriori Probabilities of the Classes in a 5 × 5 Neighborhood

(0.716, 0.284)	(0.322, 0.678)	(0.820, 0.180)	(0.669, 0.331)	(0.326, 0.674)
(0.829, 0.171)	(0.899, 0.101)	(0.897, 0.103)	(0.762, 0.238)	(0.886, 0.114)
(0.625, 0.375)	(0.158, 0.843)	(0.285, 0.715)	(0.757, 0.243)	(0.117, 0.883)
(0.087, 0.913)	(0.062, 0.938)	(0.060, 0.940)	(0.080, 0.920)	(0.090, 0.910)
(0.125, 0.875)	(0.089, 0.911)	(0.132, 0.868)	(0.157, 0.843)	(0.127, 0.873)

Note: The first entry is $p(\omega = 1 \mid X)$ and the second entry is $p(\omega = 2 \mid X)$.

*A segment is a 9 km × 11 km (5 nautical mi × 6 nautical mi) area for which the MSS image is divided into a 117 row × 196 column rectangular array of pixels.

riori probabilities of the classes of the pixel under consideration are updated using sequential context and the procedure described earlier. This procedure is repeated for five iterations, and the computational results are listed in Table 31.

The true class of the central pixel is wheat, and without using the contextual information, the central pixel will be misclassified into class "other." Table 31 shows that, using contextual information from the local neighborhood, after the third iteration the central pixel is correctly classified.

Contextual Classification Results

Comparative results with and without using contextual information in classification are presented in this subsection. It is observed from the independent test set that the classification accuracy for this segment increased by 5% with the use of contextual information from the 3 × 3 neighborhood and by 7% with the use of contextual information from the 5 × 5 neighborhood (over the accuracies obtained without using contextual information). While generally preserving the boundaries, contextual classification corrected the misclassifications of many pixels and did this more accurately with data from the 5 × 5 neighborhood than with data from the 3 × 3 neighborhood.

Accuracies in the classification of MSS images of a few segments with and without the use of contextual information are listed in Table 32.

In general, an examination of the classification maps of full images and classification accuracies on the independent test set shows considerable improvement in the classifications with the use of contextual information. The improvement is greater with the increase in size of the neighborhood. The contextual classification of a full segment with a 5 × 5 neighborhood using the methods developed here took approximately 12 minutes of total time on the Purdue University Laboratory for Applications of Remote Sensing (LARS) IBM 3031 computer system.

The variance reduction factors obtained without using contextual information and with the use of contextual information from a local neighborhood of size 5 are

TABLE 31. Computational Results of Updating the A Posteriori Probabilities of the Central Pixel in a 5 × 5 Neighborhood Using the Linear Transition Probabilities Model

Iteration	A Posteriori Probabilities before Updating	A Priori Probabilities in the Neighborhood	Estimates of Parameter θ for Different Sequential Neighborhoods				A Posteriori Probabilities after Updating
			0°	45°	90°	135°	
1	$(0.285, 0.715)$	$(0.4087, 0.5913)$	0.0	0.2904	0.4	0.4	$(0.3574, 0.6426)$
2	$(0.3574, 0.6426)$	$(0.4130, 0.5870)$	0.0	0.2656	0.4	0.4	$(0.4315, 0.5685)$
3	$(0.4315, 0.5685)$	$(0.4173, 0.5827)$	0.0	0.2416	0.4	0.4	$(0.5034, 0.4966)$
4	$(0.5304, 0.4966)$	$(0.4216, 0.5784)$	0.0	0.2194	0.4	0.4	$(0.5699, 0.4301)$
5	$(0.5699, 0.4301)$	$(0.4255, 0.5745)$	0.0	0.1995	0.5	0.4	$(0.6363, 0.3637)$

TABLE 32. Classification Accuracies (Percentages) with and without Contextual Information

Segment	Location (county, state)	Without Context	With Sequential Context[a]			With Spatially Uniform Context
			NS = 5	NS = 4	NS = 3	
1005[b]	Cheyenne, Colorado	85.88	88.46	88.46	90.38	86.54
1060[b]	Sherman, Texas	80.77	85.58	82.69	81.73	81.73
1231[b]	Jackson, Oklahoma	89.42	91.35	91.35	90.38	91.35
1520[c]	Big Stone, Minnesota	84.62	87.50	85.58	86.54	84.62
1604[c]	Renville, North Dakota	60.58	63.46	60.58	59.62	60.58
1675[c]	McPherson, South Dakota	68.27	71.15	73.08	68.27	67.31
1739[c]	Teton, Montana	68.27	75.00	72.22	73.08	70.19

[a]NS = Neighborhood size.
[b]Segments in which class 1 is winter wheat.
[c]Segments in which class 1 is spring wheat.

listed in Table 33. Table 33 shows that there is a consistent improvement in the variance reduction factor with the use of contextual information in classification.

Conclusions

In this analysis the problem of incorporating contextual or spatial information into the classification of imagery data is considered. The contextual information is introduced into classification based on the spatial dependencies between the states of nature of neighboring pixels or based on transition probabilities. The dependencies between neighboring patterns are modeled with linear and nonlinear models through a single parameter θ, which describes the transition probabilities of the classes of the neighboring patterns. An expression is developed for the likelihood function of the pattern vectors from a general local neighborhood under the following reasonable assumptions: (a) the probability density function of a pattern, given its label, is independent of other patterns and their labels; and (b) the labels of the pattern vectors are independent of the labels of their nonneighbors. Specific expressions for the likelihood function are derived for different local neighborhoods and with different transition probabilities models. The parameter θ is estimated as the one that maximizes the likelihood function.

TABLE 33. Variance Reduction Factors with and without Contextual Information

Segment	Location (county, state)	Variance Reduction Factor	
		Without Context	With Sequential Context, $NS = 5$
1005	Cheyenne, Colorado	0.5720	0.5430
1060	Sherman, Texas	0.6227	0.4717
1231	Jackson, Oklahoma	0.4407	0.4173
1520	Big Stone, Minnesota	0.6194	0.5216
1604	Renville, North Dakota	0.9865	0.9741
1675	McPherson, South Dakota	0.9985	0.9248
1739	Teton, Montana	0.9271	0.8267

Expressions are presented for updating the a posteriori probabilities of the classes of a pixel using information from a single neighbor. It is shown that these expressions can be used to update the a posteriori probabilities of a pixel under consideration for spatially uniform context and in a general sequential neighborhood. The contextual information from two-dimensional neighborhoods is introduced into the classification of imagery data as well as through a sequence of sequential neighborhoods.

The techniques presented here are applied to the classification of remotely sensed MSS imagery data. Computational results for a typical 5 × 5 neighborhood are presented.

REFERENCES

1. Hanson, A. R., E. M. Riseman, and E. Fisher, Context in word recognition, *Pattern Recognition* 8, 1976: 35–45.

2. Duda, R. O., and P. E. Hart, Experiments in the recognition of hand-printed text, part II — context analysis, 1968 Fall Joint Computer Conference, *AFIFS Conf. Proc.*, vol. 33, Thompson, Washington, D.C., 1968, pp. 1139–1149.

3. Abend, K.: Compound decision procedures for unknown distributions and for dependent states of nature, in Pattern Recognition, ed. L. Kanal, Thompson, Washington, D.C., 1968, pp. 207–249.

4. Raviv, J., Decision making in Markov chains applied to the problem of pattern recognition, *IEEE Trans. Information Theory* IT-13, 1967: 536–551.

5. Alter, R., Utilization of contextual constraints in automatic speech recognition, *IEEE Trans. Audio Electroacoustics* AU-16, 1968: 6–11.

6. Chow, C. K., A recognition method using neighbor dependence, *IRE Trans. Electronic Computers* EC-11, 1962: 683–690.

7. Kettig, R. L., and D. A. Landgrebe, Classification of multispectral image data by extraction and classification of homogeneous objects, *IEEE Trans. Geoscience Electronics* GE-14, 1976: 19–26.

8. Haralick, R. M., K. Shanmugam, and I. Dinstein, Textural features for image classification, *IEEE Trans. Systems, Man, Cyb.* SMC-3, 1973: 610–621.

9. Swain, P. H., Bayesian classification in a time-varying environment, *IEEE Trans. Systems, Man, Cyb.* SMC-8, 1978: 879–883.

10. Welch, J. R., and K. G. Salter, A context algorithm for pattern recognition and image interpretation, *IEEE Trans. Systems, Man, Cyb.* SMC-1, 1971: 24–30.

11. Chittineni, C. B., *Research plan for developing and evaluating classifiers*, Tech. Memo LEC-13300, Lockheed Electronics Co., Inc., JSC-14849, NASA/JSC, Houston, April 1979.

12. Toussaint, G. T., The use of context in pattern recognition, *Pattern Recognition*, 10, 1978: 189–204.

13. Burnside, W. S., and A. W. Panton, *Theory of Equations*, S. Chand, New Delhi, 1967.

14. Cooper, L., and D. Steinberg, Introduction to methods of optimization, Saunders, Philadelphia, 1970, pp. 87–133.

15. Kelley, L. G., *Handbook of numerical methods and its applications*, Addison-Wesley, Reading, Mass., 1967, pp. 356–359.

16. Bingham, J. A. C., An Improvement to iterative methods of polynominal factorization, *Commun. ACM*, 10, 1967: 57–60.

2.8 LAND USE MAP ACCURACY CRITERIA*

All forms of digital image processing of remotely sensed data require some form of results evaluation. Some forms of processing produce land use maps. Methods for assessing the accuracy of land use maps derived from remotely sensed imagery and related information are described in this section. The paper reproduced here broke new ground at the time it was published, and it remains valid today. A number of articles and alternative views stimulated by this article were published in *Photogrammetric Engineering and Remote Sensing* in the late 1970s.

Introduction

Remote sensing, particularly medium- and high-altitude aerial photography, is being used increasingly for the preparation of land use inventory maps by a variety of users. Until recently, this increase of use, representing an important advance for

*By R. M. Hord and W. Brooner, from *Photogrammetric Engineering and Remote Sensing*, Vol. 42, No. 5 (May), 1976, p. 671–677.

the application of aerial photographic interpretation, has been gradual, due in part to the nature of the technique and to product acceptance in user organizations.

Some organizations have been using aerial photography for planning and land use applications for several decades. Their applications have generally involved large-scale black-and-white photography, which has often served only as a base for plotting ground survey data, rather than exploiting the photography as an information source. Such users have not taken advantage of the inherent benefits of photo interpretation, such as consistency of data, timeliness, or accuracy, nor of color, color infrared, multiscale, or other remote sensing techniques conventionally available today. Ground surveys, however, almost always lead to inconsistencies in detail over space as well as to extensive requirements for collection time and resources.

Acceleration of the acceptance of photo-derived maps, however, requires addressing issues for which the remote sensing technologist is often unprepared. Included, for example, are such issues as cost/benefit in relation to traditional approaches, and accuracy of results. For the sake of discussion, it is asserted that the costs of data acquisition and analysis have been demonstrated to be markedly lower with aerial photo interpretation than with ground surveys in all but the largest-scale and most detailed, site-specific cases. Moreover, costs reduce as the scale of the data is reduced. Given the cost advantage, the discussion issue then becomes one of whether accuracy is sufficient when medium- and high-altitude photography is the principal data source. The remainder of this discussion focuses on the issue of quantifying map accuracy.

Accuracy of land use interpretation is a complex issue, both in its definition and measurement. For example, an area delineated and classified as a particular category may be in error for one or more of three reasons: (1) classification error, (2) boundary line error, and (3) control point location error. Their interrelationship will be addressed later. Each factor is of varying importance to the user, whose specification requirements differ, and each impacts the utility of the resulting products.

Classification Accuracy Estimation

Hierarchical classification schemes such as those used by the U.S. Geological Survey further complicate the first type of error. For example, two levels of detail are indicated, such as, "11" which represents "urban residential land use." The first digit (level I) is urban and built-up areas, and the second digit (level II) is residential areas. It can be seen, therefore, that an area which should be called 11 but is called 12 is in error at level II, but is correct at level I. In defining a classification accuracy estimate, either the classification must be correct at all levels reported or the level of classification must be stated for the associated accuracy estimate (i.e., x percent at level I, y percent at level II, and so on). We assert that a single accuracy estimate will have the greatest unambiguous meaning to the majority of map users; their concern for accuracy implies concern for quality and is often not supported by an understanding of the complexities which arise in attempting to quantify the accuracy of land use mapping. Therefore, a single value must refer to the lowest classification denominator: For a classification to be correct it must be correct at all levels reported.

A map is a graphic interpretation and the presentation of a complex surface that often contains abstractions. Without field-checking the total map, exact accuracy cannot be verified; hence a sampling procedure must be employed to estimate classification accuracy.

Any sampling procedure to be employed must involve field work and must be statistically valid. It is important that users understand that any accuracy estimate based on sampling requires confidence intervals which are dependent on the number of sample points selected per map.

The procedure below is suitable, as an example, for estimating accuracy for 1-acre sample points on land use maps at $1:24,000$ scale and covering $7\frac{1}{2}'$ areas (approximately 56 mi^2). The procedure involves the random selection of 1-acre points with replacement, ground-checking these points, and comparing the field observer's classification to that of the aerial photo interpreter. The estimated accuracy of classification may then be computed within specified confidence intervals.

At each level of classification (e.g., level I, level II) it is assumed that there is a "true" category for each acre (or other land surface unit) on the map. (This assumption can, of course, be challenged. Land cover and/or land use phenomena are sometimes too ambiguous to permit discrete classification, even for ground-based observers.) Each acre's assigned category is correct or incorrect (1 or 0), and the set of both is the population of concern. The mean μ of this population equals the sum of the population elements divided by the number n of these elements; that is,

$$\mu = (1/n) \sum_{i=1}^{n} x_i \tag{1}$$

Now, μ is also p, the probability that any given acre has a correctly assigned classification. Since this is a binominal distribution, the variance is

$$\sigma^2 = p(1 - p) = \mu(1 - \mu) \tag{2}$$

The confidence interval for μ can be calculated from the approximation[1]

$$P\left(-b < \left(\frac{\bar{x} - \mu}{\sigma/\sqrt{N}}\right) < b\right) = 1 - \alpha \tag{3}$$

where N is the item count of the sample, $\bar{x}$ is the sample mean, and $100(1 - \alpha)$ percent is the confidence level of the interval. For a 95% confidence interval, $\alpha = 0.05$ and, from the normal distribution tables, $b = 1.960$; hence

$$\left(\frac{\bar{x} - \mu}{\sigma/\sqrt{N}}\right)^2 < (1.96)^2$$

That is,

$$N(\bar{x}^2 - 2\bar{x}\mu + \mu^2) < (1.96)^2\mu(1 - \mu)$$

This is solved for upper and lower limits of μ.
   Example 1: If

$$N = 300 \quad \text{and} \quad \bar{x} = 0.98$$

then

$$0.9570 < \mu < 0.9908$$

That is, the true map accuracy is, with 95% confidence, in the range 0.9570–0.9908. The sample accuracy is 0.98 from a sample check of 300 points.
   Example 2: If

$$N = 300 \quad \text{and} \quad \bar{x} = 0.96$$

then

$$0.9314 < \mu < 0.9770$$

Table 34 shows the map accuracy upper and lower 95% confidence limits as a function of the number of samples and the accuracy value for these samples.
   To use this table, the user first determines the classification accuracy level to be achieved, such as 80%. Then the number of samples to be checked is determined, such as 150 points randomly selected with replacement. The table then shows by the lower 95% confidence limit that the minimum number of correct points required from the sample to achieve the specified 80% accuracy within the 95% confidence interval is, in this example, 131 correct points, so that $\bar{x}$ will be 87%.
   Table 34 also may be used to estimate classification accuracy. For example, given a sample of 250 points, of which 235 are found to be correct, then since $\bar{x}$ is 94%, the estimated lower limit of accuracy is 90% with 95% confidence. Again, with 95% confidence, the upper limit of accuracy is 96%.
   Another useful quantitative descriptor for land use maps is the spatial complexity index, defined as the average polygon ground area divided into the total area. Particularly as the digital multispectral-classification-generated land use maps find greater acceptance and hence the occurrence of single-pixel polygons increases, the spatial complexity index, which quantifies the spatial homogeneity of the map, has become an increasingly used concept. For example, on a map covering 36,000 acres with an average polygon area of 40 acres, the spatial complexity index would be expressed as 900.

TABLE 34. Map Accuracy Upper and Lower 95% Confidence Limits (μ) as a Function of the Number of Samples (N) and the Accuracy Values ($\bar{x}$) for These Samples

N	$\bar{x}$ (%)	μ (lower limit)	μ (upper limit)	N	$\bar{x}$ (%)	μ (lower limit)	μ (upper limit)
50	80	0.6896	0.8876	250	80	0.7461	0.8449
50	81	0.6808	0.8950	250	81	0.7568	0.8538
50	82	0.6920	0.9023	250	82	0.7677	0.8627
50	83	0.7034	0.9095	250	83	0.7785	0.8715
50	84	0.7149	0.9166	250	84	0.7895	0.8802
50	85	0.7624	0.9236	250	85	0.8005	0.8889
50	86	0.7381	0.9305	250	86	0.8115	0.8976
50	87	0.7500	0.9372	250	87	0.8227	0.9061
50	88	0.7620	0.9438	250	88	0.8339	0.9148
50	89	0.7741	0.9503	250	89	0.8452	0.9230
50	90	0.7864	0.9565	250	90	0.8566	0.9313
50	91	0.7989	0.9626	250	91	0.8681	0.9395
50	92	0.8116	0.9684	250	92	0.8797	0.9476
50	93	0.8246	0.9741	250	93	0.8914	0.9555
50	94	0.8378	0.9794	250	94	0.9034	0.9633
50	95	0.8514	0.9844	250	95	0.9155	0.9708
50	96	0.8654	0.9890	250	96	0.9280	0.9781
50	97	0.8799	0.9930	250	97	0.9407	0.9850
50	98	0.8950	0.9965	250	98	0.9541	0.9914
50	99	0.9111	0.9990	250	99	0.9683	0.9969
50	100	0.9287	1.0000	250	100	0.9849	1.0000
100	80	0.7112	0.8666	300	80	0.7511	0.8413
100	81	0.7222	0.8749	300	81	0.7618	0.8504
100	82	0.7333	0.8830	300	82	0.7726	0.8593
100	83	0.7445	0.8911	300	83	0.7834	0.8683
100	84	0.7558	0.8990	300	84	0.7943	0.8771
100	85	0.7672	0.9069	300	85	0.8052	0.8860
100	86	0.7786	0.9147	300	86	0.8162	0.8947
100	87	0.7902	0.9224	300	87	0.8272	0.9034
100	88	0.8019	0.9300	300	88	0.8383	0.9120
100	89	0.8137	0.9375	300	89	0.8495	0.9206
100	90	0.8256	0.9448	300	90	0.8608	0.9291
100	91	0.8377	0.9519	300	91	0.8722	0.9374
100	92	0.8500	0.9589	300	92	0.8837	0.9457
100	93	0.8625	0.9657	300	93	0.8954	0.9538
100	94	0.8752	0.9722	300	94	0.9072	0.9617
100	95	0.8882	0.9785	300	95	0.9192	0.9695
100	96	0.9016	0.9843	300	96	0.9314	0.9770
100	97	0.9155	0.9897	300	97	0.9440	0.9842
100	98	0.9300	0.9945	300	98	0.9571	0.9908
100	99	0.9455	0.9982	300	99	0.9710	0.9966
100	100	0.9630	1.0000	300	100	0.9874	1.0000
150	80	0.7289	0.8562	350	80	0.7549	0.8385

TABLE 34. — *Continued*

150	81	0.7398	0.8647	350	81	0.7656	0.8476
150	82	0.7508	0.8732	350	82	0.7763	0.8567
150	83	0.7618	0.8817	350	83	0.7871	0.8657
150	84	0.7730	0.8901	350	84	0.7979	0.8747
150	85	0.7842	0.8984	350	85	0.8088	0.8836
150	86	0.7955	0.9066	350	86	0.8197	0.8925
150	87	0.8068	0.9147	350	87	0.8307	0.9013
150	88	0.8183	0.9227	350	88	0.8418	0.9100
150	89	0.8299	0.9306	350	89	0.8529	0.9186
150	90	0.8416	0.9384	350	90	0.8641	0.9272
150	91	0.8534	0.9461	350	91	0.8754	0.9357
150	92	0.8654	0.9536	350	92	0.8868	0.9441
150	93	0.8776	0.9610	350	93	0.8983	0.9523
150	94	0.8899	0.9681	350	94	0.9100	0.9604
150	95	0.9026	0.9750	350	95	0.9219	0.9683
150	96	0.9155	0.9815	350	96	0.9340	0.9760
150	97	0.9289	0.9876	350	97	0.9464	0.9834
150	98	0.9429	0.9931	350	98	0.9593	0.9903
150	99	0.9579	0.9976	350	99	0.9730	0.9963
150	100	0.9750	1.0000	350	100	0.9891	1.0000
200	80	0.7391	0.8495	400	80	0.7580	0.8363
200	81	0.7500	0.8583	400	81	0.7687	0.8454
200	82	0.7609	0.8670	400	82	0.7794	0.8546
200	83	0.7718	0.8757	400	83	0.7901	0.8636
200	84	0.7829	0.8843	400	84	0.8009	0.8727
200	85	0.7940	0.8929	400	85	0.8117	0.8817
200	86	0.8051	0.9013	400	86	0.8226	0.8906
200	87	0.8163	0.9097	400	87	0.8335	0.8995
200	88	0.8277	0.9180	400	88	0.8445	0.9083
200	89	0.8391	0.9262	400	89	0.8555	0.9170
200	90	0.8506	0.9343	400	90	0.8667	0.9257
200	91	0.8622	0.9423	400	91	0.8779	0.9343
200	92	0.8740	0.9501	400	92	0.8892	0.9428
200	93	0.8860	0.9578	400	93	0.9007	0.9511
200	94	0.8981	0.9653	400	94	0.9123	0.9593
200	95	0.9104	0.9726	400	95	0.9240	0.9674
200	96	0.9231	0.9796	400	96	0.9360	0.9752
200	97	0.9361	0.9862	400	97	0.9483	0.9828
200	98	0.9497	0.9922	400	98	0.9610	0.9898
200	99	0.9643	0.9972	400	99	0.9745	0.9961
200	100	0.9812	1.0000	400	100	0.9905	1.0000

Boundary Errors

The second type of error, boundary lines, involves both locational tolerances and line widths relative to the scale of the final map. How wide is the map line in terms of ground distance? Line width data for standard Koh-I-Noor pen point sizes is given in Table 35. Depending on the map scale, ground distances can be considerable even when within tolerance, a fact of no surprise to those familiar with cartographic representation.

Today, with the growing access to computers and to scanning densitometers, digital cartography is finding greater acceptance. In this context maps are often represented in raster format. Each point of a map is coded as a 1 or 0, depending on whether it is or is not an element of a line. (For optimal utility the raster point diameter should approximate the line width.) If these points are spaced 10 mils between centers, the array constituting the digital codification of a 20 in. × 30 in. map would consist of 2000 lines of 3000 points each, a total of 6 million points.

A figure of merit is available[2] for specifying the degree of congruence between two such raster maps. Each point on map 1 and the corresponding point on map 2 is inspected to determine whether each is or is not an element of a line. Four totals are calculated: A, the total number of line points on map 1 for which the corresponding point on map 2 is also a line point; B, the total number of line points on map 1 for which the corresponding point on map 2 is not a line point; C, the total number of nonline points on map 1 for which the corresponding point on map 2 is a line point; and D, the total number of nonline points on map 1 for which the corresponding point on map 2 is not a line point.
Then

$$S_1 = A + C$$

$$S_2 = B + D$$

$$S_3 = A + B$$

$$S_4 = C + D$$

$$R = A + B + C + D$$

$$E_A = (S_1 S_3)/R$$

$$E_B = (S_2 S_3)/R$$

$$E_C = (S_1 S_4)/R$$

$$E_D = (S_2 S_4)/R$$

$$X = (A - E_A)^2/E_A + (B - E_B)^2/E_B + (C - E_C)^2/E_C + (D - E_D)^2/E_D$$

(4)

The larger X is, the better is the agreement between the two maps. For example, three maps in raster format each measuring 1000 × 1000 points are available. Which two are in better agreement if we arrive at the following?

In comparing maps 1 and 2, $A = 200{,}000$, $B = 500{,}000$, $C = 200{,}000$, and $D = 100{,}000$

In comparing maps 2 and 3, $A = 600{,}000$, $B = 100{,}000$, $C = 100{,}000$, and $D = 200{,}000$

TABLE 35. Ground Line Widths at Various Map Scales for Standard Koh-I-Noor Pen Point Sizes

Koh-I-Noor Line	Actual Width (in.)	Ground Distance (ft)				
		1:24,000	1:63,360	1:100,000	1:250,000	1:500,000
6 × 0	0.005	10	26.4	41.67	104.0	208.0
5 × 0	0.0075	15	39.6	62.5	156.0	312.0
4 × 0	0.012	24	63.4	100.0	250.0	500.0
3 × 0	0.014	28	74.0	116.7	291.7	583.3
00	0.016	32	84.5	113.3	333.3	666.7
0	0.017	34	89.8	141.7	354.2	708.3
1	0.021	42	110.9	175.0	437.5	875.0
2	0.023	46	121.4	191.7	479.3	958.5
$2\frac{1}{2}$	0.028	56	147.8	233.3	583.3	1,166.7
3	0.037	74	195.4	308.3	770.8	1,541.7
4	0.052	104	274.6	433.3	1,083.3	2,166.7
6	0.067	134	353.7	558.3	1,395.7	2,791.4
7	0.068	136	359.1	566.7	1,416.7	2,833.3
8	0.09	180	475.2	750.0	1,875.0	3,750.0
9	0.10	200	528.0	833.3	2,083.3	4,166.7
10	0.15	350	792.0	1,249.9	3,125.0	6,250.0

In comparing maps 1 and 3, $A = 300,000$, $B = 200,000$, $C = 300,000$, and $D = 200,000$

The three values of X are 127,000, 274,000, and 100,000. Hence, the best agreement is between maps 2 and 3, while the worst agreement is between maps 1 and 3.

An alternative measure is $(A + D)/R$, a binomial variable comparable to x of the previous subsection.

Control Point Location Accuracy

Control point accuracy deals with the absolute geometric relationship of the map with respect to a universal frame of reference such as latitude and longitude, whereas boundary errors are associated with relative geometric fidelity.

For this discussion we will use a broad definition of the term *control point:* any identifiable map representation of a landmark. By this we mean to exclude points in the middle of homogeneous areas but include field corners, buildings, and river and road intersections, that is, anything whose true ground position can be surveyed. The U.S. National Map Accuracy Standards state that for a map to be termed accurate, "for maps on publication scales larger than 1:20,000 not more than 10 per cent of the points tested shall be in error more than $\frac{1}{30}$ inch, measured on the publication scale; for maps on publication scales of 1:20,000 or smaller, $\frac{1}{50}$ inch. These limits of accuracy shall apply in all cases to well-defined points only." Since we are dealing here with land use maps that do not carry elevation information, vertical accuracy does not concern us. Hence for 1:100,000 scale maps 90% of all ground locations are claimed to be accurate to within 167 ft, whereas on 1:1,000,000 scale maps this ground location tolerance is 1667 ft. How many points need a cartographer check to have 95% confidence that the map meets this standard? We may again use the binomial distribution here, since each of the n control points on the map is either accurately placed or it is not. We want 90% of all the points to be accurately placed, and this will be so if the binomial probability p of correct placement is 0.9. Let S be the number of correct placements in m trials, that is, after checking m map points on the ground. The expected value of S is mp, and the variance of S is $mp(1 - p)$.

The variable

$$y = \frac{S - mp}{\sqrt{p(1 - p)}} \tag{5}$$

is approximately normally distributed with zero mean and unit standard deviation if m is large:

$$P(-a < y < a) = 0.95 \quad \text{for} \quad a = 1.96$$

We may assess both limits on p by solving $y^2 = a^2$ for p; that is,

$$\frac{S^2 - 2mpS + m^2p^2}{p(1 - p)} = a^2 \tag{6}$$

Example 3: $m = 5$ and $S = 5$; that is, we have ground-checked five randomly selected points and each is correctly placed. The 95% confidence interval for p is

$$0.866 < p < 1.00$$

That is, upon checking five points and finding them correct, you may be 95% confident that the probability of correct placement for any point is greater than 0.866.

Example 4: If $S = 25$ and $m = 25$, then

$$0.994 < p < 1.00$$

Here we may say with 95% confidence that the map is at least 99.4% accurate in control point placement.

Other Considerations

It is awkward to quote three values to specify map accuracy. Several composite figures of merit may warrant consideration. The authors suggest adoption of the root mean square (rms) of the lower 95% confidence limit of the 3 percent measures presented earlier. Hence if the polygon classification accuracy is at least 90%, the boundary accuracy is at least 95%, and the control point correctness is at least 97%, each at 95% confidence, then the RMS accuracy of these is 94%.

Obviously there is reason for dissatisfaction with these accuracy measures. One reason is that these three types of error are not independent. If a boundary is in error, then the adjacent polygons will be affected by having multiple land use categories in some polygons as one example of this interdependency. Another is that the field observations are taken as the standard for quality control without recognizing that ground observations are frequently in error. Despite these drawbacks and others, the authors propose these evaluation methods with the realization that no figure of merit is ever entirely appropriate.

Earlier reference was made to the cost/benefit advantage of maps generated from remotely sensed data. This can be valid only so long as the amount of ground checking for accuracy assessment is kept reasonable. In this regard the user is advised not to demand accuracy in excess of his real requirement. If the particular application for which a given map is generated need only be 80% accurate in the above rms sense, then specifying 98% accuracy is not good practice.

Finally, it is noted that the time dimension has not been addressed in this discussion. How do these error measures change with age? Is it preferable to use a map with 80% rms accuracy generated from imagery acquired 2 months ago or a map with 90% rms accuracy generated from year-old imagery?

These and similar questions may be as complex as the issue of accuracy measurement itself, and generalizations may be hazardous in their simplicity. Land use change per se is the issue here, and an aerial photograph may be regarded as a historical document which captures the landscape scene at only a single moment. The most suitable date assigned to a given land use map is the date of photo acquisition rather than photo interpretation unless field surveys can confidently update all areas of land use change.

The dynamics of land use change need examination; often the rate of change in seemingly dynamic areas is relatively low when considered in a county or regional context. From ground observations we can note (perhaps with alarm) the expansion of urban areas into adjacent hinterlands. Within central urban areas, the replacement process may be very visible to the local observer. But considered in the context of scale, new subdivisions may consume only relatively small amounts of available land per year, and many replacements may be new structures of the same previous land use. Such situations tend to reduce the necessity of 2-month-old imagery given existing imagery of comparable quality that may be a year old.

To the contrary, field surveys for interpretation verification are complicated in areas of land use change or conversion subsequent to the time of photo acquisition. What may have been an accurate interpretation of the photograph is not confirmed by field survey. The observation may lead to a recorded error in the classification accuracy estimation, penalizing the interpreter who performed his task properly and well and reducing (erroneously) the quantified quality of the map product.

Conclusion

To foster the acceptance of land use mapping from remote sensing, the cartographer must be able to specify the accuracy of his product. The three types of inaccuracy (area misclassification, boundary line error, and control point location error) have been addressed, and a procedure for quantitatively specifying each type of error has been described.

The topic is by no means closed. For example, studies to ascertain the optimal accuracy requirement specification for various applications would examine the trade-offs in the cost of field checking versus the cost of inadvertently accepting a map error.

Since the procedures are statistical in nature, some will mistrust them. Certainly counterexamples can be produced — for example, cases exhibiting good values for these figures of merit while the maps are obviously bad, at least aesthetically. Generally the cartographer's clients realize this. Maps characterized by good values for these figures of merit, even those generated from remote sensing data, should find ready acceptance. In many cases maps generated by using traditional methods may not score as well.

REFERENCES

1. Brunk, H. D., *An introduction to mathematical statistics,* Blaisdell, Waltham, Mass., 1965, pp. 177 ff.
2. Hord, R. M., *Comparison of edge and boundary detection techniques,* AIPR Symposium, Washington, D.C., December 1971.

Chapter **3**

Applications

Chapter 3 considers the use of sensors and techniques like the ones discussed earlier to derive economic benefits in four application areas. The four selected disciplines are by no means an exhaustive listing of areas in which digital image processing of remotely sensed data can have payoff. Other civilian areas include urban planning, environmental protection, search and rescue, mapping, energy conservation, wildlife preservation, and public health, to name but a few.

Meteorology is the first area visited. The economic impact of remote sensing has been larger in the realm of meteorology than in any other. The others are geology, agriculture, and maritime. Of course, the surface can only be scratched; there is not enough room to include the full range of the applications opportunities even in just these four selected areas. But perhaps some sense of the diversity and power of the digital methods can be conveyed.

3.1 METEOROLOGY

The next two subsections are devoted to the applications of digital image processing to meteorology. Section 3.1.1 deals with the use of data from the Electronically Scanned Microwave Radiometer to assess precipitation rates based on brightness temperature and the detection of severe storms. A later related paper by Rogers, Siddalingaiah, Chang, and Wilheit* extends the utility of ESMR data to

*E. Rogers, H. Siddalingaiah, A. Chang, and T. Wilheit, A statistical technique for determining rainfall over land employing Nimbus 6 ESMR measurements, *Journal of Applied Meteorology* 18(8), 978–991, 1979.

determining precipitation over land. The selected paper was chosen for its clarity to the nonspecialist. Section 3.1.2 reports on the use of VISSR (Visible/Infrared Spin Scan Radiometer) imagery to determine wind estimates from automated measurements of cloud motions. Automated cloud motion measurement has also been studied extensively at the University of Wisconsin.

3.1.1 The Nimbus 5 ESMR and its Application to Storm Detection*

Introduction

In the years which have ensued since the launch of the Tiros I meteorological satellite, a number of significant sensor advancements have been made which are contributing greatly to operational meteorological requirements. Television cameras and infrared "window" scanners made day/night cloud cover mapping a reality almost from the beginning. The Tiros medium-resolution infrared radiometer provided a first "view" of distributions of atmospheric water vapor and carbon dioxide as well as information on global albedo and long-wave radiation. The Nimbus-3 Infrared Interferometer Spectrometer (IRIS) and Satellite Infrared Spectrometer provided the first measurements of the vertical distribution of temperature and some information on the vertical and horizontal distributions of ozone, water vapor, and minor gas constituents.

Thus by Nimbus 4 (launched in April 1970), it had been proven that satellites could map cloud cover and elements of the heat and water budgets, determine atmosphere temperature profiles, and define the distribution of some of the gases which make up the atmosphere. A significant gap was, however, fairly obvious; that is, there was no way to utilize satellite observations to directly measure phenomena that could provide quantitative information on precipitation, and there was no way to obtain information on surface phenomena in the presence of clouds.

The launch of Nimbus-5 in 1972 opened a new era in meteorological satellites. Among other useful sensors, Nimbus-5 carried two microwave sensors that were designed to (*a*) provide data on cloud liquid water contents (ESMR), and (*b*) provide vertical temperature soundings in the presence of most cloud types (Nimbus Environmental Monitoring System, NEMS). In this study only the Electrically Scanned Microwave Radiometer (ESMR) data have been processed and examined.

The ESMR was designed to permit the mapping of the distribution of heavy rain clouds, for example, thunderstorm producing cells, over open ocean regions. The pictorial data displays of ESMR data dramatically reveal the proof of the design concept; for example, the rain-producing zones at frontal systems over the oceans are displayed in a manner that invites comparison with meteorological radar PPI presentations of storm systems. Since the ESMR measures emitted

*Based on R. Sabatini and E. Merritt, *The Nimbus 5 ESMR and its application to storm detection,* Earth Satellite Corp., Washington, D.C., July 1973.

radiation at 1.55 cm, a not uncommon frequency for some types of meteorological radars, the inferred parallel has a reasonably sound basis.

Description of the Nimbus 5 ESMR Experiment and Data

The ESMR is sensitive to radiation from 19.225 to 19.475 GHz (19.35-GHz center frequency). The microwave antenna scans perpendicular to the spacecraft velocity vector from 50° to the left of nadir to 50° right of nadir, encompassing 78 scan steps every 4 seconds. The beam width is 1.4° × 1.4° near nadir and degrades to 2.2° cross-track × 1.4° downtrack at the 50° extremes. At 1100 km altitude the resolution is 25 km × 25 km near nadir, degrading to 160 km cross-track × 45 km downtrack at the end of the scan. The beam width is inversely proportional to the cosine of the angle of scan, and the scan control logic is arranged to increase the angular separation between adjacent beam positions accordingly. The nominal center of the j beam position is

$$\alpha_j = \arcsin[\sin(50°)(2j - 79)/77] \qquad (1)$$

Thus

$$\alpha_1 = -50° \qquad \text{(50° to the left of spacecraft)}$$

$$\alpha_{78} = +50° \qquad \text{(50° to the right of spacecraft)}$$

There is no beam which is exactly nadir-viewing. Figure 82 graphs the distance of the center of each beam from the subsatellite track. The scanning geometry nearly compensates for the distortions introduced by the earth's curvature in the center half of the image.

The area overlap between successive orbits permits complete global coverage in 12 hours.

Brightness temperatures are measured at each scan position and are displayed in shades of gray, producing a microwave image of the earth.

ESMR data are acquired at the Rosman and Alaska data acquisition facilities and relayed to the Meteorological Data Handling System (MDHS) at Goddard Space Flight Center. The MDHS produces images and Stacked Experiment Tapes (SET) containing ESMR data. Computer programs are then utilized to produce listings and maps of brightness temperatures from the SET as described in the following section.

For a more detailed discussion of the ESMR experiment the reader is referred to the *Nimbus 5 User's Guide* (Nimbus Project, 1972).

Nimbus ESMR Data Reduction Computer Programs

The purpose of this subsection is to present an overview of the computer programs developed by NASA to process the ESMR data collected by the Nimbus 5 satellite.

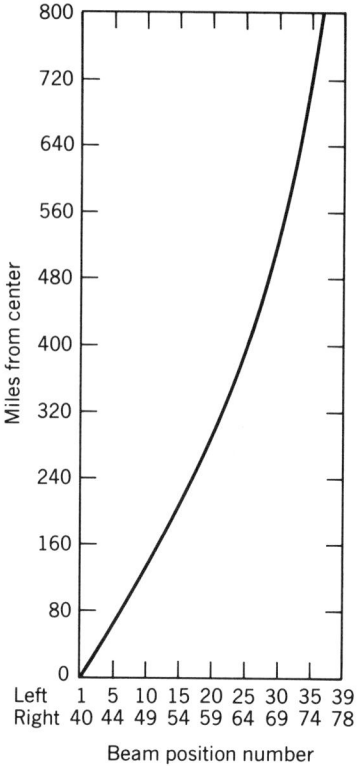

FIGURE 82. Distance in miles of center of ESMR beams from the subsatellite track.

The following three programs and associated subroutines have been obtained from NASA:

1. **ESMR Data Calibration Program**—converts the antenna data counts on the SET to brightness temperatures for the Calibrated Brightness Temperature (CBT tape).
2. **Nimbus CBT Listing Program**—provides a printout of ESMR brightness temperatures.
3. **ESMR Grid Print Map Program**—provides a printer projection map for ESMR brightness temperatures. Three projections are available:
   A polar stereographic projection
   A Mercator projection
   A horizontal stereographic projection

Additional ESMR programs are available from NASA and are described in the *Nimbus ESMR Data Reduction Program User's Guide* (1972). Figures 83 and 84 present a general flow diagram of the ESMR programs.

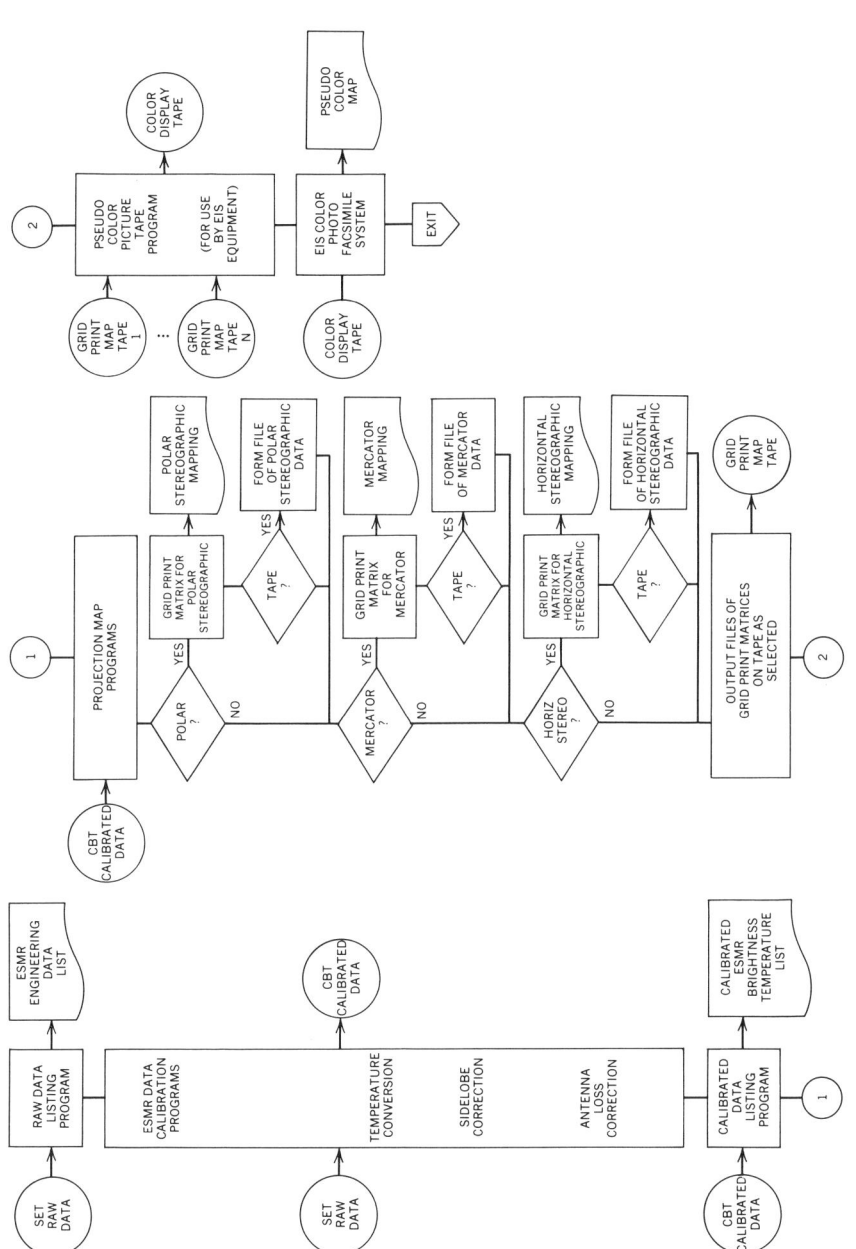

FIGURE 83. Broad flow diagram for processing ESMR tapes, I.

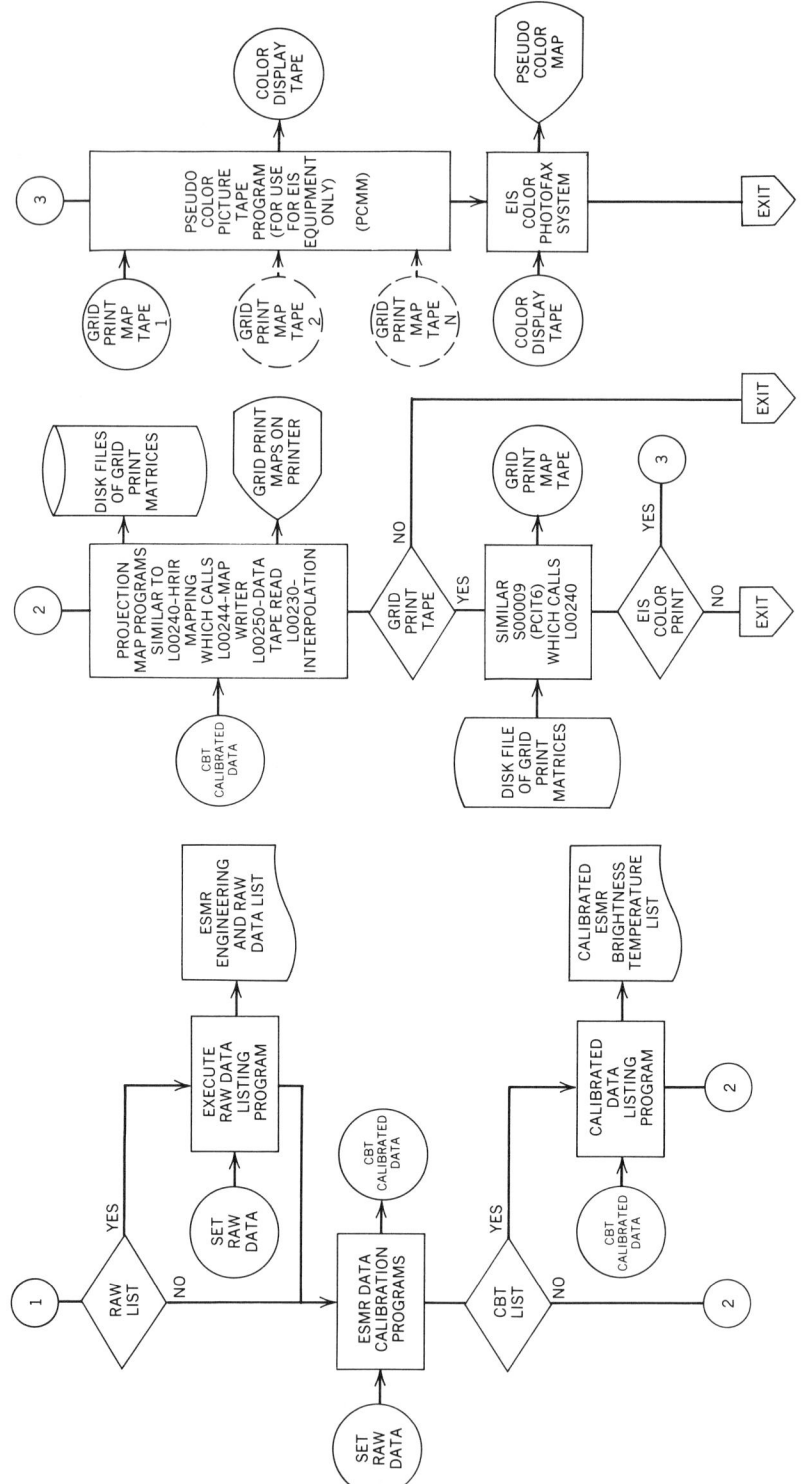

FIGURE 84. Broad flow diagram for processing ESMR tapes, II.

ESMR Data Calibration Program

The Data Calibration Program determines the latitude–longitude location for each data point and converts the antenna voltage counts to brightness temperatures. The program corrects for antenna sidelobe and antenna loss.

CBT Listing Program

The CBT Listing Program provides a printout of the ESMR calibrated brightness temperatures. The brightness temperatures for scans 1 through 78 are listed in 32 scans per page format. The starting times for each scan are also listed. Scan position 79 contains hot and cold calibration counts. Scan position 80 contains multiplexor channel data. The program also lists separately engineering data for each Nimbus major frame. These data include latitude, longitude, and altitude of subsatellite points for corresponding time.

Figure 85 presents a sample listing of the brightness temperatures recorded by the ESMR on January 12, 1973 during a nighttime Nimbus pass over the Caribbean and southeastern United States. The 200°K isotherm roughly outlines the land areas. Cuba is visible at the bottom. At the center is the elongated shape of Florida, with a band of rainclouds southeast of it and another crossing it. If this same printout is produced with a space between each line, the data would be mapped at a nearly 1:1 ratio between the vertical (along-subsatellite-track) and the horizontal (cross-track) distances. The resulting map would have 3×3 temperature values per square inch for an approximate scale of 1:2,700,000 in the center half of the scan. The instrument antenna scans cross-track nonlinearly, nearly compensating for the distortion introduced by the earth's curvature in the center half of the scan. Thus, the Nimbus CBT Listing Program can produce a map of the brightness temperatures which is able to show all data points with little distortion in the center half of the image. Such mapping was done for the rectangular area outlined in Figure 85 and is shown in Figure 89. The area shows the southern tip of Florida up to Lake Okeechobee and a precipitation band just off Florida. The distance between each point is approximately 15 miles, the resolution of the ESMR at the subsatellite point. This map and its comparison to radar data is discussed in greater detail later.

ESMR Grid Print Map Program

The ESMR Grid Print Map Program produces grid maps from the Calibrated Brightness Temperature tapes. The three projections available are (1) polar stereographic, (2) Mercator, and (3) horizontal stereographic. Each projection is defined in terms of map scale and geographic area. With each grid print map comes a data population map indicating the number of individual measurements contained in each grid point average and a latitude–longitude description for geographically locating the data. Options permit contouring of the grid print maps and use of less than the full 78 beam positions. Figure 86 shows sample output.

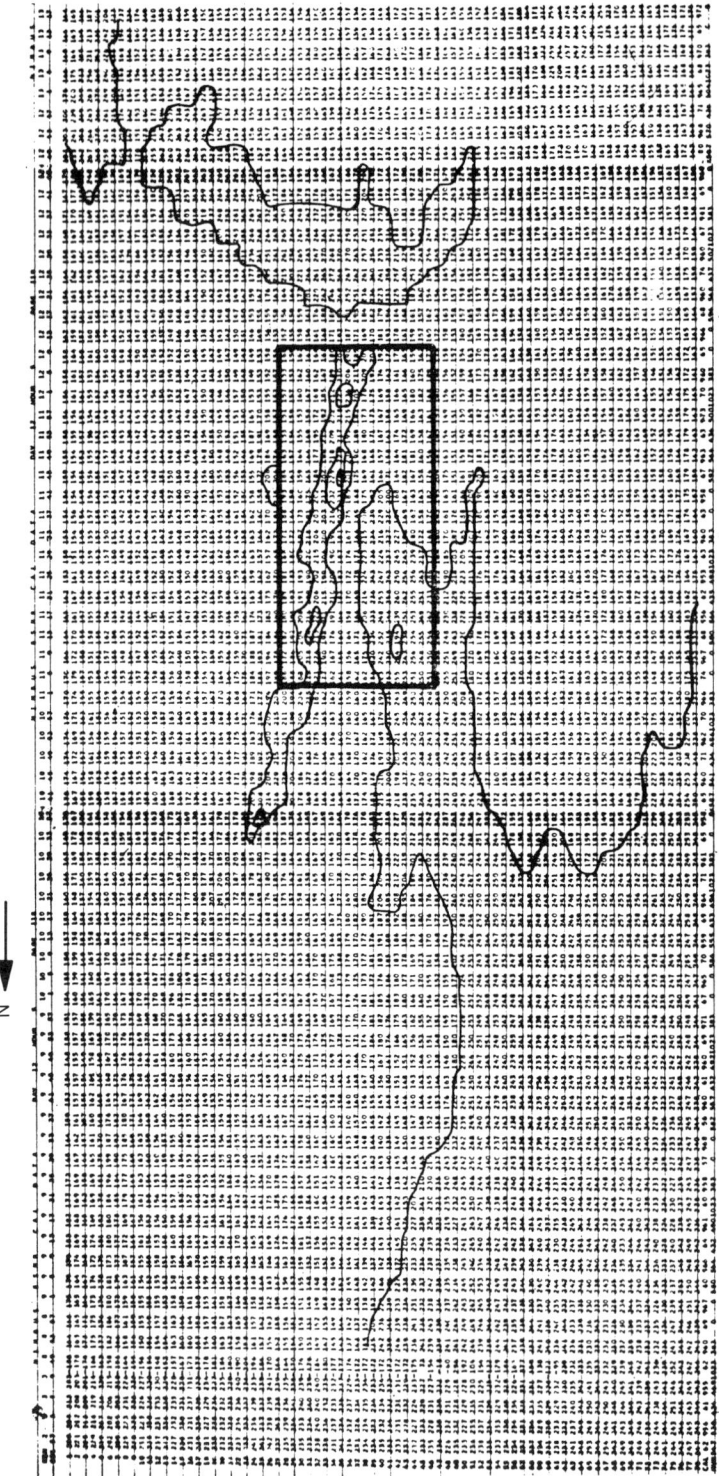

FIGURE 85. Sample output of the Nimbus CBT listing program.

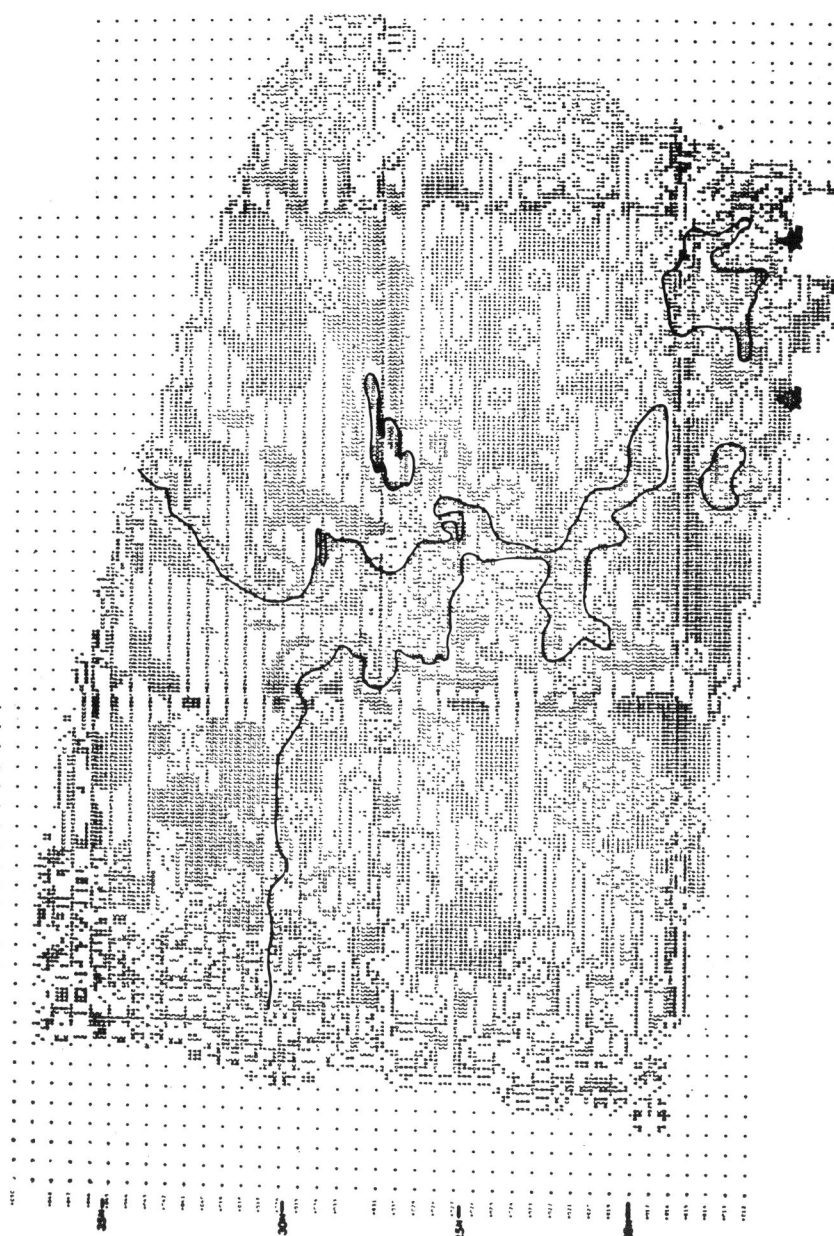

FIGURE 86. Sample output of the Nimbus ESMR grid print map program.

ESMR Data Evaluation and Applicability to Severe Storm Detection

The ESMR carried by Nimbus 5 surveys the whole earth at 19.35 GHz once every 12 hours.

When coupled with concurrent observations in the infrared and radar, the microwave data can give valuable information on precipitation distribution and intensity, especially over the oceans which have a uniform low emissivities and therefore provide a rather uniform brightness temperature background. Over land the surface emissivities are higher, and variable and brightness temperatures are in the same range as those measured over precipitating clouds; thus it is more difficult to detect precipitating clouds.

Radiation in the microwave obeys the same laws as in the infrared. In the microwave portion of the electromagnetic spectrum radiation from surfaces and the atmosphere may yield information which is additional to or nonobtainable from the visible or infrared. Because of its longer wavelength, microwave from the earth's surface is able to penetrate the atmosphere and nonprecipitating clouds with little attenuation. Atmospheric hydrometeors such as rain, hail, and snow cause strong attenuation of microwave by absorption and scattering and reemit at their temperature and emissivities.

Concept of Brightness Temperature

In the microwave region the Planck function,

$$E_\lambda = C_1\lambda^{-5}/[\exp(C_2/\lambda T) - 1] \qquad (2)$$

reduces to the Rayleigh–Jeans approximation for intensity of radiation from a blackbody,

$$E_\lambda = (C_1/C_2)\lambda^{-4}T \qquad (3)$$

since $\exp(C_2/\lambda T) \simeq 1 + C_2/\lambda T$ for large values of λT.

In the above equations E_λ is radiance at wavelength λ and temperature T; C_1 and C_2 are constants. Rather than the radiance E_λ, an equivalent brightness temperature T_B is used to describe measurements in the microwave. For a gray body,

$$T_B = \varepsilon_s T = E_\lambda(C_2/C_1)\lambda^4 \qquad (4)$$

where ε_s is the surface emissivity in the microwave and T is the actual surface temperature.

Surface Emissivities

At 19.35 GHz, variations in surface emissivities dominate over variations in temperatures. The major contributor to variations in surface emissivity is water with its extremely large dielectric constant at microwave frequencies. The emissivity of a land surface is around 0.9 and that of water surfaces is about 0.4 at 19.35 GHz, so that the brightness temperature contrast between land and water is of the order of 100°K. At 19.35 GHz, the emissivity of water varies nearly as $1/T$, making the brightness temperature of a smooth water surface virtually independent of the thermodynamic temperature of water (*Nimbus 5 User's Guide*, 1972). The emissivity of the sea surface, is however, dependent on the roughness and foam cover and, therefore, on the surface wind, as shown by Nordberg et al. (1971) and Hollinger (1971). From data given by Nordberg et al. (1971), a ΔT_B of 10°K corresponds to a wind increase of slightly less than 10 meter/sec. Using the equation presented in the *Nimbus User's Guide* (1972, p. 70),

$$T_B = 125°K + 6.8V + 300L \tag{5}$$

where V is water vapor in a vertical column in g/cm^2 and L is liquid water of nonraining clouds in a vertical column in g/cm^2. A T_B of 10°K corresponds to about 1.5-g/cm^2 change in water vapor, or 0.03 g/cm^2 of nonprecipitating clouds.

Absorption of Microwave in the Clear Atmosphere

Water vapor and molecular oxygen are the main absorbers of microwave radiation in the clear atmosphere. At 19.35 GHz, the water vapor molecule dominates the absorption process, and absorption is of the order of 0.08 dB/km at sea level for a U.S. standard atmosphere, while the absorption due to oxygen is only about $\frac{1}{10}$th of this and can be neglected. Water vapor absorption at 19.35 GHz is due to an absorption line centered at 22.235 GHz and to the far-wing absorption of infrared lines. Calculations of water vapor absorption may be performed by the expressions below as presented by Staelin (1966):

$$
\begin{aligned}
\alpha_\omega(\text{dB/km}) = {} & 140.71 \exp(-644/T)\nu^2 p\,\rho_\omega\, T^{-3.125}(1 - 0.0147\rho_\omega T/\rho) \\
& \times \{1/[(\nu - 22.234)^2 + (\Delta\nu)^2] + 1/[(\nu + 22.234)^2 + (\Delta\nu)^2]\} \\
& + 0.01107\rho_\omega^2\,\Delta\nu T^{-1.5}
\end{aligned}
\tag{6}
$$

where ν is frequency in GHz, p is total pressure in mbar, ρ_ω is the density of water vapor in gm/m^{-3}, and

$$\Delta\nu\ (\text{cm}^{-1}) = 2.58 \times 10^{-3}p(1 + 0.0147\rho_\omega T/p)\,(T/318)^{-0.625} \tag{7}$$

Attenuation of Microwave by Clouds

For wavelengths much larger than the average drop size in a cloud, scattering is negligible and the attenuation can be approximated by Rayleigh's theory. The Rayleigh model of absorption is applicable to clouds whose droplets range from a few micrometers to a few hundred micrometers. Westwater (1972) calculated the absorption coefficient for a range of microwave frequencies for water, ice, and a mixture of ice and water. He shows the following values at 19.35 GHz for the absorption coefficient (km^{-1}) per liquid water content (g/m^3):

Ice particles	1.2×10^{-3} km^{-1}/g/m^3
Water droplets	7.0×10^{-2} km^{-1}/g/m^3
Melting ice sphere	1.6×10^{-1} km^{-1}/g/m^3

Attenuation of Microwave by Precipitating Hydrometeors

When the particle size approaches the wavelength of the incident radiation, scattering is appreciable and must be considered along with the absorption, and therefore the attenuation cannot be approximated by Rayleigh theory. The much more complicated Mie model of absorption and scattering must be used. Paris (1969) used the Mie theory to calculate the volume absorption coefficient of a polydispersive cloud of hydrometeors obeying the Marshall–Palmer drop size distribution (Marshall and Palmer, 1948). The Marshall–Palmer drop size distribution is fairly representative of the drop size distribution found in most subtropical clouds and can be readily expressed in a rate of precipitation (R, mm/hr) or a liquid water content (M, g/m^3) by the relationship $M = 0.089R^{0.84}$. At 19 GHz, the absorption coefficient is nearly independent of temperature and increases almost linearly with liquid water concentration, as shown in Figure 87 from Paris (1969).

Absorption of microwave radiation by precipitation is rather high at 19.35 GHz. In fact, Paris (1969) calculated that at 19 GHz absorption is complete for liquid water content greater than 0.9 g/m^3 for a cloud 15 km thick and greater than 2.5 g/m^3 for a cloud of 5 km. This means that radiation from the surface is totally

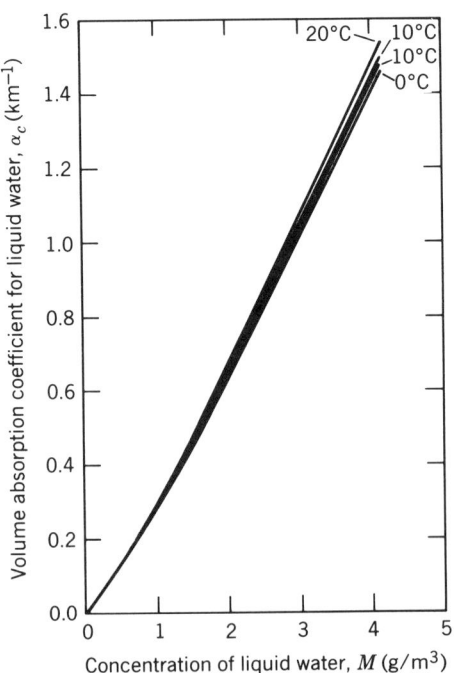

FIGURE 87. The volume absorption coefficient of a cloud of raindrops.

absorbed by a thick precipitating cloud, and what arrives at the satellite is radiation from the top layer of the cloud and the atmosphere above the cloud, the brightness temperatures measured being nearly the actual temperatures of the precipitating cloud top.

Summary of Water and Ice Content of Clouds

To correctly predict the microwave brightness temperatures of an earth–atmosphere system which includes clouds we have to correctly model the liquid water content (densities) of clouds and the concentration of drops (precipitation) in the size range which causes Mie scattering. A brief review of water content of various cloud types is therefore presented.

CIRRUS. Density ranges from less than 0.1 g/m^3 to 0.5 g/m^3 and is normally less than 0.1 g/m^3. Thickness ranges from a few hundred meters to several thousand. Cirrus are normally found at and below the tropopause. Cirrus are composed of ice and absorb and scatter one order of magnitude less than the same mass in liquid droplet form and are much less important at microwaves than liquid water clouds.

STRATUS. Table 36 gives values of liquid water densities for stratus and cumulus clouds.

Stratus clouds are low, of medium liquid water density, and not too thick (500–1000 m). Altostratus occur in the middle atmosphere, above 2000 m, contain about half as much liquid water than low stratus, and are about 500–1000 m thick.

CUMULIFORM CLOUDS. These have higher densities than stratiform clouds, more variability in water content and droplet distribution, and greater vertical extent. Densities are normally below 1 g/m^3, as shown in Table 36. Thunderstorms have the maximum liquid water concentrations and, of course, the greatest vertical extent.

The drop size distribution varies considerably in stratus clouds and cumulus clouds, but in the Rayleigh region the attenuation coefficient is independent of the drop size distribution and depends only on the liquid water content (Westwater, 1972).

TABLE 36. Cloud Liquid Water Characteristics

Cloud Type	Average Liquid Density (g/m^3)	Average Maximum Density (g/m^3)	Measured Absolute Maximum Density (g/m^3)
Stratus, stratocumulus	0.25	0.64	0.88
Altostratus, altocumulus	0.14	0.37	0.70
Cumulus	0.38	0.91	1.71
Cumulonimbus	0.51	1.72	~10.0

Source: Gaut and Reifenstein (1971).

RAIN. Rain has the greatest influence on the attenuation of microwave. The liquid water content of rain is highly variable, but can be related to rainfall rates when a raindrop distribution size is assumed. The Marshall–Palmer distribution giving the relationship between rainfall and density will be assumed in subsequent calculations of water densities and rainfall rates.

Rain from thin stratiform will have a density of about 0.1 g/m^3, from cumulus upwards to 1 g/m^3, and from cumulonimbus up to 10 g/m^3.

Microwave Brightness Temperature of the Earth's Atmosphere

The brightness temperature T_B of the earth's atmosphere as observed from a satellite above a nonscattering atmosphere may be calculated from the solution of the equation of radiative transfer. Using the Rayleigh–Jeans approximation to express the radiation intensity as a temperature, we may write

$$
T_B = \int_0^\infty T(z)\alpha(z) \, \exp\left(-\int_0^z \alpha(z) \, dz \right) dz + \varepsilon_s T_s \, \exp(-\tau)
$$

$$
+ (1 - \varepsilon_s) \, \exp(-\tau) \int_0^\infty T(z)\alpha(z) \, \exp\left(-\int_0^z \alpha(z) \, dz \right) dz
$$

(8)

and

$$
\tau = \int_0^\infty \alpha(z) \, dz
$$

where $T(z), \alpha(z)$ are the variation of atmospheric temperature and absorption coefficient with altitude z, ε_s and T_s are the surface emissivity and surface temperature, and τ is the optical depth of the atmosphere. The first term on the right in equation (8) represents the upward emission of the atmosphere between the earth surface and the satellite. The second term is the emission of the surface attenuated by the total optical depth of the atmosphere. The third term is the downward emission of the sky reflected by earth's surface toward the satellite and attenuated by the atmosphere.

Equation (8) can be integrated numerically by dividing the atmosphere into layers, each with an average temperature $\overline{T}$ and average total absorption coefficient α. The total absorption coefficient is the sum of the absorption coefficients for water vapor α_ω for small cloud droplets α_c and for precipitating hydrometeors α_r. Thus

$$
\alpha = \alpha_\omega + \alpha_c + \alpha_r
$$

(9)

At 19.35 GHz, the absorption coefficient for water vapor, α_ω, is calculated by means of equation (6). For small water droplets (nonprecipitating clouds), melting ice spheres, and ice particles, the values of α_c as calculated by Westwater (1972) can be used.

The graph in Figure 87 from Paris (1969) can be used to calculate α_r for rainfall. His values for 19 GHz will be used for the 19.35 GHz of the ESMR.

ESMR and Radar Observations of a Storm Area

A comparison of ESMR and concurrent radar observation will be presented showing the utility of the satellite microwave observation for detecting precipitation areas over the oceans.

On 12 January 1973, a quasi-stationary front with associated precipitation was located to the south of the Florida peninsula. The surface weather map for 0000 GMT, 12 January 1973 (Fig. 88) shows the front to be just south of the tip of Florida. Heavy precipitation occurred throughout southern Florida on 12 January, with Miami reporting 2.40 in. and Key West reporting 1.83 in.

Figure 85 is the output of the CBT Listing Program and presents a highly distorted map of T_B. The distortion can be partially removed by modifying the printout with a blank line between each line of brightness temperatures. The modified output of the CBT Listing Program was simulated for the area boxed and plotted at the same scale of a radar image obtained from Miami at 0513 GMT on 12 January. The T_B microwave map was overlayed on the radar image, as shown in Figure 89.

The microwave map was correctly located and oriented on the radar image by locating Miami on the Florida coastline (tight T_B gradient) and Lake Okeechobee (low T_B's) in southern Florida.

Contours of equal precipitation rates were drawn from the radar image with the aid of the calibration furnished by the National Climatic Center.

The outline of the rainband southeast of Miami coincides fairly well with brightness temperatures of 190°K and above. A point-by-point comparison of brightness temperatures and radar rainfall rates in this rainband over the ocean gives the following results:

Precipitation Rate (in./hr)	Average T_B (°K)	Standard Deviation (°K)
≤0.1–0.1	204	8.4
0.1–0.5	221	8.1
0.5–1.0	238	—

Each brightness temperature is roughly representative of an area 15 mi in diameter; thus it was often the case that a resolution circle covered more than one precipitation rate. Therefore, in calculating the average T_B for each precipitation rate range, each T_B was assigned to the precipitation ranges its resolution circle covered and weighted in the average according to the percentage area in the range.

Figure 90 shows a graph of the relationship between brightness temperatures and rainfall rates as derived from this analysis. The graph was obtained by assigning the average T_B to the middle of each precipitation range values and assigning $T_B \pm 1$ standard deviation to the limits of the precipitation range. Thus,

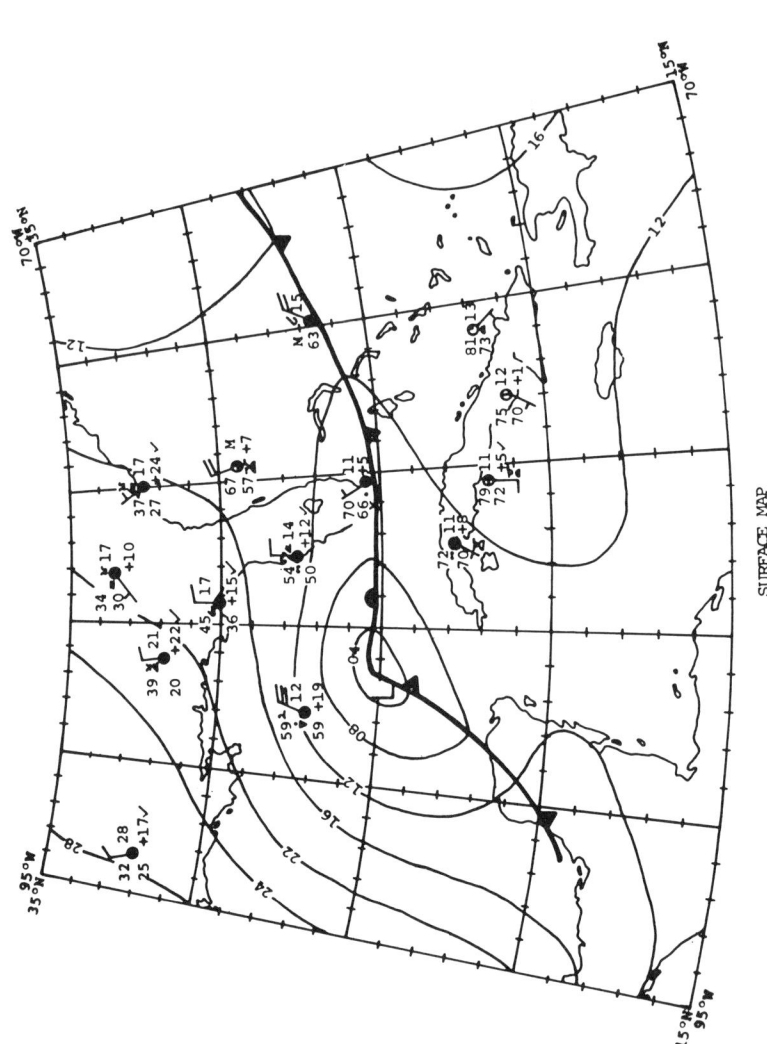

SURFACE MAP

0000Z FRIDAY JANUARY 12, 1973

FIGURE 88.   Surface weather map for 0000 GMT, 12 January 1973.

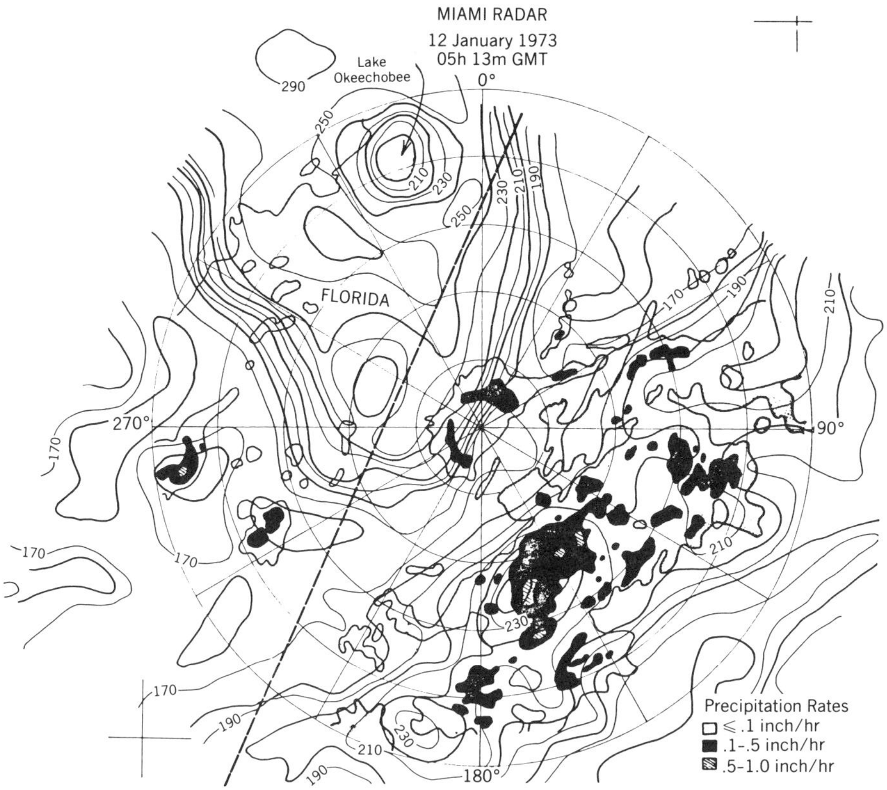

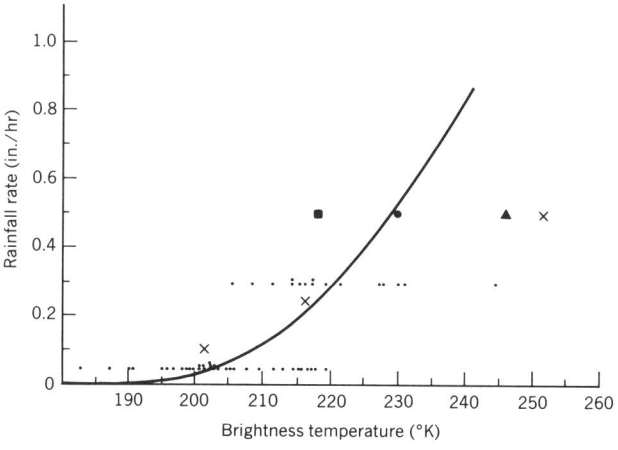

MIAMI RADAR
12 January 1973
05h 13m GMT
Lake
Okeechobee
FLORIDA
0°
90°
180°
270°
Precipitation Rates
≤ .1 inch/hr
.1-.5 inch/hr
.5-1.0 inch/hr

FIGURE 89. Radar image obtained from Miami.

Rainfall rate (in./hr)
Brightness temperature (°K)
1.0
0.8
0.6
0.4
0.2
0
190
200
210
220
230
240
250
260

FIGURE 90. Relationship between ESMR and radar.

$$0.05 \text{ in./hr} = 204°K$$

$$0.1 \text{ in./hr} = 212°K$$

$$0.3 \text{ in./hr} = 221°K$$

$$0.5 \text{ in./hr} = 229°K$$

$$0.75 \text{ in./hr} = 238°K$$

The curve was then extrapolated toward higher temperatures. The relationship is based on a very limited data sample and may not be valid when extended to higher precipitation rates and to different storms. Nevertheless, qualitatively the brightness temperature patterns agree fairly well with the radar echoes, and one can safely surmise that the highest precipitation rates coincide with the highest brightness temperatures. For comparison purposes, a clear area (as determined by the concurrent thermal infrared imagery) over the ocean just north of Cuba was characterized by an average T_B of 150°K. A cloudy nonprecipitating Atlantic area (as determined by the Waycross Radar and THIR image) was characterized by an average T_B of 164°K.

Precipitation areas are not as clearly distinguishable over land as they are over oceans. The precipitation area crossing central Florida as detected by an uncalibrated radar from Tampa is characterized by a T_B of 249°K over land, while an adjacent area of nonprecipitating clouds over land has a T_B of 243°K.

REFERENCES

Gaut, N. E., and E. D. Reifenstein III, *Interaction model of microwave energy and atmospheric variables,* Final Report, Contract NAS8-26275, February 1971.

Hollinger, J. P., Passive microwave measurements of sea surface roughness, *Trans. IEEE Geosci. Electron.* GE-9, 1971: 165–169.

Marshall, J. S., and W. McK. Palmer, The distribution of raindrops with size, *J. Meteor 5,* 1948: 165–166.

Nimbus ESMR data reduction program user's guide, prepared by Programming Methods Incorporated, Federal Systems Division, for the Laboratory for Meteorology and Earth Sciences, Goddard Space Flight Center, Greenbelt, Md., 1972.

Nimbus Project, *The Nimbus 5 user's guide,* ed. R. R. Sabatini, NASA, Goddard Space Flight Center, Greenbelt, Md., 1972.

Nordberg, W., J. Conaway, D. B. Ross, and T. Wilheit, Measurements of microwave emission from a foam-covered wind driven sea, *J. Atmos. Sci.* 28, 1971: 429–435.

Paris, J. F., Microwave studies at Texas A&M University, in *Microwave observations of the ocean surface,* SP-152, Naval Oceanographic Office, Analysis of the NASA/Navy Review, 11–12 June 1969.

Staelin, D. H., Measurements and interpretation of the microwave spectrum of the terrestrial atmosphere near *L*-cm wavelength, *J. Geophys. Res.* 71, 1966: 2875–2881.

Westwater, E. R., *Microwave emission from clouds,* NOAA Technical Report ERL 219-WPL 18, 1972.

Wilheit, T., ESMR experimenter, NASA/GSFC, personal communication, June 1973.

3.1.2 Automated Cloud Detection and Displacement Measurement*

A computer model used operationally at the National Environmental Satellite Service (NESS) provides for the automated detection and displacement measurement of selected cloud imagery to determine wind estimates. Infrared (IR) data (8-km resolution) acquired from two geostationary satellites, SMS-2 and GOES-2, are used as input to the model, and displacement measurements are computed from two digital IR picture images which are 30 or 60 minutes apart in time.

Availability of IR data (which are calibrated to be directly converted to temperatures) motivated the development of the current model, which computes displacement of cloud patterns defined by a specific range of temperatures. The limits of this temperature range are converted to IR digital counts, and a 16 × 16 target array at a prescheduled grid point is examined to determine the number of counts within range. A predetermined threshold value is used to determine whether the 16 × 16 array contains a sufficient number of samples for displacement measurement.

Once the target array has passed the detection/selection criteria, it is compared with a collocated search area from a picture image 30 or 60 minutes later in time using a refinement of the basic cross-correlation algorithm.

Automated measurements of cloud motions have been made operationally at NESS since 1972. The data used are digital images obtained from geostationary satellites. These spacecraft, whose orbits are synchronous with the earth's, scan the same area of the earth at regular intervals, making it possible to compare successive images.

The current operational spacecraft are GOES-2 (subpoint, 0°N, 75°W) and SMS-2 (0°N, 135°W), both at an altitude of approximately 36,000 km. The sensor aboard each is the Visible/Infrared Spin Scan Radiometer (VISSR) which views the earth in the visible (0.5–0.7 μm) and IR (10.5–12.6 μm) spectra with a cycle of about 18 minutes per image. An IR image (to which we will confine our remarks here) consists of approximately 1800 lines of 8-bit samples. The images are begun every 30 minutes and are stored on computer disk.

Cloud Motion Measurements

Successive images 30 or 60 minutes apart are used in the motion computations. The first step is to register or align the two images with respect to each other. Both the registration process and the subsequent cloud motion measurements rely on the classical formula for correlation:

$$\rho = \frac{\sigma_{AB}}{\sigma_A \sigma_B}$$

where A and B are $m \times n$ arrays, σ_{AB} is the covariance between A and B, and σ_A

*Based on G. Hughes, C. Novak, and R. Schreitz, Automated techniques for the detection and displacement measurement of selected cloud imagery observed in geostationary satellite data, 8th Applied Imagery Pattern Recognition Conference, Gaithersburg, Md., April 1978.

and σ_B are the respective standard deviations of A and B. A 16×16 target array, centered on a predefined point on the surface of the earth, is extracted from image 1; and a companion 32×32 search array, centered on the same point, is extracted from image 2. Correlations are computed for all possible positions of the smaller array within the larger array. These positions are identified by their horizontal and vertical lag values, which range from -8 to $+8$. The displacement of the maximum correlation, relative to the center of the 17×17 array of correlations, is interpreted as the displacement of the target. In the registration process, a preselected set of 50–100 distinctive geographic features of "landmarks" serve as targets, and the modal offset is used to register the two images. This process, augmented by a two-dimensional interpolation technique, results in a registration error of less than one-eighth of an image element. In the cloud motion measurements, the targets are centered on points in a predefined grid, and the computed displacements are converted to direction and speed. Two cloud images and their corresponding correlation matrix are given in Figure 91.

The 64×64 arrays are alphanumeric displays of IR data observed 30 minutes apart and centered at the same latitude/longitude position. The imagery feature being measured is bounded by the solid line in the upper left-hand array. Thirty minutes later, the upper right-hand array shows that the feature has been displaced 7 elements, the distance between its initial position (dashed line) and its final position (solid line). Cross-correlation applied to these two arrays results in a correlation matrix (lower figure) where the location of the maximum correlation coefficient is found 7 elements away from the origin.

In practice, the so-called first guess/fast displacement algorithm is used for computational efficiency. The algorithm uses a first guess vector and an allowable deviation from this guess to define an optimal search area.

Only the coefficients within the larger rectangle are computed initially. If a peak value occurs within the smaller rectangle, its position (relative to the center of the matrix) is taken to be the displacement. If the peak occurs outside the smaller rectangle, the coefficients for all 17×17 lag positions are computed and searched for a peak. Operationally, about 75% of the calculations make use of the shortened method.

The displacements so obtained are then converted to a meteorological direction (degrees from north) and speed (knots). The vectors are used by the National Weather Service (NWS) as estimates of the winds at an altitude of 1 km. An example of the output of one such computer run is given in Figure 92. The approximate time for the program is 1.5 CPU minutes to process approximately 150 wind estimates from one spacecraft.

Current Model

The development of the current model was motivated by the fact that the digital IR values can be calibrated and converted to temperatures. Thus, the motions (at 1000 m) could be computed using only the digital values corresponding to a specific temperature range. The original model used the complete range of values (0–255) in the computation, although an individual target does not necessarily

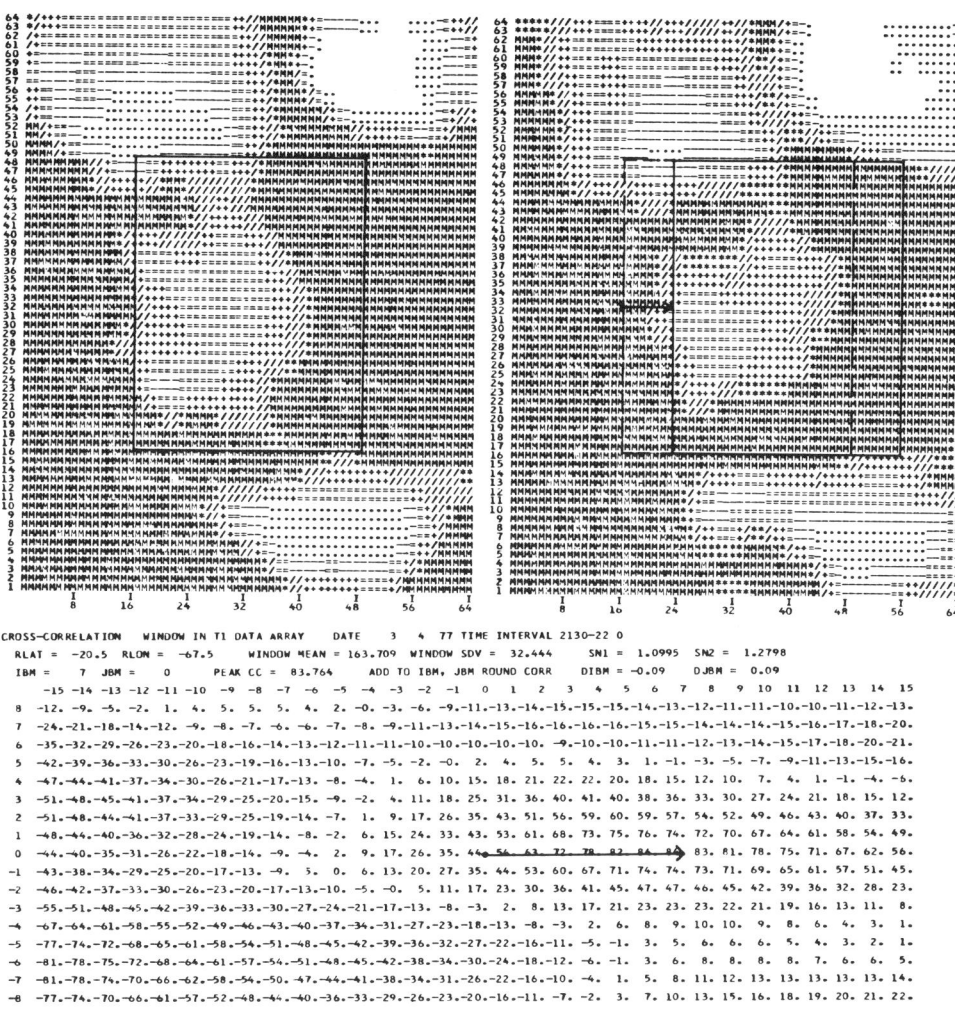

FIGURE 91. Two cloud images.

contain clouds of the proper altitude. These invalid measurements were removed after the fact by a meteorologist.

This so called "temperature slice" model uses a global temperature analysis provided by NWS to define a temperature range corresponding to clouds having maximum tops of 3000 m. The limits of this temperature range are converted to digital counts and the target array is examined for the number of samples within the range. If a threshold number of samples is exceeded, the array is "sliced"; that is, samples outside the range are set equal to the mean and the cross-correlation proceeds.

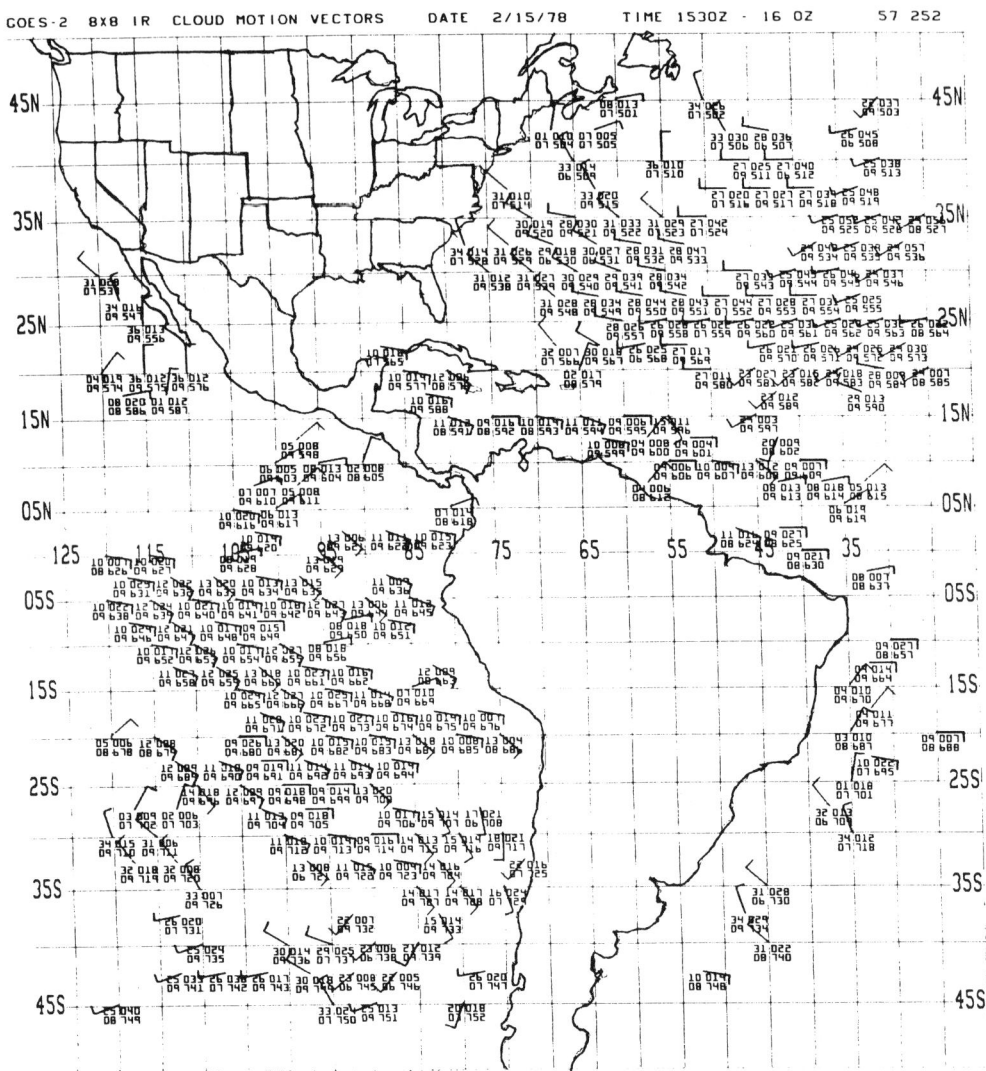

FIGURE 92. Output of computer program for cloud motions from GOES-2 spacecraft.

The resulting model is both more efficient and theoretically more meaningful. Targets with insufficient samples are not subjected to further computations. Targets which exceed the threshold number of samples but which contain samples outside the range that is, higher clouds, have the effect of the higher clouds eliminated. Without "slicing," the displacement measurement would be a composite of the motions at perhaps several levels. An expanded form of the correlation equation,

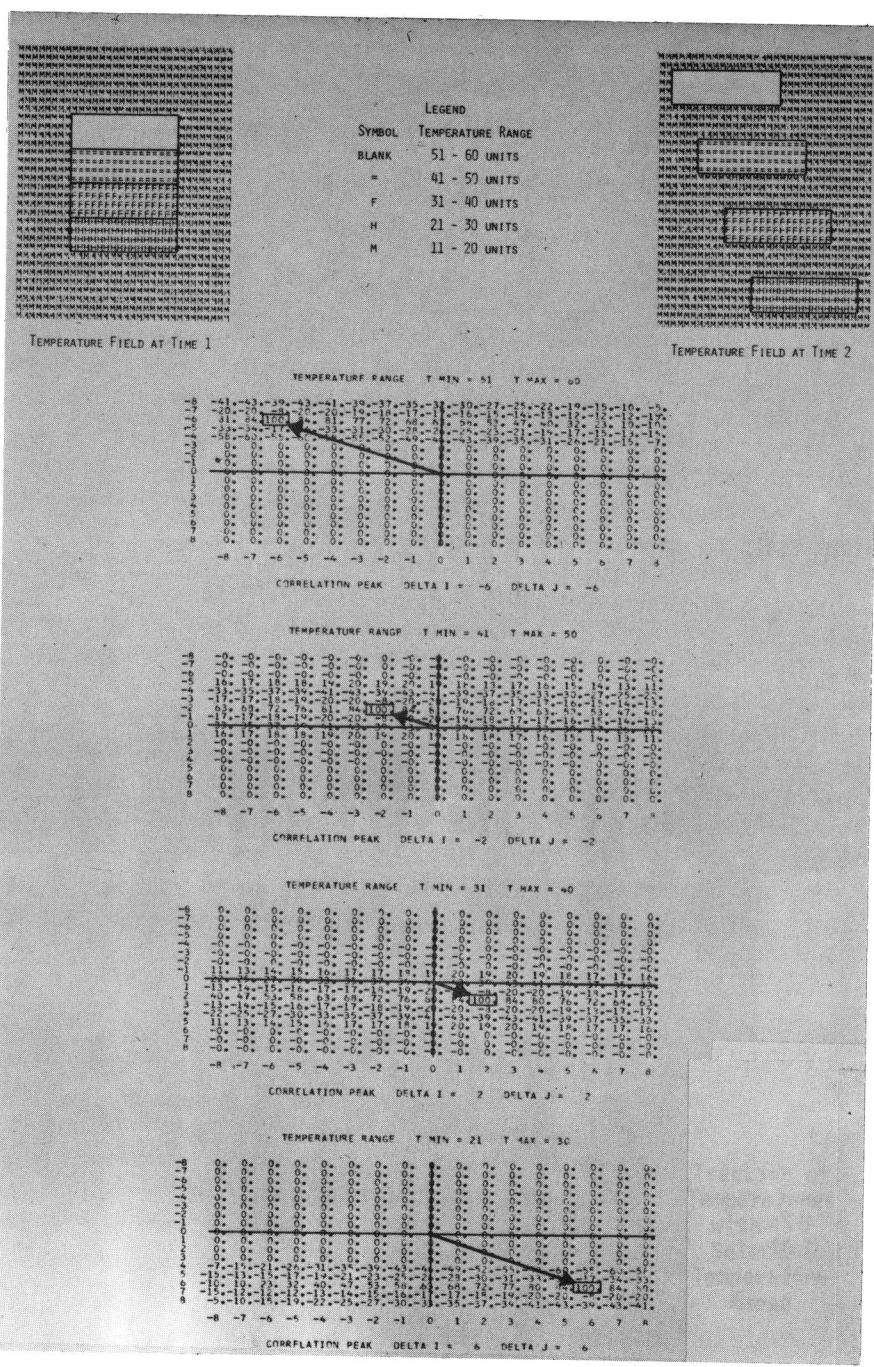

FIGURE 93. Tracking of temperatures within a selected temperature range.

$$\rho = \frac{1}{mn} \frac{\Sigma ij \, (a_{ij} - \overline{a}) \, (b_{ij} - \overline{b})}{\sigma_A \sigma_B}$$

better demonstrates the mechanics of the temperature slicing. If the values outside the specified range (in the target/search array) are set equal to the mean, they will be eliminated from the correlation computation and, hence, from the displacement measurement. Figure 93 illustrates a hypothetical case in which the target array is divided into four rectangles, each of which contains "data samples" in a unique temperature range. By defining appropriate slice limits in each of four successive computations, these rectangles were individually detected and tracked.

Summary

A measure of the success of the temperature slice model can be seen in the number of erroneous wind estimates. The implementation of the new model resulted in a 40% reduction in the number of wind estimates edited by the meteorologist who reviews the output before it is sent to NWS.

There is an inherent physical problem which greatly increases the difficulty of computing cloud motion in a specific altitude range. This problem is responsible for a high percentage of the erroneous low-level wind estimates and, at the present time, almost precludes the computation of high-level wind estimates. The relationship between IR digital counts and altitude involves a significant degree of uncertainty. IR digital counts are proportional to radiation detected in a specific wavelength band. The counts are converted to temperature, the temperature to pressure, and the pressure to altitude. The IR sensor responds to radiation not only from the highest cloud, but also to radiation from lower clouds and the surface of the earth. Booth (1972) and Parikh (1976) have explored various pattern classification schemes as cloud discriminators. When such an approach is successful in identifying the higher clouds, the temperature slice method could produce wind estimates at all levels.

REFERENCES

Booth, A. L., *Objective cloud type classification using visual and infrared satellite data*, Technical Note BN768, Institute for Fluid Dynamics and Applied Mathematics, University of Maryland, College Park, 1973.

Leese, J. A., C. S. Novak, and B. B. Clark, An automated technique for obtaining cloud motion from geosynchronous satellite data using cross correlation, *Journal of Applied Meteorology* 10(1), 1971: 118–132.

Novak, C. S., and M. T. Young, The operational processing of wind estimates from cloud motions, *Proceedings of the Symposium on Meteorological Observations from Space: their contribution to the first GARP Global Experiment*, 1976, pp. 214–221.

Parikh, J., Cloud pattern classification from visible and infrared data, Technical Report TR-442, Computer Science Center, University of Maryland, College Park, 1976.

3.2 GEOLOGY

Two sections cannot do justice to the applications of digital image processing of remotely sensed data to geology. The intention, rather, is to provide some sense of the richness and variety of the opportunities offered to the geologist by these and related methods. Section 3.2.1 discusses the manipulation of Landsat MSS color composites to discriminate rock types in the Goldfield mining district of Nevada. The digital algorithms employed are primitive in this early paper,* but the demonstration of handsome payoff in terms of detailed geological interpretations underscores the power of even simple digital processes when applied to suitable data.

Section 3.2.2 is an anomaly for this book. Geologic exploration with high-resolution radar is not an example of digital image processing; rather it applies qualitative visual image interpretation. However, no comparably thorough exposition of the utility of airborne imaging radar is available that uses digital processing. The growing importance of radar as a source of image data in all application areas compels the practitioner to become acquainted with the comparative advantages and disadvantages of radar with respect to passive image sources. As images, the illustrations in this section lend themselves to digital image enhancement and analysis.

It should not be inferred from this selection, however, that all interpretation of radar images is performed visually. Digital processing is used for quantitative studies not only for geology, but also for soil moisture, sea ice, forestry, and so forth.

3.2.1 Discrimination of Rock Types in Nevada[†]

Introduction

Landsat provides an important tool for geologic exploration in the form of small-scale multispectral visible and near infrared images. These images, which show large areas under nearly constant lighting conditions, have been applied to a wide variety of geologic problems ranging from detection and delineation of fault zones and volcanic centers to studies of coastal erosion and sediment transport. Most applications, however, have relied mainly on photo interpretation of these synoptic views rather than on multispectral reflectance analysis. Regional morphologic features are commonly quite conspicuous, but visible and near infrared spectral reflectance differences among rocks are usually small and therefore not readily apparent through visual examination of the images.

A technique which combines digital computer processing and color compositing has been devised for enhancing subtle spectral reflectivity differences. This

---

*For a more recent related paper, see A. Goetz and L. Rowan, Geologic remote sensing, *Science* 221, 1981: 781.
[†]Based on L. C. Rowan, P. Wetlaufer, A. Goetz, F. Billingsley, and J. Stewart, *Discrimination of rock types and altered areas in Nevada by the use of ERTS images,* Professional Paper 883, U.S. Geological Survey, Washington, D.C., 1974.

technique has been applied to part of a Landsat image of south–central Nevada with emphasis on the Goldfield mining district. Analysis has focused on discrimination of the geologic materials, especially in mineralized areas, on the basis of visible and near infrared spectral reflectivity differences. This discussion presents the geologic interpretation and evaluation of the resulting processed images. Brief descriptions are given of the general nature of the visible and near infrared spectral reflectivity of rocks and the techniques used.

Visible and Near Infrared Reflectance of Rocks and Minerals

Our understanding of the relationships between mineralogical composition and visible and near infrared spectral reflectivity is based mainly on published laboratory measurements (Hunt and Salisbury, 1970; Ross et al., 1969; Hunt et al., 1971a, b, 1973a, b, 1974a, b). All these measurements were made on crushed, monogeneous, and generally unaltered samples. Although application of these data to analysis of MSS images of large geologically complex areas is made difficult by several factors, especially scale differences and surface state conditions, the laboratory data do provide a framework for initial investigations.

Analysis of these laboratory spectra show that electronic transitions in constituent metal ions result in broad optical absorption bands in the ultraviolet, visible, and near infrared wavelengths. For example, iron, the most common transitional metal ion, has conspicuous ferrous and ferric absorption bands centered at about 1.0 and 0.92 μm, respectively, and several closely spaced weaker bands between 0.40 and 0.55 μm (Hunt and Salisbury, 1970). Other transitional metal ions that give rise to absorption bands include copper, titanium, chromium, and manganese (Hunt et al., 1971a, b). Absorption bands due to vibrational processes in water and hydroxyl molecules also occur in the near infrared wavelengths, but the only one of these bands within the response range of the MSS is centered at about 0.95 μm.

Absorption bands are commonly quite intense and therefore result in conspicuous reflectance minima, predominantly in ferromagnesian and hydrous mineral spectra. Diagnostic individual spectral features, however, are generally subdued beyond recognition by addition of anhydrous nonferrous minerals such as quartz and feldspar to form polymineralic rocks. The shape of rock spectra are nonetheless still affected by the absorption bands. For example, the reflectance of mafic and ultramafic rocks changes very little between 0.4 and 1.1 μm (Fig. 94B). In contrast, felsic rock reflectance, which is generally affected less by absorption bands, increases continuously throughout this range, although at a slower rate between 0.70 and 1.1 μm than at shorter wavelengths (Fig. 94A). The slopes of reflectance curves for rocks with intermediate composition are typically between those of felsic and mafic rocks (Ross et al., 1969). As we will discuss later, the spectra for unaltered rocks (Fig. 94) and altered rocks generally have significantly different shapes.

Differences in spectral shape can be used to distinguish among geologic materials and, in some cases, to place bounds on their bulk composition. Several fac-

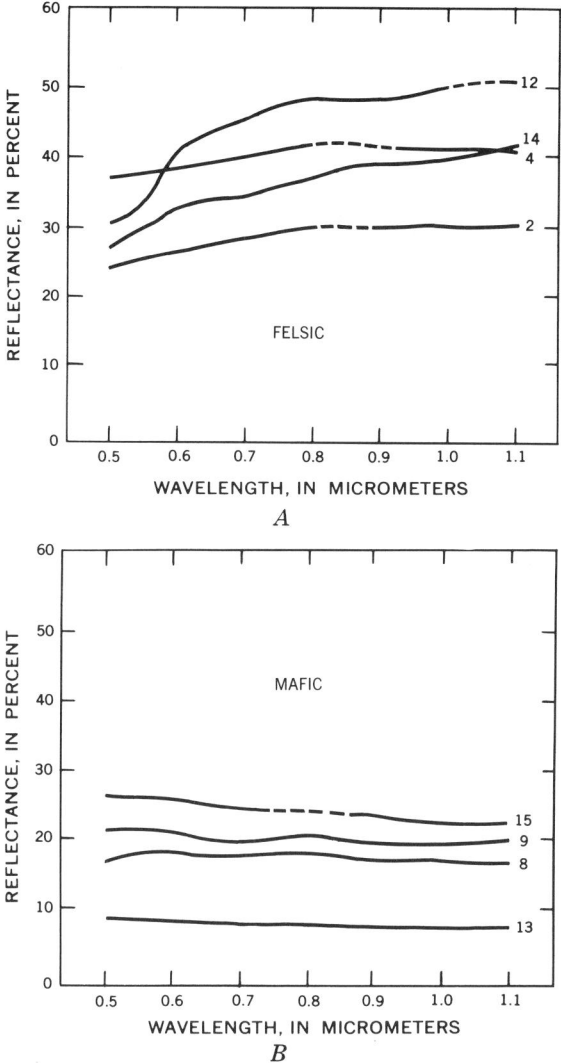

FIGURE 94. Visible and infrared reflectance spectra.

tors — including impurities both in the crystal lattice and on the surface of the rocks, atmospheric conditions, and system calibration — complicate both these efforts. Surficial weathering products such as limonite and clay minerals are especially important, as they obscure the original rock surface.

The characteristic spectral shapes for the rock (and soil) units can be derived from the MSS data to provide the basis for compositional estimates, although spectral details are somewhat subdued by the breadth of the MSS bands, especially band 7. Unknown atmospheric effects and system calibration complications preclude obtaining absolute spectral reflectivities at this time. On the other hand, absolute calibration is not required for simple discrimination that depends mainly on relative reflectance differences among rock types. Rock (and soil) units can be distinguished on the basis of very subtle reflectance differences, even though the absolute spectral reflectivities and gross lithologies cannot yet be determined from the MSS data.

Computer Image Enhancement

Development of image-processing techniques has been greatly stimulated during the last 20 years by the widespread use of imaging devices in planetary and terrestrial remote sensing. Enhancement techniques range from the relatively simple single-band contrast enhancement to two-dimensional filtering in spatial or Fourier domain to complex cluster analysis, sometimes coupled with ratioing of spectral bands.

The picture data can be taken from film images by scanning densitometry or, as in the case of the MSS, taken directly from computer-compatible tapes (CCTs). Tape-recorded digital data are preferable because on film, accuracy is lost in recording radiometric information and is further degraded by duplication, and scanning introduces additional noise into the signal. Other advantages of using the tapes for analysis include considerable analytical flexibility, reproducible results, and relatively reasonable costs.

Film methods were used in one of the earliest successful attempts (Whitaker, 1965) to discriminate rock units on lunar photographs on the basis of spectral reflectivity. In the apparently uniform regolith of Mare Imbrium, two basaltic lava flows were distinguished in a black-and-white composite produced by masking blue and infrared wavelength telescopic photographs. Using a similar approach, but manipulating digitized multiband telescopic photographs in the computer, a method was devised for analyzing Apollo orbital multiband photographs (Billingsley et al., 1970; Goetz et al., 1971). An extension of this technique, using a photographic ratio method, was developed by Yost, Anderson, and Goetz (1973). During this same general period, Vincent and Thomson (1972) were ratioing thermal–infrared spectral images to detect emissivity variations related to chemical and mineralogical differences. These results, along with the evaluation of laboratory spectra described in the previous section, made clear the considerable potential of ratioing for multispectral analysis of Landsat data (Rowan and Vincent, 1971).

Ratioing is an effective method for distinguishing among rock types because the main spectral differences in the visible and near infrared spectral regions are found in the slopes of the reflectivity curves; individual absorption bands are broad and weak and therefore cannot be used in most cases for discrimination of rock type on the standard MSS images. In addition, areas of interest geologically generally have some vertical relief. The ratioing process removes first-order brightness effects due to topographic slope and allows attainment of higher image contrast through additional processing. On the other hand, terrain effects are highly disturbing in color-additive displays or in analysis by clustering methods based solely on brightness.

Although digital computer processing ultimately proved to be necessary for this study, attempts were made to use more rapid visual and optically assisted techniques. Visual comparison of the MSS bands of many Nevada scenes resulted in only a few places where band-to-band differences could be related to rock type. For example, widespread volcanic rocks in the northern Antelope Range and Schell Creek Range in White Pine and Elko counties are conspicuously darker in the near infrared bands than in the visible bands (frame No. E-1053-17533, not shown). Enhancement by color-additive viewing is mainly useful for determining vegetation distribution. Although discrimination of rock units in the study area is not substantially improved by this method, color-additive techniques have proved more useful than simple comparison of individual black-and-white MSS images in other areas.

Attempts to enhance spectral reflectance differences by compositing a negative of one spectral band and a positive of another, as described by Whitaker (1965), were also generally unsuccessful. Of the many problems, the most serious were the general lack of enhancement actually achieved and the introduction of photographic processing errors that were of the same order of magnitude as the spectral reflectance differences. In order to make the spectral reflectance differences visible, a very high film contrast is necessary. At high-contrast levels, however, some important information is lost in the nonlinear part of the film response curve. Therefore, in general, purely optically assisted methods of analysis appear to have a somewhat limited value for geologic multispectral analysis.

Digital Methods

Various techniques have been developed for digital processing of images. Only those relevant to discrimination among rock materials will be discussed here. The steps in the enhancement of the MSS images are outlined in the flow diagram in Figure 95.

The dynamic range of the MSS is encoded to 64 brightness levels. Application of system calibration to take care of nonlinearities results in approximately 80 brightness levels, coded to 7-bit accuracy. The digital numbers (DNs) on the tape represent data values that are linear with brightness and range from 0 to 127. For convenience in using existing computer programs, the MSS data have been expanded into 8 bits, resulting in a DN range of 0–255. Future references to DN values will refer to the 8-bit range.

Contrast Stretching

The MSS system is designed to cover a large dynamic range in scene brightness to respond to the effects of sun angle and albedo variation as the spacecraft covers the globe. Consequently, the brightness range of any one image will generally occupy only part of the available dynamic range, resulting in a low-contrast image. In reconstructing an image from the digital data, it is therefore desirable to stretch the DN range to increase the contrast. Stretching begins by forming a histogram plot of the number of pixels per DN value (Fig. 96). The brightness values above

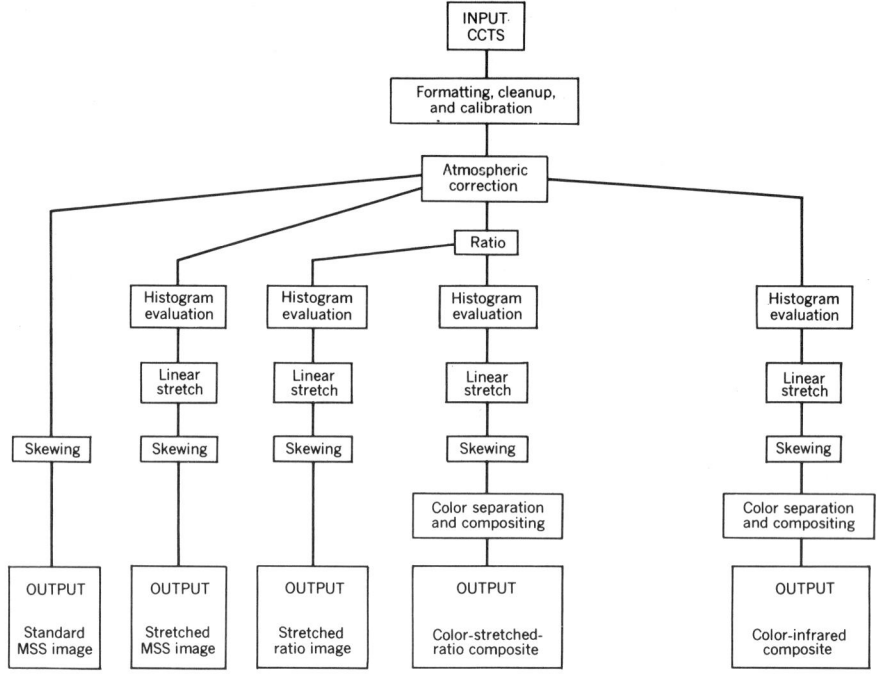

FIGURE 95. Flow diagram.

and below which no appreciable data exist can then be located and used as stretch limits. The stretch may be linear or nonlinear.

A linear stretch increases the scene contrast uniformly over the dynamic range of the output product. The stretch limits determined from the histogram are placed at the extreme points of the dynamic range (that is, 0–255 DN values), and the other points are spaced linearly between these end points (Fig. 96).

In a nonlinear stretch, such as a cube root stretch, the cube root of each DN value is taken. The resulting DN range is linearly stretched as above. This proce-

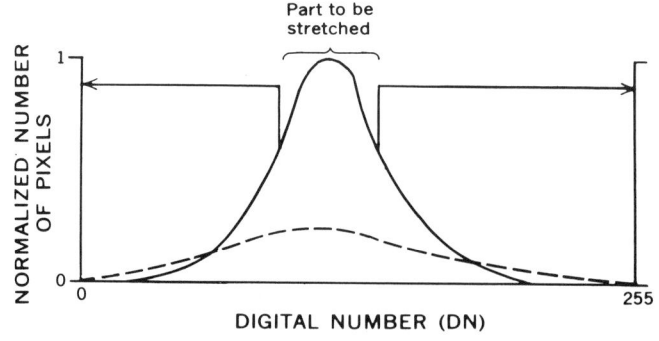

FIGURE 96. Schematic histogram.

dure increases local scene contrast in the dark areas at the expense of contrast in the brightest areas. In an exponential stretch, the inverse occurs. There does not seem to be a general rule of thumb that can be applied to all images in determining the required stretch parameters. Care must be taken to see that useful raw data at the extremes of the DN range are not saturated and lost.

Atmospheric Effects

The effects of absorption and scattering in the atmosphere vary among the different types of enhanced images. When the images are simply stretched to increase contrast, no effect is noted. If color composites are made from the stretched images, however, the relative color balance will be affected by the atmospheric scattering. The greatest effect is seen in the formation of ratios. Because

$$\frac{A + \varepsilon_1}{B + \varepsilon_2} \neq \frac{A}{B}$$

where ε_1 and ε_2 are the additive atmospheric scattering components in bands A and B, the scattering term must be removed.

The value for the atmospheric scattering component can be determined by locating the lowest DN values in the image. These values will normally occur over water or in cloud shadows. In a study area in Arizona (Goetz, 1974), measurements of ground-reflected radiation made before, during, and after the passage of a small (about 300-m-diameter) fair-weather cumulus cloud showed a reduction in reflected intensity by a factor of 5–6 during cloud passage. From these data we can anticipate that, to first order, most of the light received from the dark shadows is in fact atmospherically scattered sunlight. Therefore, corrections are made by subtracting the appropriate band-dependent values, determined from cloud shadows in each band, from each pixel DN value. Other methods of determining the band-to-band values may be used. The atmospheric corrections are applied to the data before skewing, stretching, and ratioing.

Ratioing

In the ratioing process, two spectral band images that have been corrected for atmospheric effects are divided pixel by pixel. The resultant image will show the variations in the slopes of the spectral reflectivity curves between the two wavelength bands. Differences in albedo are suppressed, however, and very dissimilar materials easily separable on a standard photographic image may become inseparable on a ratio image because their spectral reflectivity slopes are similar. On the other hand, a distinct advantage of this method is that one type of material will appear the same or similar in a ratio image regardless of the local topographic slope angle.

The ratio image generally shows a narrow histogram of DN values and may therefore be contrast stretched to enhance the visibility of the spectral differences. The type of stretch used, whether linear or nonlinear, as discussed above, will determine which areas in the image are most strongly enhanced. For instance, a linear stretch will result in the visual enhancement of dark areas (low DN values in the original ratio image) and light areas (high DN values in the original ratio image). Other types of stretches will enhance other DN value ranges in the original ratio image.

The error introduced by scattering is greatest for low-reflectance targets. It is highest in band 4, where the scattering component may make up as much as 50% or more of the recorded brightness level. Band 7 has the lowest scattering component.

Although atmospheric absorption can play an important role in the form of apparent interband radiance variations, we are not able to separate these effects from variations in the MSS absolute calibration, done before we received the tapes. Band 7, spanning the 0.95-μm water band, might be expected to be affected most severely by atmospheric absorption.

Residual errors, most easily detected on ratio images in highly sloped and shadowed regions, may result either from improper atmospheric correction or from the fact that such areas are illuminated mainly by the blue sky and reflections from surrounding terrain; if this is the case, the atmospheric absorption and scattering calculated from other regions are not appropriate for these areas.

Display Products

The computer-enhanced images are recorded on black-and-white film with a flying-spot CRT recorder. These 70-mm transparencies can be printed on paper or combined in a color-additive process, using either a viewer, color-negative stock, or diazo transparencies. The color combination of ratio images provides the photo interpreter with a vivid display akin to a classification map.

Geology of the Study Area

The surficial character of the study area (Fig. 97) makes it an ideal choice for analysis of Landsat spectral data. The topography of the test site is varied, ranging from smooth-textured alluvial basins to rugged ranges and a large mesa. Vegetation is sparse, covering 10–20% of the desert valleys where sagebrush is dominant, and is substantially denser only in the higher ranges where piñon pine, juniper, and grasses are predominant. Although vegetation type and distribution can sometimes reflect geology, minimal vegetation was considered preferable for these initial evaluations so that surficial features and rock units would not be obscured. The most important characteristics of the terrain are the widespread hydrothermal alteration and the broad compositional range of Tertiary igneous rocks, which provide an excellent opportunity for testing the discrimination potential of the MSS images.

Rock Units

Widespread Tertiary volcanic and intrusive rocks cover approximately 95% of the surface of the study area not covered by alluvium. Precambrian and Paleozoic rocks are exposed only locally; Mesozoic plutonic rocks, mainly quartz monzonite in composition, are of limited distribution in the study area.

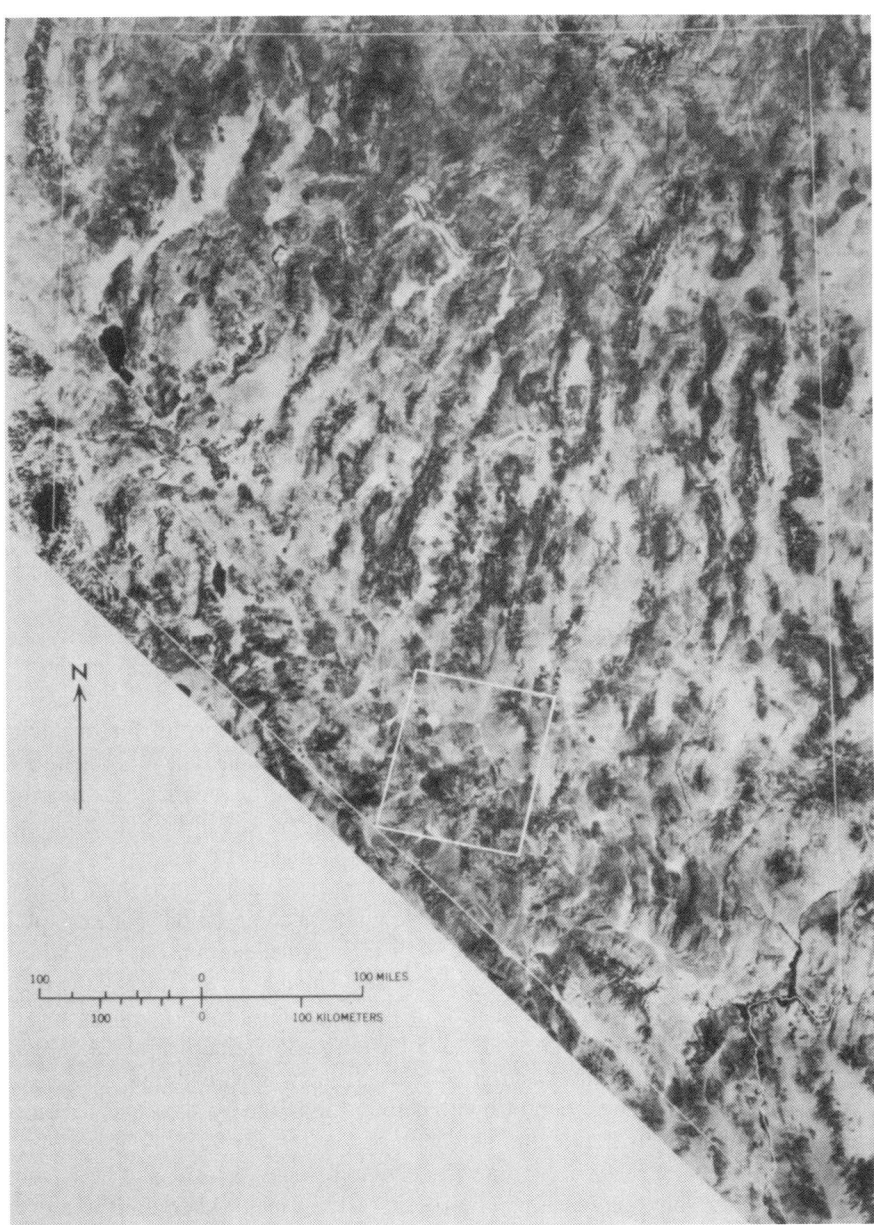

FIGURE 97. State of Nevada.

The Tertiary units exceed 6000 m in thickness in a composite section; tuffs of rhyolitic, dacitic, and quartz latitic composition are the most common. Lava flows and intrusive rocks of similar compositions and of andesite and basalt are also widespread (Cornwall, 1972). Although sedimentary rocks are subordinate in the Tertiary sequence, Miocene tuffaceous sedimentary rocks are common in the central and east–central parts of the area. The sources for the tuffs and flows are thought to have been as many as 9 or 10 volcanic centers, several of which are within the study area (Ekren et al., 1971). Especially noteworthy are the Black Mountain and Timber Mountain calderas. The Black Mountain caldera, in the southeastern corner of the study area, was the source for the Thirsty Canyon ash-flow and ashfall tuff which underlies much of the southern half of the area. Tuffs derived from the larger Timber Mountain caldera southeast of the study area are prominent in the west–central and southwestern parts of the study area. In addition, the Goldfield mining district is a volcanic center (Albers and Cornwall, 1968) and is thought to be a resurgent caldera (Albers and Kleinhampl, 1970). The volcanic and intrusive rocks of the Goldfield mining district are restricted to that district.

The Quaternary units consist of surficial deposits and sporadic outcrops of basalt flows and cinder cones. The basaltic deposits are conspicuous because of their low reflectivity relative to the adjacent materials. The surficial units, however, include complexly related alluvium, colluvium, desert wash and landslide deposits, playa materials, and, on the western edge of the area, a single exposure of bedded clay and silt. With the exception of the playa deposits, which are characterized by reasonably uniform composition and high reflectivity, these surficial deposits are lithologically heterogeneous. This compositional heterogeneity, as well as grain size, surface coatings, and vegetation cover variations, gives rise to large spectral reflectance differences. Although the processing techniques described in the previous subsection appear to be potentially useful for mapping these materials, the surficial deposits present a formidable analytical problem because of their large extent and the transitional nature of boundaries. The rest of this discussion will exclude the surficial deposits; emphasis will be placed on the Tertiary units, especially where they have been hydrothermally altered.

The unaltered Tertiary volcanic and intrusive rocks have from very high to low albedos, but these variations are not a reliable guide to composition. For example, although most of the silicic rocks have high to intermediate albedos, the widespread rhyolitic Thirsty Canyon tuff appears dark enough on the images, as well as in the field, to be mistaken for a rock of intermediate or even mafic composition. The unaltered rocks in the study area have a wide variety of muted colors, the most common being brown and gray, with green, yellow, and pink tints.

Alteration by the introduction of hydrothermal fluids and by subsequent weathering has resulted in hydration and oxidation of the Tertiary volcanic host rocks. The end products range from high-albedo clay-rich rocks, which may be locally silicified, to variably colored limonitic rocks. All the major alterations products are present in the Goldfield mining district, which is the most productive and best known altered zone in the study area.

Geology of the Goldfield Mining District

The Goldfield mining district is principally composed of Miocene volcanic rocks overlying Ordovician shale and chert and Mesozoic granitic rocks (Albers and Stewart, 1972). The middle Tertiary units include ashfall and ashflow tuff along with flows and intrusive bodies of andesite, dacite, rhyodacite, quartz latite, and rhyolite (Cornwall, 1972). Upper Tertiary basalt and welded tuff locally cap these units (Ashley, 1970). Alteration and mineralization are extensive, especially in the lower Miocene andesite and dacite, the primary ore-bearing rocks.

Ashley (1970) described two types of alteration. The older deuteric or propylitic alteration varies considerably, each variation characterizing a single volcanic unit. According to Ashley and Keith (1973), however, the chemical changes in most of these rocks have probably been quite limited. The younger intense hydrothermal alteration is more conspicuous and has a similar character in all rock units. Harvey and Vitaliano (1964) and Ashley and Keith (1973) described three mineralogically distinct zones of hydrothermal alteration. In order of decreasing alteration, silicified rocks with associated alunite, and kaolinite give way to illite-kaolinite-bearing argillized rocks, which grade into montmorillonite-bearing argillized rocks having surficial coatings of limonite and jarosite resulting from oxidation. Limonite is also common in the first two zones. The silicified and argillized rocks have a bleached appearance except where stained red and yellow by limonite and jarosite. Figure 98 shows the general limits of the alteration zones which cover more than 38.4 km^2 (Ashley, 1970).

Geologic Interpretation of Images

Using the procedures previously described, five processed image sets (Figure 95) have been produced from the basic digital MSS tape of the south–central Nevada study area. Listed in order of increasing enhancement of spectral reflectance information, the MSS image products are (1) standard unenhanced images, (2) stretched images, (3) stretched color–infrared composites, (4) stretched ratio images, and (5) color–stretched ratio composites (hereafter referred to as color–ratio composites).

Standard MSS Images

The standard MSS images are of low to moderate contrast; consequently, the level of detail is low for both bright and dark objects. In the northeastern part of the 3 October 1973 frame, No. E-1072-18001, materials with low and high albedos, such as mafic rocks and playa deposits, respectively, are distinguished easily, but little detail can be discerned within these areas. Furthermore, few differences are detectable through band-to-band comparison, except in vegetated areas, which are dark in the visible bands and light in the near infrared bands. This general lack of band-to-band contrast testifies to the subtlety of the spectral differences among the rock types. The small magnitude of the spectral differences, along with the low scene contrast of the standard images, seriously limits their value for rock-type discrimination.

Image contrast for the rocks and soils of the study area, as well as for most areas examined in Nevada and southern California, appears to be highest in the MSS image for bands 6 and 7, at a scale of 1:500,000. However, although most of the playas are easily distinguished from clouds and other features in the area on the basis of shape and texture, little confidence can be placed in most other dis-

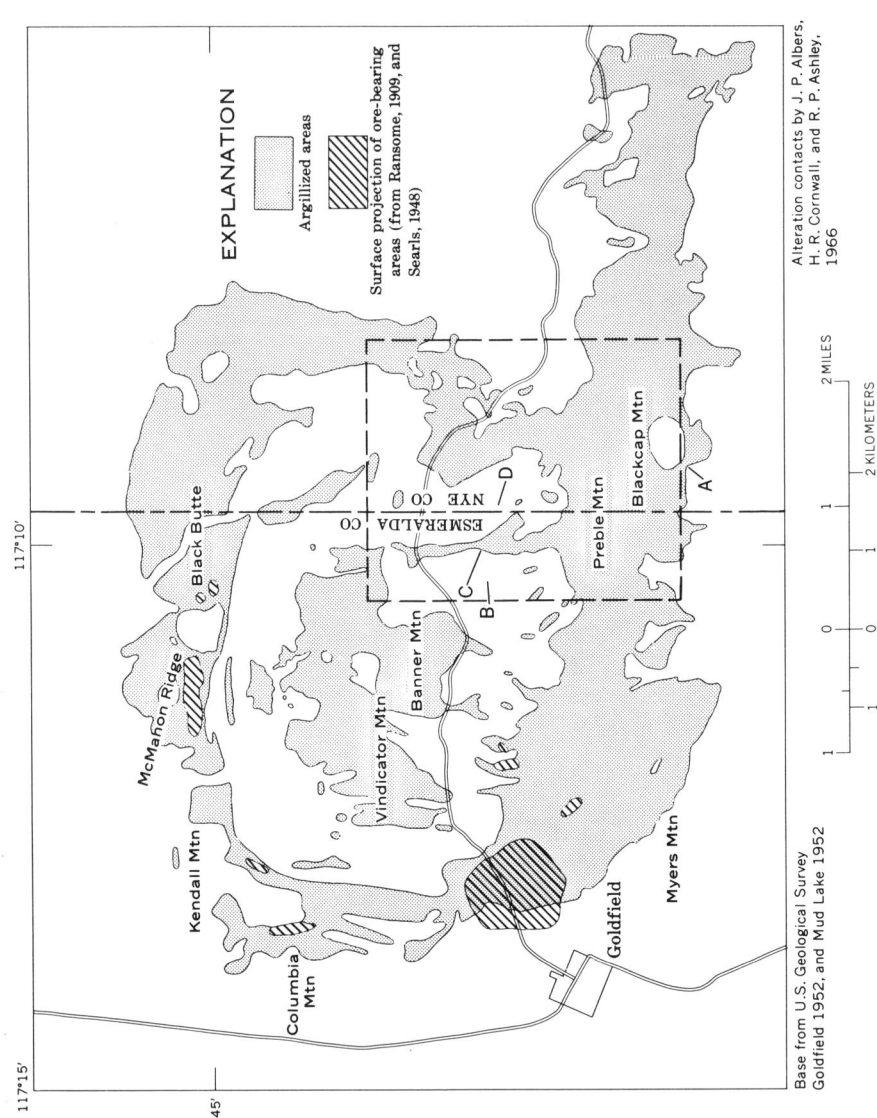

FIGURE 98. Map showing areas of alteration.

tinctions. For example, several dark areas are quite prominent, but the compositions of the rocks in those areas range from rhyolitic to andesitic to basaltic. The two large dark patches south–southwest of Mud Lake are basaltic and andesitic in composition, but three minor outcrops of tuffaceous sedimentary rocks occur in the northernmost dark area. The three small dark spots on the northeastern margin of the Cactus Range are basalts, whereas the larger dark area slightly to the south represents part of a mafic intrusive body. East of the Cactus Range, andesite makes up the low hills that are dark on the image. The areal extent of these andesitic outcrops is exaggerated on the image because of the presence of a talus apron around the outcrop. Approximately 12 km south of Tolicha Peak is a series of basalt flows, whose boundaries do agree well with the geologic map. Although some subtle reflectivity differences among the mafic rocks exist on the image, variations are not consistent enough to allow discrimination, for example, among andesites, basalts, and the mafic intrusive body with any degree of confidence.

Not all dark areas are indicative of mafic rocks, however. The most prominent dark area on the standard MSS image in the southeastern corner represents the previously mentioned Thirsty Canyon tuff and a rhyolite. Only the southernmost circular area is basalt (Black Butte), and it is indistinguishable from the dark-appearing tuff and rhyolite north of it. Tuffaceous sedimentary rocks that crop out within the dark area southwest of Mud Lake are also indistinguishable from the neighboring andesite and basalt. In addition, a few small exposures of Precambrian and Cambrian rocks appear dark on the image. Therefore, rock-type discrimination on this standard MSS image is severely limited even if only a two-component classification system of mafic and felsic rocks is used.

In general, felsic rocks vary in tone on the image from medium to light gray. Discrimination of rock units within this tonal range of gray is rarely possible in any band, especially as the alluvium in the image appears as very similar gray levels. Isolated outcrops can be discriminated, however, as in the rhyolitic rocks and tuffaceous sediments on the westernmost margin of the Cactus Range. Nevertheless, on Pahute Mesa, a dark pattern on the image correlates only locally with the mapped distribution of tuff and rhyolite.

Major mineralized areas, as indicated by the locations of mines and mining districts (Fig. 99), are not distinguishable on the image. The largest, the Goldfield district, is a uniform light gray, and the image gives no indication of the extensive alteration zone there. Other altered areas are also indistinct.

Stretched MSS Images

The stretched MSS images of the study area show substantially more scene contrast and spatial detail than do the standard images. The increased scene contrast, a direct result of the stretching process, allows slightly improved discrimination of rock types. The stretched images have been generated directly from the digital tapes, with the result of an apparent increase in spatial resolution. The standard images, on the other hand, have passed through several photographic reproductive processes after their generation from the tapes, processes that have caused a loss of some resolution.

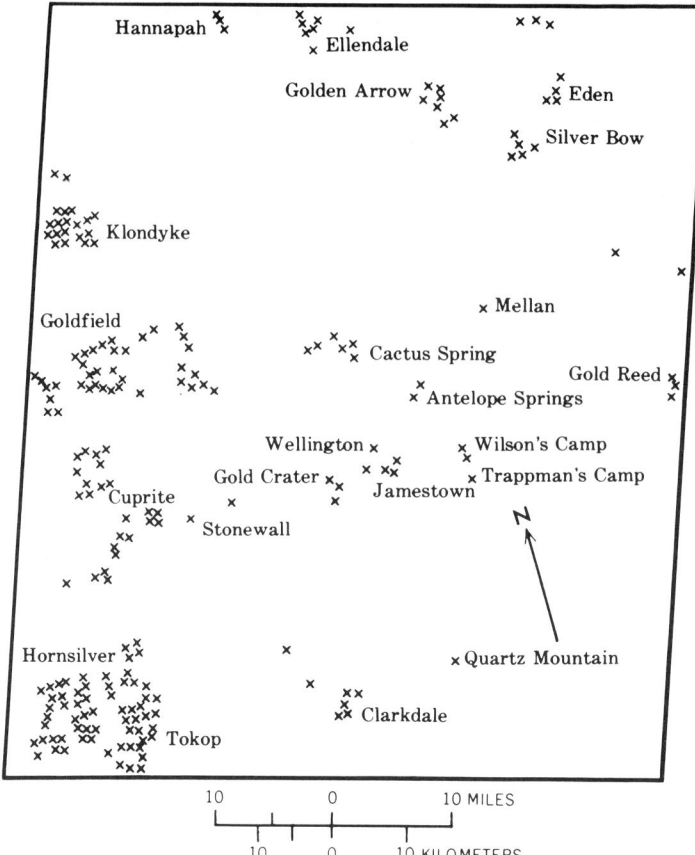

FIGURE 99. Major mining districts in the study area.

The most apparent improvement of rock-type discrimination is the separation of the basalt and the mafic intrusive in the Cactus Range on the basis of the generally lower albedo of the basalt. Other limitations on discrimination of mafic rock types, however, as described for the standard images, also apply to the stretched images. In addition, discrimination among the felsic rock units, particularly in the Pahute Mesa area, is not improved through stretching the images. Although the Goldfield district stands out better in the stretched than in the standard images because of the increased contrast, most of the mineralized areas are not prominent in the stretched images. In general, although scene contrast and apparent spatial resolution are increased, stretching of the radiance data without additional enhancement results in only slightly better discrimination of the geologic materials of this area.

Color–Infrared Composites
Color composites can be prepared either by transmitting filtered light through the positive film transparencies or by combining color separates. For example, a

color–infrared composite was prepared using blue, green, and red filters and positive transparencies for stretched MSS bands 4, 5, and 7, respectively. Colors in this composite are directly related to the film densities in the positive transparencies. Hence, vegetation is red in this composite because the high reflectivity of vigorous vegetation in MSS band 7 compared with MSS bands 4 and 5 results in a relatively low density in the MSS band 7 transparency.

In this study area, the color–infrared composite is most useful for discrimination of vegetated areas. The darkest red, and therefore the densest vegetation, occurs in the Kawich Range and on Stonewall Mountain. Additional red tinges are apparent on Gold Mountain, on parts of Pahute Mesa, and north of Monitor Peak. These sparsely vegetated areas cannot be detected easily on the black-and-white stretched MSS images.

Color–infrared composites appear to offer little improvement over the stretched MSS images for discrimination of rock types. All the points of confusion among the felsic, intermediate, and mafic rock types described for the standard and stretched images are also present in this color composite. Although some of the known limonitic areas, such as southwest of Quartz Mountain and north of Monitor Peak, are light orange–brown, the other altered areas are not distinctive. The light orange–brown color, suggesting high reflectance in the MSS bands 5 and 7 compared with the MSS band 4, is consistent with limonite spectra discussed later. Nonetheless, color–infrared composites do not appear to offer a reliable means for detecting hydrothermally altered areas.

Stretched Ratio Images

From the previous discussions, it is clear that the spectral reflectance differences among rock types are between altered and unaltered rocks are generally too small to be detected by visual comparison of the MSS images or through analysis of color–infrared composites. Ratioing of the spectral bands, however, provides additional means for enhancing spectral differences. Although ratioing alone is adequate to show large differences, such as those between the visible and near infrared bands for vigorous vegetation, stretching is also necessary for adequate enhancement of the typically subtle spectral reflectance differences found among most geologic materials. The resultant images represent a visual display of the differences between the bands in slope of the spectral reflectance curve of each geologic unit. Used in combination, ratioing and stretching offer a powerful means for discriminating rock types and alteration zones.

In the stretched ratio images, the linear stretches used have been selected to show maximum scene contrast rather than to relate film density to the DN values. Hence, the ratio range selected for stretching and the amount of stretch applied to each ratio image are different. Relative spectral reflectivity, therefore, cannot be determined directly from these particular stretched ratio images because corresponding gray levels among the images do not represent equal DN ratio values.

Within each of the images, the extremes of the 16-step gray scale represent the largest spectral reflectivity differences. The darkest areas in each stretched ratio image are those for which the denominator of the ratio is greater than the numerator. Conversely, the numerator is greater than the denominator for the lightest areas. For example, the largest differences in reflectance between bands for vegetation are in the images for ratios 5/6 and 5/7 (vegetation very dark) and for ratio 4/5 (vegetation extremely light).

More variation within the playas is present in the stretched ratio images than in the previous single-band enhanced images in which no variation is apparent. Mud Lake and the southernmost playa are most notable. Within all the stretched ratio images, however, the playas can be confused with other geologic materials.

Mafic rocks appear very light and distinctive in the images for MSS 4/5, 4/6, and 4/7, but discrimination is still problematic. For example, basaltic and andesitic rocks are indistinguishable, and in the image for MSS 4/5 the mafic rocks can be confused with vegetation and with the felsic rocks on Pahute Mesa. Felsic rocks do not stand out in the stretched ratio images. Most appear as medium tones of gray, although the felsic rocks on Pahute Mesa are dark in the image for MSS 5/7 and can, in this image, be discriminated from the lighter mafic rocks.

Most of the hydrothermally altered areas are indistinct in the standard MSS and stretched MSS images and in the stretched color–infrared composites. The Goldfield mining district (Fig. 99), although the largest producing district in the study area is especially inconspicuous in these image products. In the stretched ratio images, however, the Goldfield district shows a pattern that, as we will discuss next, is nearly identical with the altered area mapped by Jensen et al. (1971).

Although the Goldfield district and many other known altered areas are apparent on the stretched ratio images, they cannot be discriminated from other areas with similar gray tones. The gray levels of the playas especially appear to be nearly identical with those of the altered areas.

Color–Ratio Composites

Color-compositing techniques offer an efficient means for combining black-and-white stretched ratio images for discrimination of rock types. Whereas two spectrally different areas may be nearly indistinguishable in a black-and-white stretched ratio image, proper color combination of two or more images permits discrimination on the basis of color differences. Discrimination is increased not only because information from several ratio images is combined, but also because the human eye is capable of discriminating two orders of magnitude more hue values than values of gray. Color–ratio composites constitute the most useful image product for geologic analysis generated during this study.

A large number of color and ratio image combinations is possible. Ideally, selection of stretched ratio images for compositing should be based on the spectral reflectives of the materials of interest, but spectral data for the study area are too limited to provide an adequate basis for specific selection. Therefore, 70-mm positive transparencies of the six stretched ratio images were combined in a color-additive viewer to determine the combinations most useful for discriminating the main rock types and altered areas.

Analysis of combinations of two stretched ratio images and two color filters, from a choice of red, blue, and green, showed that no single two-component composite examined could provide adequate discrimination among major rock types, altered areas, vegetation, and playas. The most common problem involved difficulty in distinguishing the altered zones from the playas and some alluvial areas. Furthermore, in combinations in which the altered areas are detectable, some of the major rock types are not distinctive. Three component composites (three stretched ratio images and three color filters) have more overall discrimination potential than do the two-component composites, but discrimination between the altered areas and the playas and alluvium remains a problem.

Another color-compositing technique was tried in an effort to circumvent these problems. Color–ratio composites were prepared using diazo transparencies. Because in the diazo process the most intensely exposed parts of the image are "burned off" upon development, no color is contributed by areas that are clear (for example, DN = 0) in the transparency. Conversely, high film density results in intense color in the diazo color separates. It is important to understand that the end product of this technique does not represent the reverse of the color-additive method. That is, a composite of negative transparencies made in a color-additive viewer would not be the same as a positive–transparency composite produced using the diazo process.

An optimum combination for geologic analysis of the study area was determined using the diazo process. Although all the innumerable possible color combinations and stretched ratio image combinations have not been evaluated, the most effective three-component color–ratio composite for discriminating between altered and unaltered areas and among the regional rock units was prepared using the following color and stretched ratio image combination: blue for MSS 4/5, yellow for MSS 5/6, and magenta for MSS 6/7. Note that these ratios involve all four MSS bands.

Analysis of Color–Ratio Composite

A color–ratio composite was analyzed initially by studying available geologic maps and 1:250,000 scale black-and-white photomosaics to determine which rock units should be distinguishable and to identify problem areas to be checked in the field. Several regional geologic studies have been conducted previously in this area, most notably those of Cornwall (1972), Albers and Stewart (1972), and Ekren et al. (1971), but a map based on a compilation by John H. Stewart and J. E. Carlson, U.S. Geological Survey, of the State of Nevada is the most useful single map because it integrates all these previous studies. In addition, the original compilation scale of 1:500,000 is especially compatible with analysis of Landsat images. Three field evaluations, each 1 week long, were aided appreciably by overflying the area before working on the ground. Black-and-white aerial photographs at a scale of 1:62,500 also proved to be nearly indispensable for orientation and for distinguishing outcrops and surficial deposits.

In the color–ratio composite, mafic rocks are generally white, felsic rocks are pink, playas are blue, and vigorous vegetation is orange. Limonitic areas are green to yellow–green, and essentially limonite-free hydrothermally altered areas range from green to brown and, more rarely, to red–brown. Clouds are a dark pink–brown, and cloud shadows are white; topographic shadows are also white.

Most mafic rocks can be discerned easily on the image. They appear white, usually accompanied by small patches of pink. In places, such as immediately southwest of Mud Lake, some of the pink patches associated with the white area are clearly due to the presence of tuffs and tuffaceous sediments, but in other areas the pink is probably related to residual soil formed on mafic rocks. These light-pink-colored soils are exemplified by Malpais Mesa southwest of Goldfield, the basalt approximately 6 km north of Tolicha Peak, and the basalt northeast of the Kawich Range; they are probably felsic in composition either because the mafic minerals have been decomposed and leached, leaving a residuum consisting mainly of slightly weathered feldspar, or because a veneer of felsic eolian material overlies the basalt. Some discrepancies between the map units and the color–ratio composite may be accounted for by talus deposits, as mentioned previously.

Unfortunately, several other rock units besides mafic rocks also appear white, although not consistently so. The two white spots northeast of Mud Lake are dark gray to black Paleozoic limestone and chert (Cornwall, 1972), and a white strip north of Gold Mountain is dark green–gray siltstone and very fine grained quartzite of Precambrian and Cambrian age (Albers and Stewart, 1972). The eastern part of the white square area west of Stonewall Mountain is part of an undifferentiated Paleozoic unit. The western part is presumably talus.

Additional problems in discriminating mafic rock types are the white-appearing cloud shadows and topographic effects — for example, thin white areas northeast of Stonewall Mountain. Examination of the stretched MSS images, however, makes identification and discrimination of these nongeologic features straightforward.

Dark pink, brown–pink, and orange–pink are characteristic of felsic rocks on the image. Light vegetation cover causes the orange tone but it appears that rock reflectance, and therefore rock-type variation, can be seen if the vegetation cover is light, as on Pahute Mesa. On the other hand, where vegetation is dense, as on the Kawich Range and on Stonewall Mountain, spectral differences reflecting rock type are not recorded. These areas appear a very dark orange.

In general, the agreement between pink hues and the distribution of felsic rocks is quite striking. Most impressive is the orange–pink hue of the Thirsty Canyon tuffs and flows. Because the surface of these rocks has a low albedo, they are easily confused with the mafic volcanic rocks both on the other images evaluated and in the field. Ratioing minimizes albedo effects, however, so that discrimination of these felsic rocks from mafic rocks is possible on the color–ratio composite. The general shape of the visible and near infrared reflectance spectra of these felsic rocks must be very similar to that of the other felsic rocks. In general, however, the older tuffs and flows are less uniform on the color–ratio composite than newer ones.

Playas are easily discriminated from the altered rocks. The only blue areas that are not playas are two small patches representing mine tailings, north and northeast of Goldfield. In addition, variation within and among playas is much more pronounced in the color–ratio composite than in any of the previously enhanced products. The southernmost playa and Mud Lake show the most variation, presumably because of compositional and perhaps grain size differences.

One of the principal objectives of this study was to evaluate the potential of the MSS data for detecting hydrothermally altered areas. Although none of the previous image products seems to have much potential for discriminating altered areas, these areas can be identified with a high degree of confidence in the three-component color–ratio composite.

Discussion

One of the ultimate objectives of this investigation is to derive quantitative spectral reflectivity information from the MSS images and to use these data for making estimates of the bulk composition of surface materials. Although this objective is not presently feasible because of inadequate calibration data, the spectral differences seen as color variations in the color–ratio composite can be evaluated qualitatively by considering laboratory spectra that appear to be reasonably representative of the surface materials.

As discussed previously, visible and near infrared reflectance spectra for mafic and felsic rocks (Fig. 94) have generally different shapes. Although data for only a few of the rock types present in the study area are included in Figure 94, a general idea of the anticipated differences can be gained through comparison of these spectra. A useful method for making such comparisons is examining calculated ratio values based on the width of the MSS bands used in processing the color–ratio composite (MSS 4/5, 5/6, and 6/7). These ratio values are not representative of the absolute MSS values because of undetermined effects of the solar spectrum and of atmospheric variations, but the calculated ratio values can be useful for comparative purposes. In general, ratios for the felsic rocks are less than unity, whereas the mafic rock ratios are greater than unity (Table 37). Comparison of the basalt with the two rhyolite spectra shows that the ratios for basalt are approximately 7–12% higher, except in the MSS 4/5 ratio for the pink rhyolite (Table 37), where the difference is about 25%. These low percentage values attest to the subtlety of the spectral differences among even very different rock types. It is not surprising, therefore, that some form of computer processing is necessary to enhance the spectral signatures of mafic and felsic rocks in the standard Landsat images. Additional computer processing, perhaps using cube root stretch of the ratio values to increase scene contrast in the dark areas in the stretched ratio images, might permit further discrimination between, for example, basalts and andesites.

Evaluation of the altered areas is more difficult because spectroradiometric studies have not been conducted on altered rocks. A few laboratory spectra, however, have been published for most of the main minerals that constitute the Goldfield alteration zone, including limonite, montmorillonite, alunite, kaolinite, and jarosite (Fig. 100). Comparison of the ratios calculated for these spectra and for the mafic rocks (Table 37) shows that the MSS 4/5 values for the alteration products are significantly lower than those for mafic rocks and that the MSS 5/6 ratios are slightly lower. Therefore, these alteration products should be very distinctive where they occur in areas underlain by mafic rocks. Comparison with the felsic rock ratios indicates that geothitic and hematitic limonite, alunite, and montmorillonite should be distinguishable from felsic rocks because of their lower MSS 4/5 ratios and somewhat lower MSS 5/6 ratios. The jarosite MSS 4/5 ratio is only slightly lower than the felsic rock ratios; however, the MSS 6/7 ratio for jarosite is markedly higher than the largest value for the felsic rocks, so this ratio could be

TABLE 37. Ratios Calculated for MSS Bands for Selected Mafic and Felsic Rocks and Alteration Minerals

Sample	Number	MSS 4/5	MSS 5/6	MSS 6/7
Mafic Rocks				
Basalt	13	1.05	1.03	1.06
Gabbro	9	1.07	1.01	0.99
Peridotite	8	1.01	1.01	1.05
Serpentinite	15	1.04	1.05	1.07
Felsic Rocks				
Rhyolite (pink)	12	0.79	0.92	0.96
Rhyolite (gray)	4	0.96	0.96	0.98
Granite	14	0.89	0.94	0.91
Granite	5	0.97	0.98	0.92
Granodiorite	2	0.92	0.94	0.97
Alteration Minerals				
Limonite (goethitic)	—	0.61	0.72	0.94
Limonite (hematitic)	—	0.49	0.85	0.95
Jarosite	—	0.78	0.99	1.27
Montmorillonite	—	0.65	0.84	1.07
Alunite	—	0.66	0.88	0.97
Kaolinite	—	0.93	0.98	0.99

used for discrimination purposes. The ratios shown in Table 37 for kaolinite resemble those for the felsic rocks of Figure 94, and it is doubtful that such small differences could be distinguished in a color–ratio composite. However, as Ashley (1970) pointed out, illite and alunite are associated with kaolinite in the more intensely altered zone at Goldfield. Both of these minerals might subdue the effects of the kaolinitic spectral reflectivity and result in low ratios.

Although altered and unaltered rocks are discriminated with a high level of confidence in the color–ratio composite, limonitic and limonite-free hydrothermally altered areas do not appear to be distinguishable at this stage of the analysis. Although this distinction does not appear to be economically important in the study area because metallization occurs in both types of areas, it could be important in areas where limonite is not particularly diagnostic of hydrothermal alteration. The calculated ratios in Table 37 suggest that the MSS 4/5 ratio should provide the best opportunity for distinguishing bleached argillized and limonitic altered rocks, especially where the limonite is hematitic rather than goethitic. In addition, higher spectral resolution and bands at longer wavelengths should significantly aid in discriminating these materials and in estimating the mineralogical makeup of the surface materials.

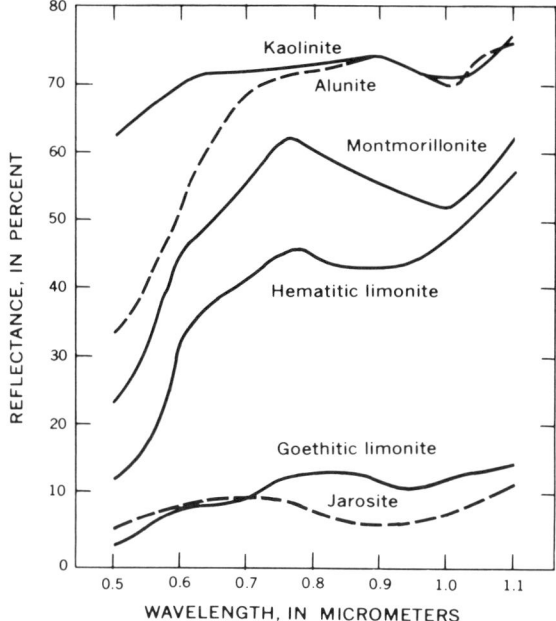

FIGURE 100. Reflectance spectra for alteration minerals.

The above discussion illustrates the critical need for a clearer definition of the relationship between mineralogical composition and visible and near infrared spectral reflectivity of altered and unaltered rocks. Ideally, in situ spectroradiometric measurements, at several scales and coordinated with detailed sampling for subsequent mineralogical analysis, should be made in a variety of geologic settings and environmental conditions. An important use for these in situ measurements is formulation of spectral reflectivity standards that take into account varying surface state conditions. The standard areas should be of different albedos but nearly uniform over large areas, and the spectral information should be collected at the same time as the satellite overflight so that atmospheric conditions are similar. Spectral reflectance standards acquired in this fashion could be used to normalize all the DN values in the MSS images, provided that atmospheric transmission and path radiance are uniform over the run. Then quantitative spectral reflectivity information could be extracted directly from the DN values for any surface unit.

Summary and Conclusions

The results of this study show that most of the major rock units and altered and unaltered areas in the study area can be separated on the basis of visible and near infrared spectral reflectivity differences recorded from satellite altitude. Because these differences are mainly due to slight variations in the slopes of the spectra, they are not detectable through visual or optically assisted techniques. Digital ratioing of the MSS bands and subsequent stretching to increase the contrast are necessary to enhance these differences. Although the basic spectral information is contained in the stretched ratio images, color–ratio composites, especially combinations of three ratio images, appear to be the best means of display for geologic visual interpretation as an aid in mineral resources exploration and regional geologic mapping.

For discrimination of hydrothermally altered areas and of regional rock (and soil) units in the study areas, the optimum color–ratio composite appears to be a combination of diazo color transparencies having the following ratio images and colors: blue for MSS 4/5, yellow for MSS 5/6, and magenta for MSS 6/7. In this composite, mafic rocks are generally distinguishable from felsic rocks despite their similar albedos in some places. Details within playas are best seen in the color–ratio composite. Some of the most important practical limitations, however, include erroneous identification of basalt as felsic rocks where soil overlies areas of the basalt and lack of discrimination thus far of basaltic and andesitic rocks.

Hydrothermally altered areas appear as anomalous color patterns within the volcanic rocks on the color–ratio composite. Comparison of known mining district locations and clusters of anomalous colors, ranging from green and dark green to brown and red–brown, shows good agreement in the widespread Tertiary volcanic rocks. In the Goldfield mining district, for example, very striking agreement was found between the anomalous green pattern and the map of the altered area by Jensen et al. (1971). However, the density of anomalies compared with that of the mines is not as high in the southwestern part of the study area, where the alteration and mineralization are associated with intrusion of pre-Mesozoic rocks rather than with the Tertiary volcanic rocks.

In general, the green to dark green color pattern in this color–ratio composite represents predominantly limonitic areas, which except for two cases are hydrothermally altered. All green areas in the composite are not limonitic, however. Several green areas, as is well illustrated in the Gold Crater area, are essentially limonite-free argillized volcanic rocks. Red–brown areas in the color–ratio composite commonly appear to be silica-rich light-colored volcanic rocks. In the Cuprite district, the red–brown color correlates very well with the silicified zone in which hydrothermal alteration has taken place, but the red–brown areas near Tolicha Peak are very glassy tuffs without any obvious hydrothermal alteration. Brown areas, though studied in less detail than the other color anomalies, generally seem to represent light-colored volcanic rocks, with hydrothermal alteration present at least in the Silver Bow district. The presence of hydrothermal activity in other brown areas could not be determined at this time.

Although adequate spectroradiometric measurements are not yet available for defining the relationships between the mineralogical composition and spectral reflectivity of the surface materials, analysis of laboratory spectra suggests that ratio differences of 7–12% between the felsic and mafic rocks may be shown in the color–ratio composite. Comparison of the ratios calculated for unaltered rocks and alteration minerals common to the Goldfield district shows that in general the altered rocks should be discriminable, especially in the MSS 4/5 ratio. The inability to distinguish between limonitic and limonite-free hydrothermally altered areas is believed to be due to the similar general shapes of limonite and nonferrous alteration minerals and to the low spectral resolution and absence of spectral bands beyond 1.1 μm in the MSS. Higher-spectral-resolution images of the area, especially recorded beyond 1.1 μm, should provide adequate discrimination between limonite and the nonferrous alteration minerals. Additional computer processing of the MSS data may also aid in distinction of materials.

Because the studies described here have been directed towards areas of outcrop, surficial deposits have received very little attention; even a cursory examination of the color–ratio composite, however, shows considerable spectral information for these materials, which should be useful for study of these areas. An interesting result pertaining to these surficial deposits is that a thin coating of ferric iron apparently forms on the fragments derived through weathering of some felsic rocks, thereby causing the anomalous green color on the color–ratio composite in some alluvial areas.

Refinement and further testing of these computer-processing and geologic interpretation techniques are necessary for realizing the full potential in geologic exploration. An especially critical need at this stage is for in situ spectroradiometric measurements coordinated with detailed sampling for subsequent mineralogical analysis in a variety of geologic and environmental conditions.

Additional computer-processing techniques that deserve consideration include cluster analysis and automatic classification. However, the advantage of interpreting a color–ratio composite over using a pure classification scheme, such as the LARSYS method (Landgrebe, 1971), is that the analyst can take into account many other factors, such as the distinction between outcrop and surficial deposits, before delineating boundaries among units. An additional advantage of the ratio method used in this study is that computer-processing time is reduced by a factor of 100 or more below the time for the LARSYS classification scheme, with an accompanying substantial cost saving. Nevertheless, consideration should be given to such classification schemes as well as to both supervised and unsupervised cluster analysis techniques.

Although many questions have been left unanswered by this report, the results indicate that geologic exploration can benefit substantially by the use of digital computer processing of visible and near infrared MSS images. Limitations imposed by the low spatial and spectral resolutions of the MSS must be overcome in subsequent satellite systems so that the results can be applied to larger-scale problems. In the meantime, in order to define the effects of spatial resolution and intervening atmosphere, visible and near infrared spectral data should be collected from aircraft platforms, processed in a manner similar to that applied to these MSS images, and analyzed in conjunction with existing geologic maps, field study, and spectral reflectivity information.

REFERENCES

Albers, J. P., and H. R. Cornwall, *Revised interpretation of the stratigraphy and structure of the Goldfield district, Esmeralda and Nye Counties, Nevada* [abs.], Spec. Paper 101, Geological Society of America 1968, p. 285.

Albers, J. P., and F. J. Kleinhampi, *Spatial relation of mineral deposits to Tertiary volcanic centers in Nevada,* Prof. Paper 700-C, U.S. Geological Survey, 1970, pp. C1–C10.

Albers, J. P., and J. H. Stewart, 1972, *Geology and mineral deposits of Esmeralda County, Nevada,* Bull. 78 Nevada Bureau of Mines and Geology 78, 1972.

Ashley, R. P., *Evaluation of color and color infrared photography from the Goldfield mining district, Esmeralda and Nye Counties, Nevada,* U.S. Geological Survey Open-File Report, 1970.

Ashley, R. P., and W. J. Keith, *Geochemistry of the altered area at Goldfield, Nevada, including anomalous and background values for gold and other ore metals,* U.S. Geological Survey Open-File Report, 1973.

Bancroft, G. M., and R. G. Burns, Interpretation of the electronic spectra of iron in pyroxenes, *Am. Mineralogist* 52(9–0), 1967: 1278–1287.

Billingsley, F. C., and A. F. H. Goetz, 1973, Computer techniques used for some enhancement of ERTS images, in *Symposium on significant results obtained from the Earth Resources Technology Satellite-1,* vol. I, *Technical Presentations,* comp., ed., S. C. Friden et al., sec. 13, NASA SP-327, 1973, pp. 1159–1168.

Billingsley, F. C., A. F. H. Goetz, and J. N. Lindsley, Color differentiation by computer image processing, *Photog. Sci. Eng.* 14(1), 1970: 28–35.

Cornwall, H. R., *Geology and mineral deposits of southern Nye County, Nevada,* Bull. 77, Nevada Bureau of Mines and Geology, 1972.

Ekren, E. B., R. E. Anderson, C. L. Rogers, and D. C. Noble, *Geology of northern Nellis Air Force Base Bombing and Gunnery Range, Nye County, Nevada,* Prof. Paper 651, U.S. Geological Survey, 1971.

Goetz, A. F. H., Quality and use of ERTS radiometric information in geologic applications, in *Proc. 4th Ann. Conf. Application Remote Sensing Arid Land Resources and Environment, Tucson, Ariz., Nov. 14–16, 1973,* 19 U.S. Geological Survey.

Goetz, A. F. H., F. C. Billingsley, E. Yost, and T. B. McCord, Apollo 12 multispectral photography experiment, in *Proc. 2nd Lunar Sci. Conf., Houston, Tex., January 11–14, 1971,* ed. A. A. Levinson, MIT Press, Cambridge, Mass., 1971, vol. 3, pp. 2301–2310 (*Geochim. et Cosmochim. Acta,* Supp. 2).

Harvey, R. D., and C. J. Vitaliano, Wall-rock alteration in the Goldfield district, Nevada, *J. Geology,* 72(5), 1964: 564–579.

Hunt, G. R., and J. W. Salisbury, Visible and near-infrared spectra of minerals and rocks, I, *Silicate Min. Mod. Geol.* 1(4), 1970: 238–300.

Hunt, G. R., J. W. Salisbury, and C. J. Lenhoff, Visible and near-infrared spectra of minerals and rocks, III: oxides and hydroxides, *Mod. Geol.* 2(3), 1971a: 195–205.

———, Visible and near-infrared spectra of minerals and rocks, IV: sulphides and sulphates, *Mod. Geol.* 3(1), 1971b: 1–14.

———, Visible and near-infrared spectra of minerals and rocks, VII: acidic igneous rocks, *Mod. Geol.* 4, 1973a: 217–224.

———, Visible and near-infrared spectra of minerals and rocks, VI: additional silicates, *Mod. Geol.* 4, 1973b: 85–106.

———, Visible and near-infrared spectra of minerals and rocks, VIII: Intermediate igneous rocks, *Mod. Geol.* 1974a (In press).

———, Visible and near-infrared spectra of minerals and rocks, IX: Basic and ultrabasic igneous rocks, *Mod. Geol.* 1974b.

Jensen, M. L., R. P. Ashley, and J. P. Albers, Primary and secondary sulfates at Goldfield, Nevada, *Econ. Geol.* 66, 1971: 618–626.

Kral, V. E., Mineral resources of Nye County, Nevada, *Nevada Univ. Bull.* 45(3) (Geology and Mining Ser. No. 50), 1951.

Landgrebe, D. A., *Systems approach to the use of remote sensing,* LARS Inf. Note 041571, Purdue University, Lab. Application Remote Sensing, 1971.

Ransome, F. L., *The geology and ore deposits of Goldfield, Nevada,* Prof. Paper 66, U.S. Geological Survey, 1909.

Ross, H. P., J. E. M. Alder, and G. R. Hunt, A statistical analysis of the reflectance of igneous rocks from 0.2 to 2.65 microns, *Icarus* 11(1), 1969: 46–54.

Rowan, L. C., Near-infrared iron absorption bands: applications to geologic mapping and mineral exploration, in *Proc. 4th Annual Earth Resources Program Rev.,* Houston, Tex., 1972, pp. 60-1–60-18.

————, *Iron-absorption band analysis for the discrimination of iron-rich zones:* U.S. Geological Survey Open-File Report, 1973.

Rowan, L. C., and R. K. Vincent, Discrimination of iron-rich zones using visible and near-infrared spectral analysis [abs.], *Geol. Soc. Am. Abs. Prog.* 3(7), 1971: 691.

Searls, Fred, Jr., 1948, A contribution to the published information on the geology and ore deposits of Goldfield, Nevada, *Nevada Univ. Bull.,* 42(5) (Geology and Mining Ser. No. 48), 1948.

U.S. National Aeronautics and Space Administration, Goddard Space Flight Center, *Data users handbook* [for Earth Resources Technology Satellite], Doc. 71SD4249, 1971.

Vincent, R. K., and F. J. Thompson, Discrimination of basic silicate rocks by recognition maps processed from aerial infrared data, in *Proc. 7th Int. Symp. Remote Sensing Envir.* 1: Ann Arbor, Mich., University of Michigan, Institute of Science and Technology, Willow Run Laboratories, 1972, pp. 247–252.

[Whitaker, E. A.], Colors and the meso-structure of the maria, in *Ranger VII, pt. II, Experimenters' analyses and interpretations,* ed. R. L. Heacock et al., Tech. Rept. 32-700, California Institute of Technology, Jet Propulsion Lab. 1965, pp. 29–39.

White, W. B., and K. L. Keester, Optical absorption spectra of iron in rock-forming silicates, *Am. Mineralogist,* 51(5–6), 1966: 774–791.

Yost, E., R. Anderson, and A. F. H. Goetz, Isoluminous additive color method for the detection of small spectral reflectivity differences, *Photog. Sci. Eng.* 17, 1973: 117–182.

3.2.2 Geologic Exploration with High-Resolution Radar*

High-resolution, side-looking radar imagery has many qualities giving it the appearance of aerial photography. However, the similarities between the two media are more apparent than real. The principal reason for this is that radar wavelength, illumination, and geometry are different from those of an aerial camera. The unique characteristics of radar imaging resulting from these differences cause radar to have advantages as well as limitations. These must be understood in order

*Based on Hubert O. Rydstrom, *Geologic exploration with high resolution radar,* Goodyear Aerospace Corp., Litchfield Park, Ariz., 1966.

to perform accurate analysis and interpretation. When they are not understood, an incorrect conclusion may be formed, or a proper conclusion may be made but for an incorrect or a fallacious reason. Solutions reached in this latter way, although satisfying a momentary requirement, soon lead to incorrect conclusions or a failure to recognize the true data obtainable from the imagery.

Adapting to a Radar World

The investigator faced with the task of using imagery from an unfamiliar sensor has a natural tendency to base analyses or interpretations on information or data relating to his specialty. Thus, an interest in geology causes him to see radar images of rocks in terms of those rock parameters which he commonly employs in a field investigation. Unfortunately, the radar does not sense rocks in the same way that the geologist does. Since the radar cannot change, the only alternative is for the investigator to revise his thinking and methods, making them compatible with the radar world.

By performing analysis and interpretation according to radar concepts, the way to greater and more meaningful exploitation of radar as an investigative tool may be realized. At the same time, overly optimistic generalizations or, conversely, underestimation of capabilities may be avoided.

The analysis and interpretation of radar imagery presented here was performed without the benefit of digital processing. This material is included primarily because the power of this unique data source is often overlooked; it shouldn't be. As imagery, these data are susceptible to the same enhancement and manipulation algorithms as photography, but their distinctive character provides the analyst with another dimension of information to exploit.

Interpretation Problem Areas

When the geologist is not familiar with radar, he may attribute an image anomaly to rock mineral constituents rather than to a difference in surface vegetation. That the particular rock may support a specialized plant community, which in turn causes a variation in radar reflectivity, can certainly be an interpretative clue — geobotanical analysis from radar imagery — but to ascribe the image directly to rock type is misleading at best.

As another example, it is well known that the dielectric constant of a material is a factor in the transfer or propagation of electromagnetic energy. It is also well known that interstitial water in rock material changes its overall dielectric constant. Thus it has been said that a porous fracture zone, in which the material is moister at the surface than the surrounding rock, should be evident as a variation of tone on a radar image. Commonly, however, this moist surficial zone supports more (or different) vegetation than does the surrounding rock. This vegetation, being the predominant surface presented to the radar beam, is also the predominant backscatterer and can effectively mask or override contributions attributable to such parameters as dielectric constant.

Radar Imaging

Illumination.   Radar is called an active sensor because it supplies its own illumination. Not requiring sunlight, it is capable of gathering information any time of the day or night. This illumination, propagated by the antenna, is essentially unidirectional compared to the multidirectional illumination provided by atmospherically diffused sunlight. The radar thus illuminates directly, and receives reflections from, only those surfaces on a line of sight from the antenna, as shown in Figure 101. The shadows or tonal changes which can be produced by this illumination are helpful in defining elevated features; and by decreasing the depression angle of the beam, shadowing may be enlarged or intensified to assist in the analysis of low-relief terrain. However, radar shadowing need cause no loss of information because flight parameters can be organized to illuminate any side or dimension of a feature.

In northern latitudes when weak solar illumination and adverse weather conditions exist, radar can provide a continuous flow of information on ship and ice movements, for example, which is unattainable by other means.

Wavelength and Reflection.   A surface with roughness on the order of one-half wavelength of the impinging electromagnetic energy appears smooth to that energy and acts as a specular or mirror-like reflector. Rougher surfaces act as diffuse reflectors, scattering energy in numerous directions. Since radar energy is of much longer wavelength than that of the visible spectrum, many more surfaces appear smooth to a radar than to a camera. The relationship of wavelength and reflection to surface roughness is illustrated in Figure 102.

Specular Reflection

On radar imagery, the strongest returns and the no-returns (other than shadow) are generally caused by specular reflectors. The strong returns from these reflectors

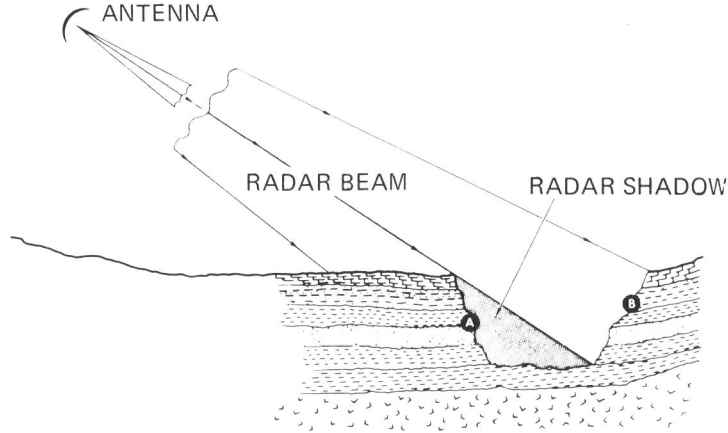

FIGURE 101.   Radar illumination and shadow.

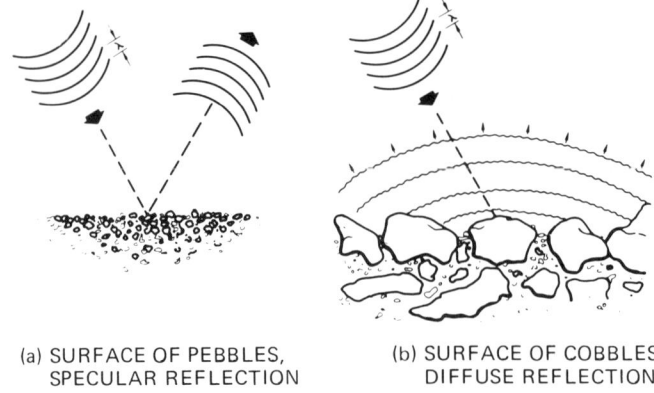

(a) SURFACE OF PEBBLES,   (b) SURFACE OF COBBLES,
    SPECULAR REFLECTION   DIFFUSE REFLECTION

FIGURE 102. Specular and diffuse reflection.

occur when a surface is so oriented or so configured, as illustrated in Figure 103, that the major part of the energy is returned to the antenna. The no-returns occur when the reflection is away from the antenna, as in Figure 102a. The larger number of specular reflections recorded on radar imagery causes feature images or feature emphasis to differ from those recorded by photography. Because of the reflective differences, features difficult to discern on photography may stand out boldly on radar imagery; conversely, features with commonly detectable photographic tonal differences may be a no-return on radar imagery.

Diffuse Reflection

The broad range of tones between the strong return and no-return extremes of specular reflections are caused by diffuse reflections from a variety of surfaces, such as vegetation. The essential depiction by imaging radars is of the principal surface being illuminated. It is in this connection that major misunderstandings of radar images can occur. A predominant covering of the earth's surface, varying in density from place to place, is vegetation, which forms a primary diffuse reflector of radar energy.

As was previously suggested, there are probably contributions to the total return caused by dielectric or other properties of the rock or soil beneath the vegetation, but at the present time there is no way of evaluating these contributions in meaningful terms. For this reason, the geologist must take care not to attribute solely to rock constituents the return reflections caused for the most part by an overlying vegetation. On the other hand, the association of particular rocks and vegetation communities could permit geobotanical identification from the radar imagery.

Aerosol Penetration

The energy of imaging radars, particularly at X-band wavelength, is attenuated very little, if at all, by aerosols. Thus, radar sensors can gather data through dust,

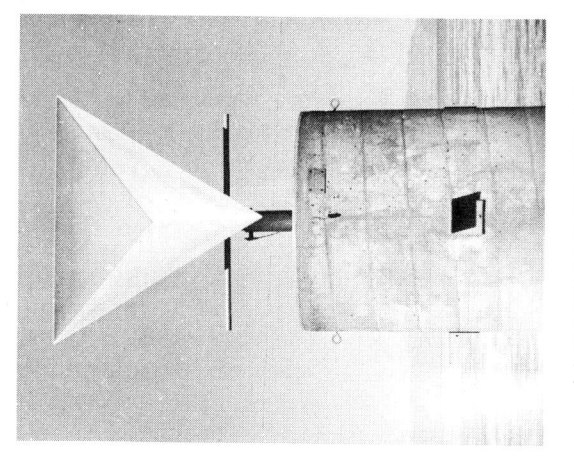

(c) MANMADE TRIHEDRAL REFLECTOR

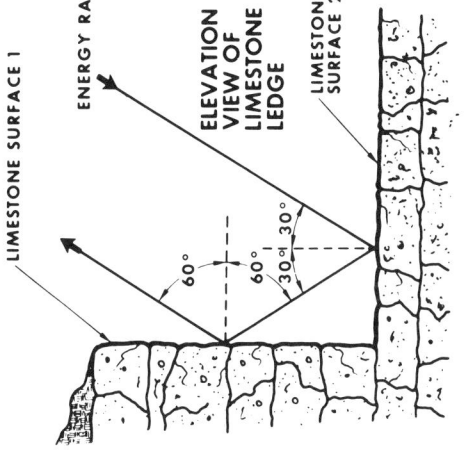

(b) DIHEDRAL REFLECTOR ORIENTED AT 90 DEGREES TO RADAR BEAM

(a) ANGLE OF INCIDENCE EQUAL TO ZERO DEGREES

FIGURE 103. Specular return reflectors.

clouds, fog, haze, rain, or snowfall when other sensors are ineffective. It is evident that this capability is advantageous for monitoring ice floes and ship traffic on a continuing basis during foul weather when the information is most acutely needed.

Imaging Snow-Covered Areas

In areas blanketed by snow, a camera may record a vast white sameness because it is sensitive to the diffuse reflections from the minute roughnesses of the snow surface. Radar energy, unaffected by this minor roughness or the white color, records the larger variations in roughness and configuration representative of old ice as opposed to new ice, for example.

Selected Examples of Natural Features

Following are examples of imagery from an X-band Synthetic Aperture* Radar. A few salient natural features, as well as several cultural features, on each image are annotated, and when discussed in the accompanying text are emphasized by italicizing the annotation words.

The annotated features and the brief text are not assumed to be complete or indicative of the extent to which interpretation could be carried. Many additional analyses and discussions could be prepared from these images, depending on the interests or requirements of the investigator.

Conway, Arkansas

Figure 104 illustrates a portion of the Arkansas Valley structural province, where differential erosion of sandstone and shale has developed a grand pattern of sweeping ridge and valley curves.

The width and relationship of the tonal variations of these curves provide the means for differentiating the gentle back slopes from the steeper scarp slopes and determining dip and strike directions. For example, at the left-hand arrow of *low hills,* a narrow band of return on the concave side of the curve is bordered toward the convex side by a broader band of weaker returns. This indicates an outward radiating dip slope and an anticlinal structure. At the annotation *dip (obscure),* the broad return bordered at the top by a narrow no-return indicates a dip slope in the opposite direction — the flank of a second anticline. This is separated from the first anticline by a broad valley developed on a younger shale in the associated syncline. Using the same approach, the large, roughly circular feature at *dip and strike of rock* can be interpreted as a synclinal basin.

The identification of a *marsh* is based on the intensification of returns common to the vegetation community associated with this type of terrain. The *drainage course* and related small bodies of water below the *dried-up reservoir* also support the conclusion that the environment is suited to marshland.

A *town, interstate highway,* and *spur dikes* also are noted.

*For a brief discussion of this type of radar, the reader is referred to *Basic concepts of synthetic aperture side-looking radar,* Bulletin GIB 9167, Goodyear Aerospace Corporation, Litchfield Park, Ariz., 1969.

RADAR IMAGE OF CONWAY, ARKANSAS, AREA

FAR RANGE

NEAR RANGE

TOWN

DIP (OBSCURE)

INTERSTATE HIGHWAY

DRAINAGE COURSE

LOW HILLS

DRIED UP RESERVOIR

MARSH

SPUR DIKES

DIP AND STRIKE OF ROCK

N

FIGURE 104. Conway, Arkansas.

Mohawk, Arizona

The Basin and Range province of southwestern United States is characterized by *fault block mountain(s)*, such as the Mohawk in Figure 105, separated by extensive, flat valleys or alluvial plains. The well-defined ridgeline and the radar shadow show that this mountain has been highly eroded into a sharp, serrate ridge. The ridge-to-toe lineations are typical of the weathering of schistose rock in this area. The nearby detritus from this erosion forms a *piedmont slope* rather than a series of alluvial fans, indicating an extended period of erosion. The former, much larger size of the mountain block is attested to by inliers (*visible pediment*) representative of the eroded toe of the mountain block.

In this arid subtropical climate, brief but torrential storms cause heavy runoff which forms an intricate drainage network (*distributaries, parallel drainage*). The strong returns from these drainageways are caused by vegetation that thrives on the occasional water they provide and from the large boulders that predominate because the fines are washed away. A notable change occurs at the *sand dunes*. Here the drainage pattern is replaced by that of windblown sand, for the rainfall immediately percolates into the porous material. The dunes are somewhat more irregularly arranged toward the north, possibly because of more changeable wind currents at the pass through the mountains.

A number of cultural features of interest also are shown with descriptive annotations.

Olton, Texas

The strong influence of natural phenomena on human ecology is illustrated by Figure 106. The *sand hills area,* with its distinctive tone and texture, adjoins the *agricultural area* along a sharp line of demarcation. The field patterns are well defined and show a variety of tones. The very dark or black patches are fallow fields or those prepared for planting; the lighter tones are fields in crop of several types. These tones will vary from season to season through different stages of growth or land preparation.

The arrows from *sinkholes* point out several of the depressions common to parts of the plains region. Their origin is controversial (e.g., blowout, solution) but their images are characteristic of limestone sinks. Although origin cannot be determined from the radar imagery either, the presence of these depressions restricts the possible choices for rock and soil types in the area.

A *transmission line* serves a *power plant*, which is located in the marginal land of the sand hills area. The spot returns of a transmission line are usually returns from the towers, although at certain orientation angles to the radar beam, cables will return energy.

Port Arthur, Texas

In Figure 107, the area along the seaward edge of a low coastal plain is strikingly patterned by the long, sweeping curves of *beach ridges*. These were developed during formation of the bar by deposition from alongshore currents sweeping across the mouth of the *bay*.

NEAR RANGE

INTERCHANGE

HIGHWAY REST STOP

DISTRIBUTARIES

FAULT BLOCK MOUNTAIN

PIEDMONT SLOPE

FAR RANGE

CANAL

IRRIGATED FIELDS

WIND MACHINES

TRAINS

VISIBLE PEDIMENT

SAND DUNES

N

HIGHWAY CONSTRUCTION

BARREN FIELD

PARALLEL DRAINAGE

RADAR IMAGE OF MOHAWK, ARIZONA AREA

FIGURE 105. Mohawk, Arizona.

FIGURE 106. Olton, Texas.

254

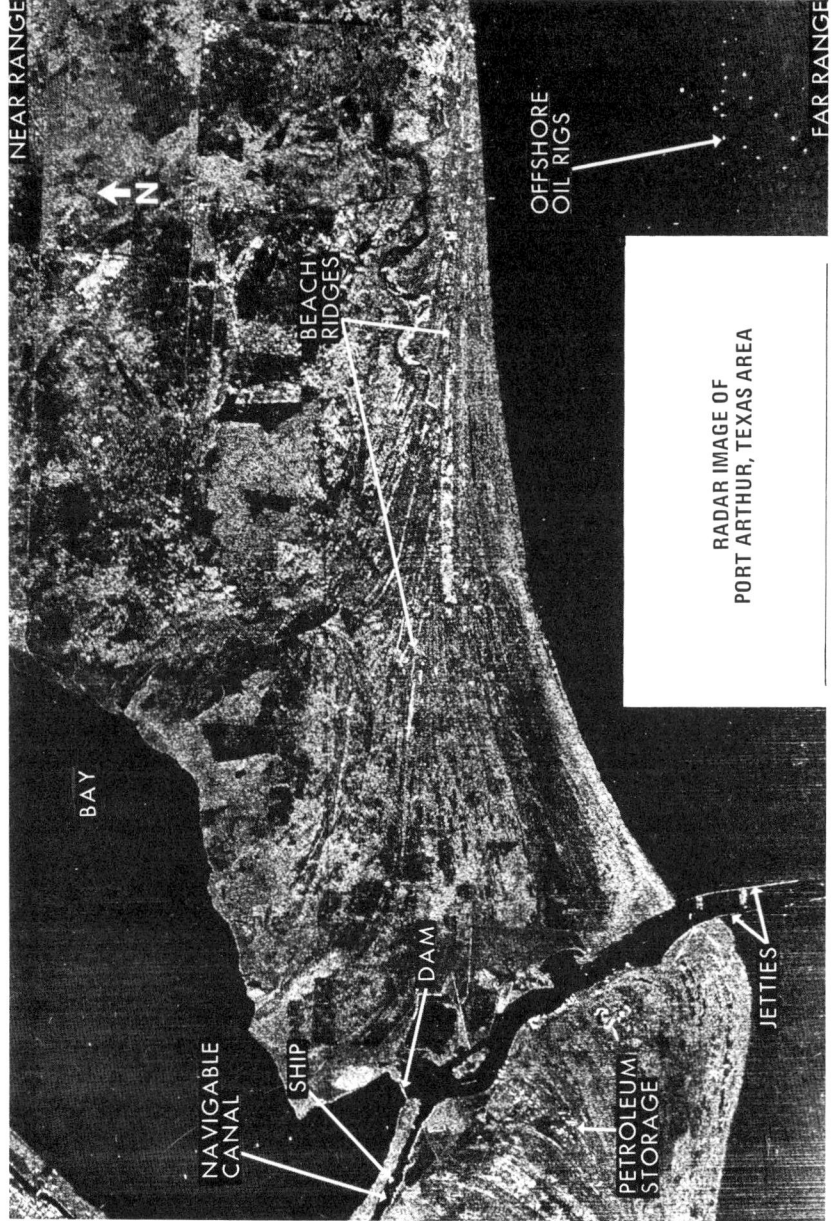

NEAR RANGE

N

OFFSHORE OIL RIGS

FAR RANGE

BEACH RIDGES

BAY

RADAR IMAGE OF
PORT ARTHUR, TEXAS AREA

DAM

NAVIGABLE CANAL

SHIP

PETROLEUM STORAGE

JETTIES

FIGURE 107. Port Arthur, Texas.

255

So that *ship(s)* may use the *navigable canal, jetties* have been constructed to force a stronger current flow, thus maintaining an adequate channel depth. Controlled flow of water from the bay, probably through the canal, is evident from the *dam*. Presence of the *offshore oil rigs* and a *petroleum storage* facility indicates one of the economic reasons for maintaining the channel.

Moriarity, New Mexico

The most striking geological feature in Figure 108 is the large dike at the right-hand arrow of *dike–igneous rock*. The strong returns from the side facing the radar (near range side) are typical of those from the numerous and variously oriented rock faces produced by weathering of resistant, jointed rock in this relatively dry climate. The no-return of a radar shadow occurs on the far range side of the dike. The dikes radiate away from a central point (*plug*) which may be the remnant of a volcanic neck or the denuded portion of a larger igneous mass. The *wind gap* and *water gap* are evidence of uncovering of this dike by erosion.

A mountain showing erosional characteristics of *granitoid rocks* is shown in the upper left. The base of these mountains is becoming engulfed in alluvial deposits which form a piedmont slope. The absence of well-defined alluvial fans attests to an advanced stage of erosion and deposition. The opposing stream patterns in the alluvium clearly delineate a *drainage divide*.

Another divide occurs at the *erosional scarp*. The north-flowing drainage is working headward into poorly consolidated valley fill materials, probably silty clay, as indicated by the subpinnate drainage pattern. A somewhat less silty and more resistant rock apparently occurs north of the divide, where the drainage density is low but the channels are sharply imaged. This is probably a shale overlying the even more resistant sedimentary rocks to the right. A prominent "V," which indicates westerly dipping beds, can be seen near the upper part of these outcrops.

Cairo, Illinois

Figure 109 shows the confluence of the Ohio River (right) with the Mississippi River.

The *anomalous drainage pattern* is composed of the large meandering stream and the tributary dendritic drainage. Drainageways compatible in size with the resolution of the radar are imaged in fine detail, approaching that of topographic maps at a scale about 1 inch to the mile. This type of imagery can be used in identification of soils, rock, or geologic structure from drainage pattern analysis. Comparison of radar imagery with maps can reveal course changes that have occurred, suggesting the use of radar to provide updating information.

The *river bottom vegetation* is imaged as very strong returns typical of verdant growth and contrasts sharply with the adjacent cultivated fields. Vegetated areas such as these can be very precisely mapped from radar imagery information. By comparative methods, variations in return intensity have been used to differentiate between different kinds of field crops. Strong vegetation returns may also be observed along the *levee* and the *drainage ditches*. Sometimes a ditch will be wide enough to show a no-return from the water, as at the upper arrow of this annotation. These ditches may be part of a *stream course* which has been canalized.

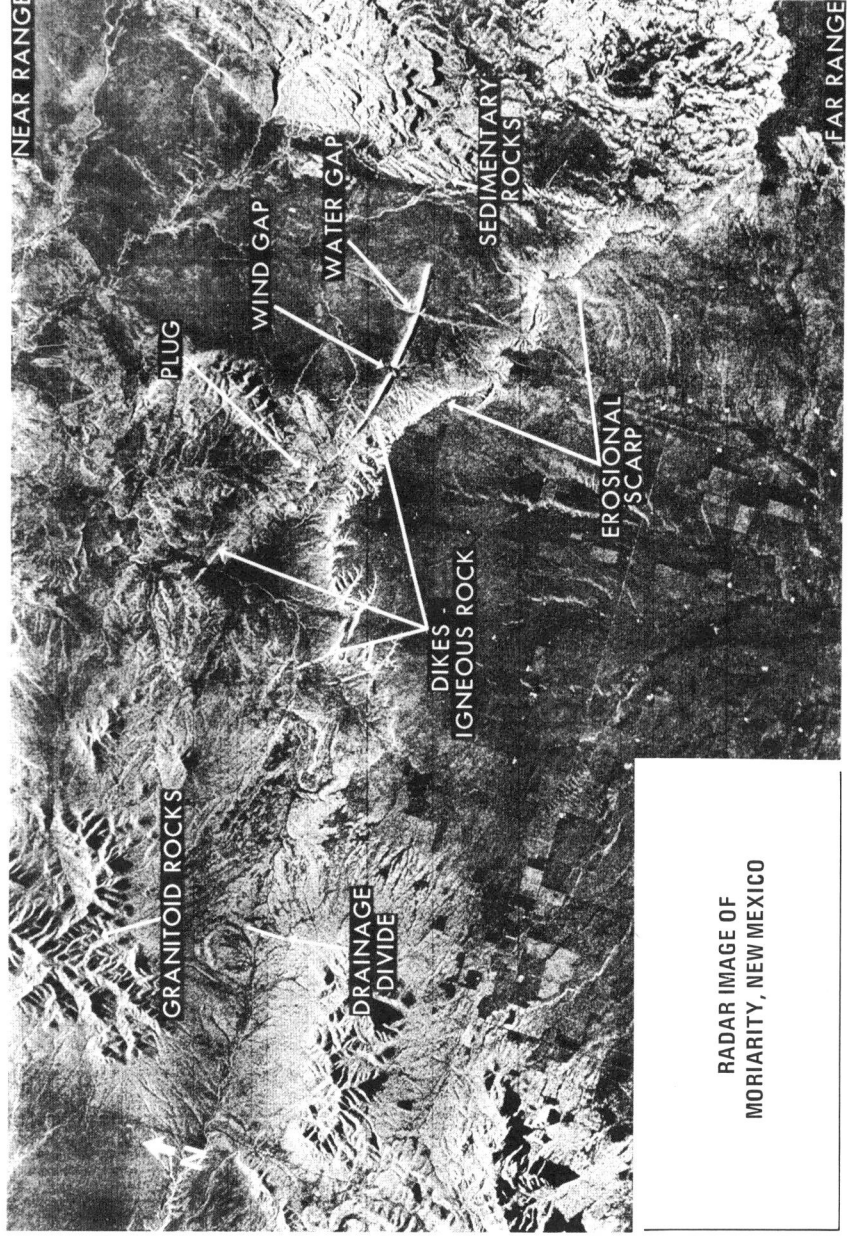

FIGURE 108. Moriarity, New Mexico.

RADAR IMAGE OF
MORIARITY, NEW MEXICO

257

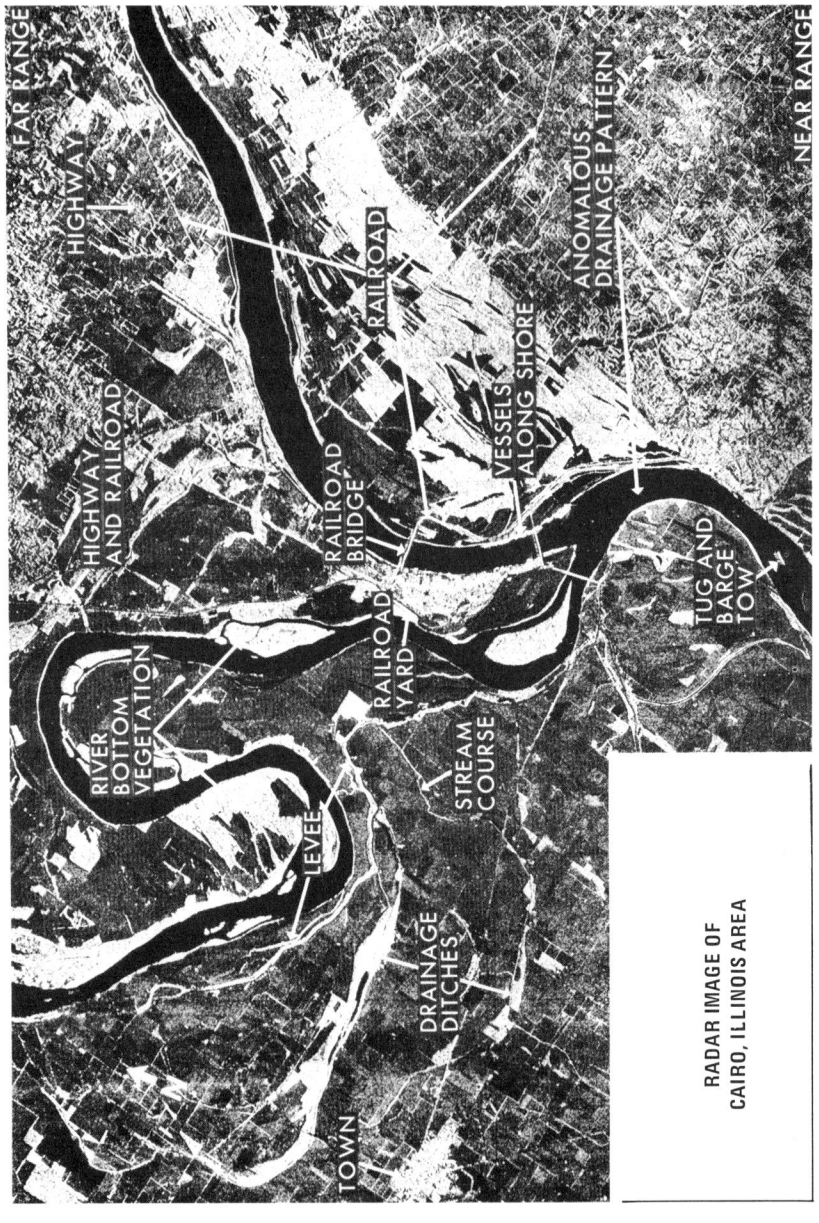

FAR RANGE

HIGHWAY

HIGHWAY AND RAILROAD

RIVER BOTTOM VEGETATION

LEVEE

TOWN

DRAINAGE DITCHES

RAILROAD BRIDGE

RAILROAD YARD

STREAM COURSE

RAILROAD

VESSELS ALONG SHORE

ANOMALOUS DRAINAGE PATTERN

NEAR RANGE

TUG AND BARGE TOW

RADAR IMAGE OF
CAIRO, ILLINOIS AREA

FIGURE 109. Cairo, Illinois.

The *town* has a rectangular pattern and an image texture which distinguish it from the heavy vegetation. Several other cultural features are indicated by appropriate annotation.

Quincy, Illinois

Figure 110 illustrates the application of radar to flooding assessment. Because radar is a continuous capability sensor, imagery such as this can be obtained at night or in rainy, overcast weather, providing immediate information useful for flood control or evacuation methods.

The no-return of the *flooded fields* is interrupted by the rectangular pattern of returns from *hedgerows* projecting above the water. Preflood information on the heights of representative vegetation could be useful in estimating the depth of water in the area. This depth would range from zero at the right-hand side (near *pair of dikes*) to a maximum left of *school*.

Higher ground at *edge of floodplain* is marked by a prominent rise and a change in topography from very flat and unbroken to gently rolling and somewhat dissected, evident in the smaller, more irregular fields, woodlots, and stream course vegetation.

The point at which the break in the principal *dike or levee* occurred is not apparent. Comparison with preflood imagery, however, would be one method which could be helpful in locating it. The dikes in this area are imaged as a return from associated vegetation. If they were smooth, mounded earth lacking in vegetation, they would not usually be so detectable because the radar energy would seldom strike them at an incident angle (0°) favorable for energy return.

Navigability of the Mississippi River is maintained in part by structures such as the *dam and lock*.

Radar Imagery Analysis Applied to Geologic Problems

This subsection provides examples of more specific and detailed applications of radar imagery analysis to problems in the geology of water, minerals, and oil. Because radar is a tool for surface geologic investigation, geomorphic features contribute heavily to the analyses, aided by the tonal differences provided by radar illumination. In addition, the synoptic view of radar imagery provides a means for relating local detail to regional trends in a manner difficult to achieve by other means.

Applications of Radar Imagery Analysis to Ground Water
Supply Investigations

Many scores of governmental agencies are involved in some aspect or another of water supply — thousands, if municipalities are included. The problems of coordinating their efforts, developing an inclusive, usable system of records, and ensuring proper well-drilling methods are but a few of the weak links apparent to knowledgeable and concerned people in and out of government.

Problems related to the development and use of ground water supplies are unquestionably more insidious than for surface water simply because of much slower

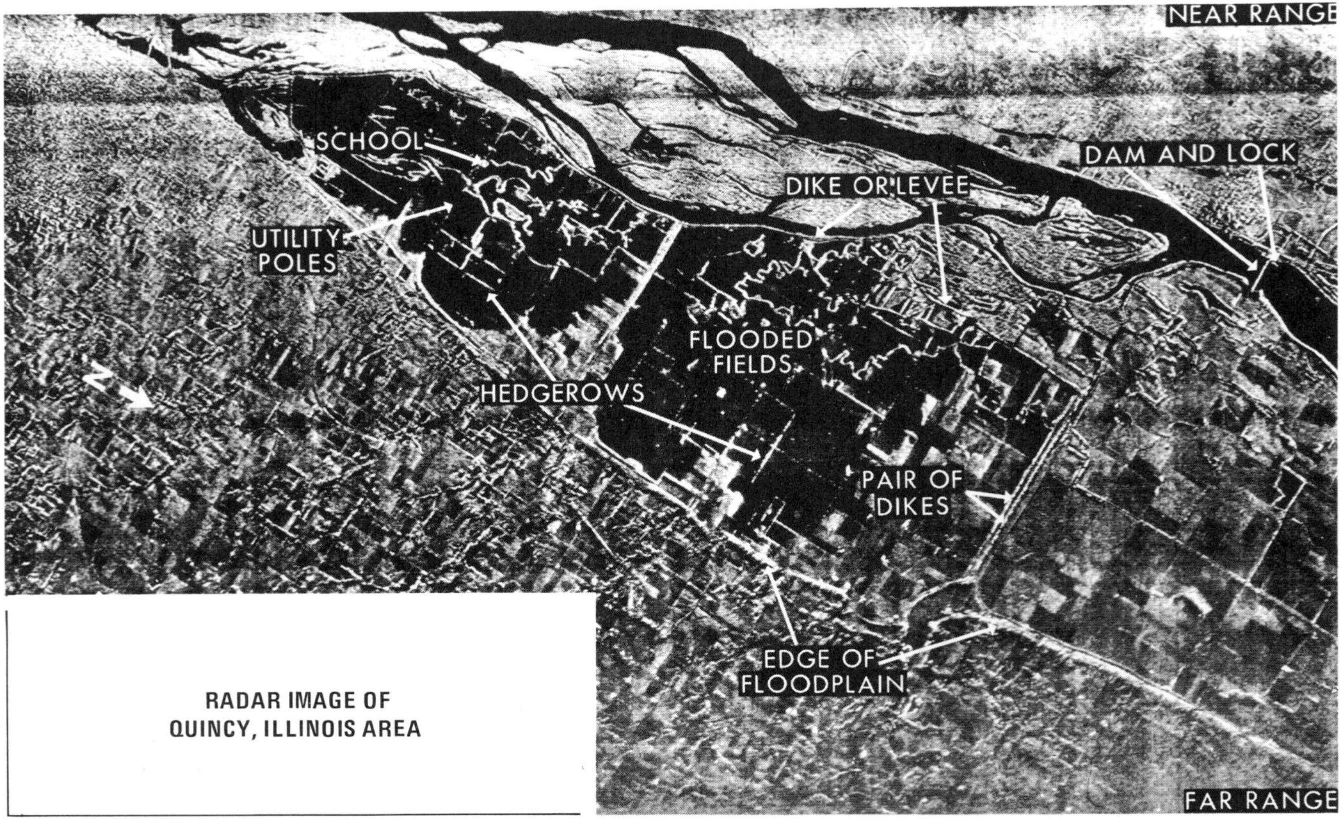

NEAR RANGE

FAR RANGE

DAM AND LOCK

DIKE OR LEVEE

FLOODED FIELDS

PAIR OF DIKES

EDGE OF FLOODPLAIN

HEDGEROWS

SCHOOL

UTILITY POLES

N

RADAR IMAGE OF
QUINCY, ILLINOIS AREA

FIGURE 110. Quincy, Illinois.

movement underground. Thus, an improvident action, an improper procedure, or faulty workmanship may set the stage for a consequence which will not become apparent for years.

Ground Water Basin Studies
Proper control and use of precious ground water supplies must start with a program of regional scope. One of the primary objectives must be a greatly expanded understanding of the geology, hydrogeochemistry, and interrelationships of ground water provinces.

The U.S. Geological Survey has undertaken a program which studies models of ground water basins. With these models, which electrically simulate the ground water basin, such things as flow patterns and water levels can be studied, and the effect of drilling new wells can be predicted. However, these models are reliable only to the degree of accuracy and detail of the information supplied for them.

One kind of information that is difficult to obtain for these basins is the configuration of the bedrock upon which the water-bearing valley fill is deposited.

Determining Basin Configuration
To determine the configuration of a ground water basin is an enormous and discouragingly expensive task. The large majority of water wells do not penetrate to bedrock and therefore supply no usable information on the basin shape. Subsurface geophysical methods could be employed, but the tremendous cost of blanketing an area of several hundred square miles or more, multiplied by the number of basins, staggers both the imagination and the budget.

The Role of Radar in Basin Studies
By providing a synoptic view of a basin and its environs (e.g., mountains) on a continuous strip or on several easily mosaicked strips, radar imagery can provide valuable additional information for basin studies. This information is basically derived from analysis of lineaments which are prominently displayed and readily related to each other on the imagery.

For example, lineaments occurring in mountain rocks are frequently traceable into the more recent sediments on the opposite side of the valley. Projection of these lineaments across the valley localizes probable anomalies that may be pertinent to the ground water situation in the basin. The localization provides reasonable limits to areas for investigation by other means with a large savings in both time and money.

Radar Imagery Lineaments in Sulphur Springs Valley, Arizona
The north half of Sulphur Springs Valley in Arizona is a large intermontane basin which is characterized by interior drainage into Willcox Playa. For this reason, it has an independent water supply but is consequently dependent upon how well this water supply is managed. Mismanagement can cause serious depletion or dissemination of poor water into good water areas. Undesirable water is common in parts of the area, particularly around the town of Willcox, where shallow wells

produce a distasteful water of high hydrogen sulfide content because of proximity to the playa. The deeper wells are often extremely high in tooth-mottling fluoride, as noted in the report on water quality studies conducted here in 1960.

In recent studies of radar imagery of the area, shown in Figure 111, it appeared that a number of lineaments could correspond to deep basin configurations (structures) which related to water quality. These lineaments, examples of which are shown at A, B, C, E, F, and G on the imagery, are most prominent in the basin-rimming outcrops but may be discernible on the piedmont slopes at as C and G. The lineament at H is not readily apparent but shows more clearly beyond the area included on the imagery. Two of the lineaments, E and F, appear on the Arizona geologic map as faults. The lineament at A is in alignment with, and a few miles from, a fault mapped at J. On the radar image, a reasonable alignment or correspondence can be ascertained between lineaments A and C, between B and G, and between E and H.

Validation of the Projected Fault

The theory of radar lineament projection to establish probable basin structure buried at depth appears to have some merit. But other data were considered necessary during the study to substantiate it as a tool with reliability for basin analysis.

Investigation disclosed an electrical resistivity probe run in the northwest corner of Section 18, T. 13S., R.24E.*

The orientation of the probe stations was N15°W, very nearly at right angles to the lineament. By the method employed, two curves are developed, one each from readings taken to the left and right of the center electrode.

The slope of these curves is related to resistivity increments of the valley fill or rock material from the surface to a depth similar to station distance from center. Bedrock, which is much more resistant than valley fill, is recognizable as a sharp upswing in the curve and a very steep curve slope. For the probe run in Section 18, the northwest half placed the basin bedrock at 1100 ft below the surface, the southeast half at 550 ft. The geophysical interpretation by Mr. Manera, made without regard for the radar imagery lineaments or the mapped faults, was that the bedrock is cut by a northeasterly trending fault with a downthrow of 550 ft on the northwest side.

Further evidence of a bedrock anomaly was obtained from an aerial gravity survey conducted by the U.S. Geological Survey.[†] The gradients indicated by this survey show a considerable increase in depth to bedrock commencing several miles north of Willcox. The line of demarcation obtainable from this survey is not so precise as that from the radar lineaments or the resistivity probe, but provides substantiation for them both.

Although the correlation of these separately obtained data is satisfying, a great deal of additional validity testing should be performed to arrive at a more substantial weight of evidence factor or percentage of reliability figure for the radar imagery interpretations.

---

*P. A. Manera, ER Geophysical Company, Phoenix, Arizona.
[†]H. Schumman, U.S. Geological Survey, Phoenix, Arizona.

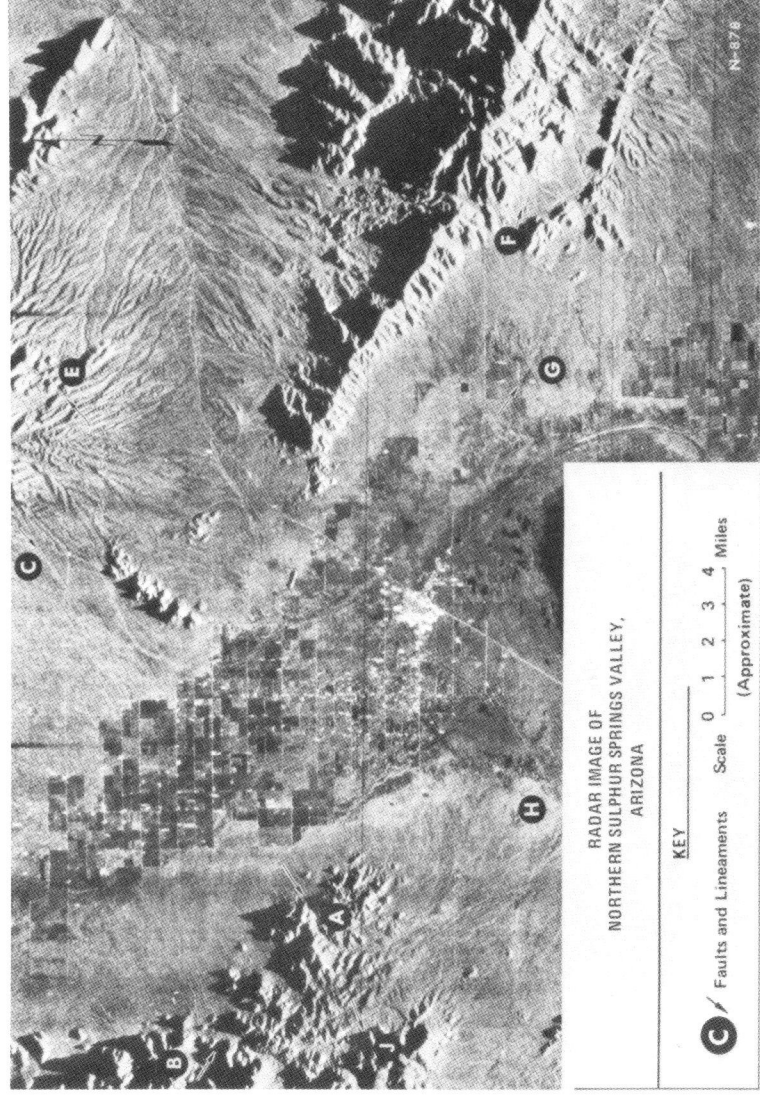

RADAR IMAGE OF
NORTHERN SULPHUR SPRINGS VALLEY,
ARIZONA

KEY

Faults and Lineaments   Scale 0 1 2 3 4 Miles
                                (Approximate)

FIGURE 111. Cochise County, Arizona (Northern Sulphur Springs Valley).

Application to Ground Water Problems

The combination of information on water quality analyses, pumpage data, water levels, and basin configuration can be applied to the problem of basin water usage. In the strictest sense, water conservation and the prevention of poor water and good water intermingling — in effect, containment of pollution — implies close control and supervision of drilling, well construction methods, and usage.

But assuming such controls to be effective, the effort in this part of the Sulphur Springs Valley would be directed toward ensuring continued separation of the waters of different quality and applying each quality to its best use. For example, the hydrogen sulfide water and high-fluoride-content water is satisfactory for most irrigation purposes. When necessary, this water could be piped to cropland. Conversely, municipal supplies would be piped from wells meeting Public Health Service standards. In addition, pipelines would deliver acceptable domestic water to farms with unsatisfactory local supply.

The fault in the basin basement has a considerable influence on the ground water distribution patterns, the depths from which water can be obtained, and the water quality. These can be complex relationships. North of the fault, water is yielded from many more valley fill aquifers than south of it because the fill is more than 500 ft thicker. However, the deeper wells here are also high in fluoride, whereas shallower wells are low in fluoride. Thus, an obvious but perhaps not exclusive control concept would be to pump irrigation water from depth north of the fault and conserve the shallower wells for domestic supplies. To maintain a water balance, shallow hydrogen sulfide wells south of the fault would be used for irrigation only, as has been suggested previously.

A complete basin study to conserve supplies, maintain the status quo on water quality zones to prevent intrabasin pollution, and provide a master plan for best use would take into account many additional factors beyond those briefly presented above. Among such factors would be other hydrogeochemical considerations such as sodium content of irrigation waters, the effect of other possible faults (BG and EH, for example), and the problem of the many improperly built wells in the area.

Because the aquifers are generally very fine sand, each well should be gravel-packed and perforated to exacting specifications to prevent both sanding and eventual collapse of strata. Otherwise, continued withdrawal could lead ultimately to settling and destruction of the arable surface. This means that proper shutdown and cementing of some existing wells must be undertaken, followed by construction of new wells of correct design in the same or other locations.

Conclusion

This brief analysis demonstrates that the control which will be demanded if irreplaceable water supplies are to be conserved and protected will require a massive effort supported by data from every available source. Radar can make an important contribution to these data by providing synoptic imagery from which basin structures can be analyzed and field work organized and directed to accomplish the task in a cohesive and economical manner.

Metallic Minerals Prospecting from Radar Imagery

Early prospectors of gold, silver, and the base metals relied for the most part on surface indications of mineralization in initial evaluation and location of claims. Realization of the size of the bonanza or of the disappointment of a pinchout occurred after actual mining had commenced. The common surface indicators signaling mineralization to the prospector were the same as those often visible on radar imagery today: fracture zones, veins, and associations of rock types.

At the grosser resolutions of common imaging radars, the veins, normally only a few feet wide, may not be separately resolvable in most cases. However, when the fractures in which the veins usually occur are large enough to be resolved and have an influence on the topographic expression, they are prominently displayed on the radar imagery. In addition, the imagery can be analyzed — again within system capability — for probable rock types. Thus, a radar system can provide a means of initially prospecting for areas, often inaccessible without great expense, which have indicators of potential mineralization. The economic advantages of thus delimiting the areas for primary ground investigation are obvious.

Johnson Mining Camp Area, Arizona

To test the possibilities of radar for metallic minerals prospecting in this manner, the Johnson Mining Camp area in the Little Dragoon Mountains of Arizona, shown at A within the striped box on the imagery of Figure 112, was compared to adjacent areas. Although not large by some standards, the operations at Johnson have produced many millions of dollars worth of copper, zinc, lead, tungsten, silver, and gold.

Analysis of lineaments — probable fractures — in Figure 112 show that the Johnson area has a large number of closely spaced, intricate, patterned linear trends and an association of rock outcrops unique to the imaged area. This particular combination of lineaments and rock does not appear in the Gunnison Hills (annotation 1), Red Bird Hills (2), Steele Hills (3), Winchester Mountains (4), Galiuro Mountains (5), or Johnny Lyon Hills (6).

A separate view of the Johnson Camp area is given in Figure 113A. In Figure 113B, the camp is again shown at A and the lineaments are delineated in black, using a solid line for fractures which appear on the imagery and on geologic maps of the area and a dotted line for related lineaments which appear on the radar imagery. Outcroppings of granitic rocks, intensely jointed to the northeast and the only such occurrence in the entire area, are disposed around and between the annotations B and C. Differentiates from rocks of this type are commonly the carriers of ore minerals, and their association with adjacent highly fractured rock strongly suggests possible mineralization.

Conclusion

The Johnson Mining Camp area is the only location for miles around which has produced economic quantities of metallic ores. Its potential as such a producer is demonstrable from radar imagery, which shows it to have characteristics unique from those of adjacent areas.

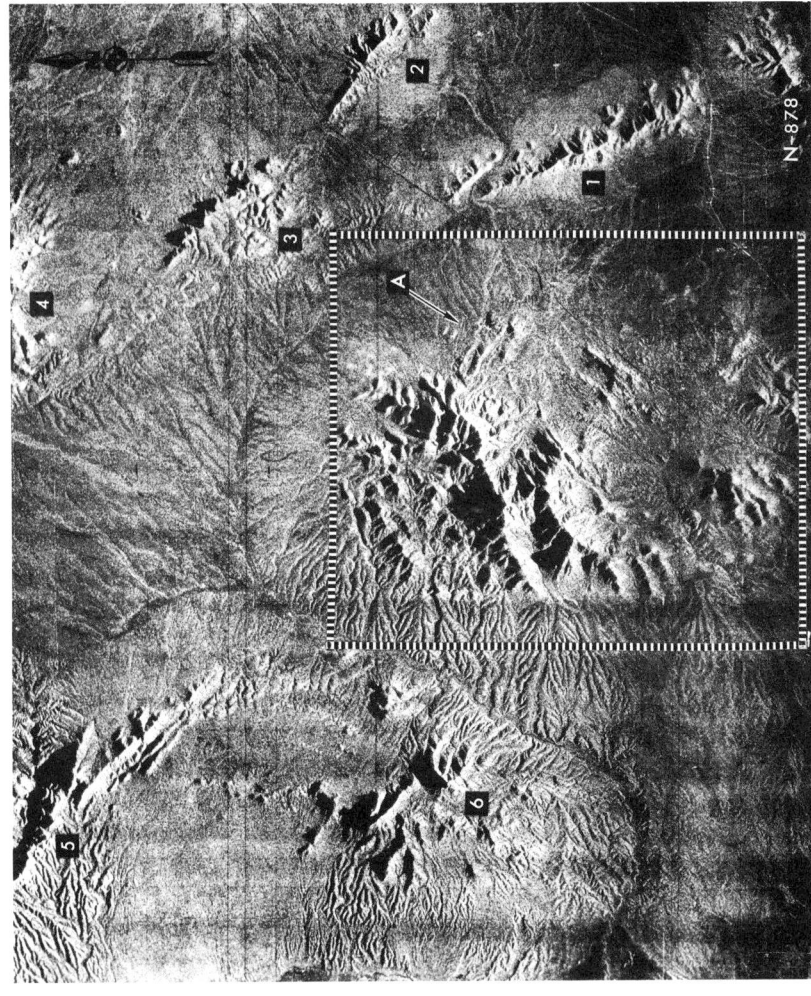

FIGURE 112. Cochise County, Arizona (Johnson Mining Camp).

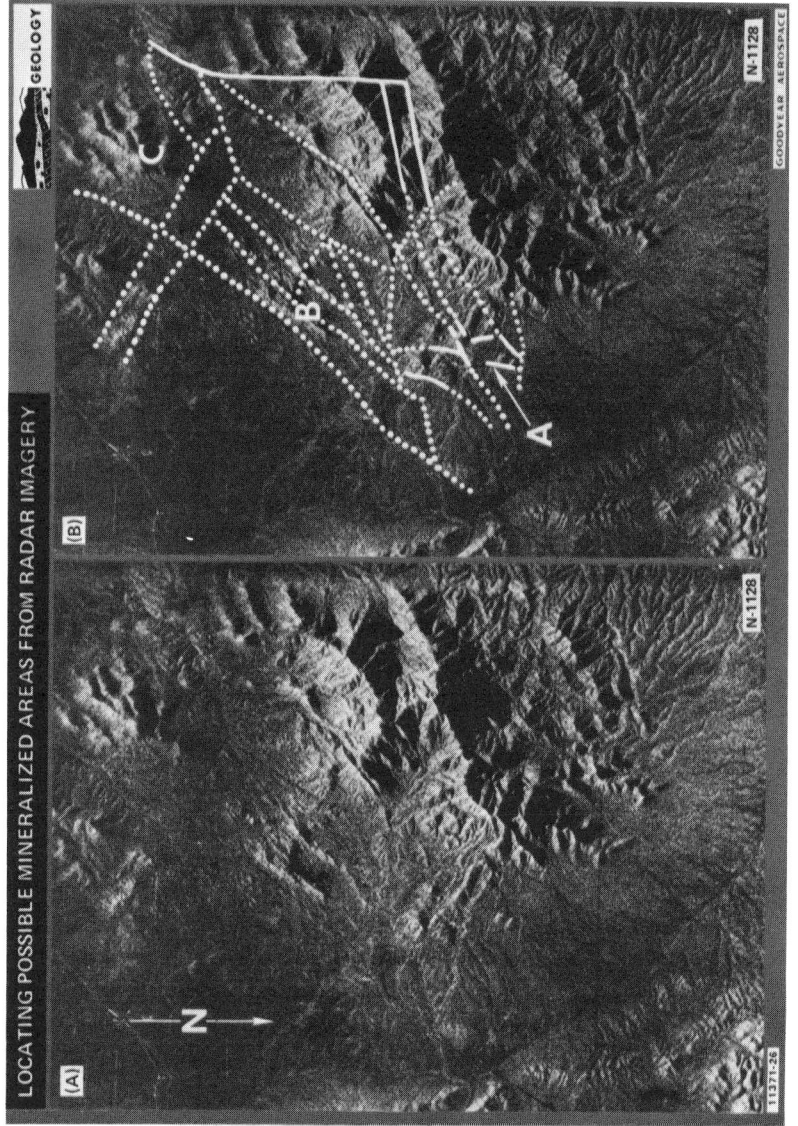

FIGURE 113. Unannotated vs. annotated Johnson Mining Camp.

267

Interpretation of Geologic Structure, Meteor Crater Area, Arizona

Radar imagery of the Meteor Crater area in northern Arizona contains numerous lineaments which appear significant of geologic structure. Most of these lineaments are not included on the generalized Arizona geologic map, but a number of them appear in a part of the area mapped in detail by Shoemaker.* These structural signatures were analyzed in light of known structures to produce a revised map containing more detail than that of the existing geologic maps.

Lineaments on the Radar Imagery

Figure 114 is the radar imagery which is the basis for this analysis. The two prominent lineament trends, northeasterly and northwesterly, are exemplified by Canyon Diablo, annotated A, which leaves the area in a direction opposed at 90° to that of entering. The general dip slope of the rocks in the imaged area is northeasterly and follows regional tendencies toward monoclinal flexure. That this dip slope is far from flat and locally subject to a variety of interruptions is apparent at numerous places, such as B, C, D, E, F, G, and H. (Between B and D of the figure, the left-to-right discontinuous, sinuous white patches resulted from improper film handling and should be disregarded.)

Joints

These large trends are repeated continuously in the detail of the rectangular drainage pattern. Figure 115, an enlargement from Figure 114, is a section of Canyon Diablo, D, which illustrates a complexity of 90° turns. A previous field investigation of the joint system showed that it follows the regional trends and exerts control over the drainageways. This extremely dense joint system occurs in the brittle Kaibab limestone, illustrated by the photographic inset of Figure 115. The strong radar energy returns from the canyon at every orientation are caused by the surface roughnesses, the variously oriented limestone faces, and particularly by the trihedral corner reflectors (triangular-shaped outline in the photograph) created by the orthogonal joint system and bedding planes. These reflectors return entering energy regardless of the angle of entry. The continuity of returns, as a corollary, is evidence of the blocky form of the exposed rock. This same joint system control and strong imaging is apparent in other canyons, C, F, G, and H in Figure 114, cut into the limestone.

Mapped and Imaged Faults

The major faulting in the area follows the northwesterly trend, but the northeasterly trend is notably lacking. Also, the fault lines and zones are curved or sinuous and may deviate from the jointing, indicating variable localized stresses that probably originated after setting of the joint framework. The most prominent faulting shown in the imagery occurs along the fault lines and zones are curved or sinuous and may deviate from the jointing, indicating variable localized stresses that proba-

---

*E. M. Shoemaker, *Impact Mechanics at Meteor Crater*, U.S. Geological Survey Open-File Report, 1966.

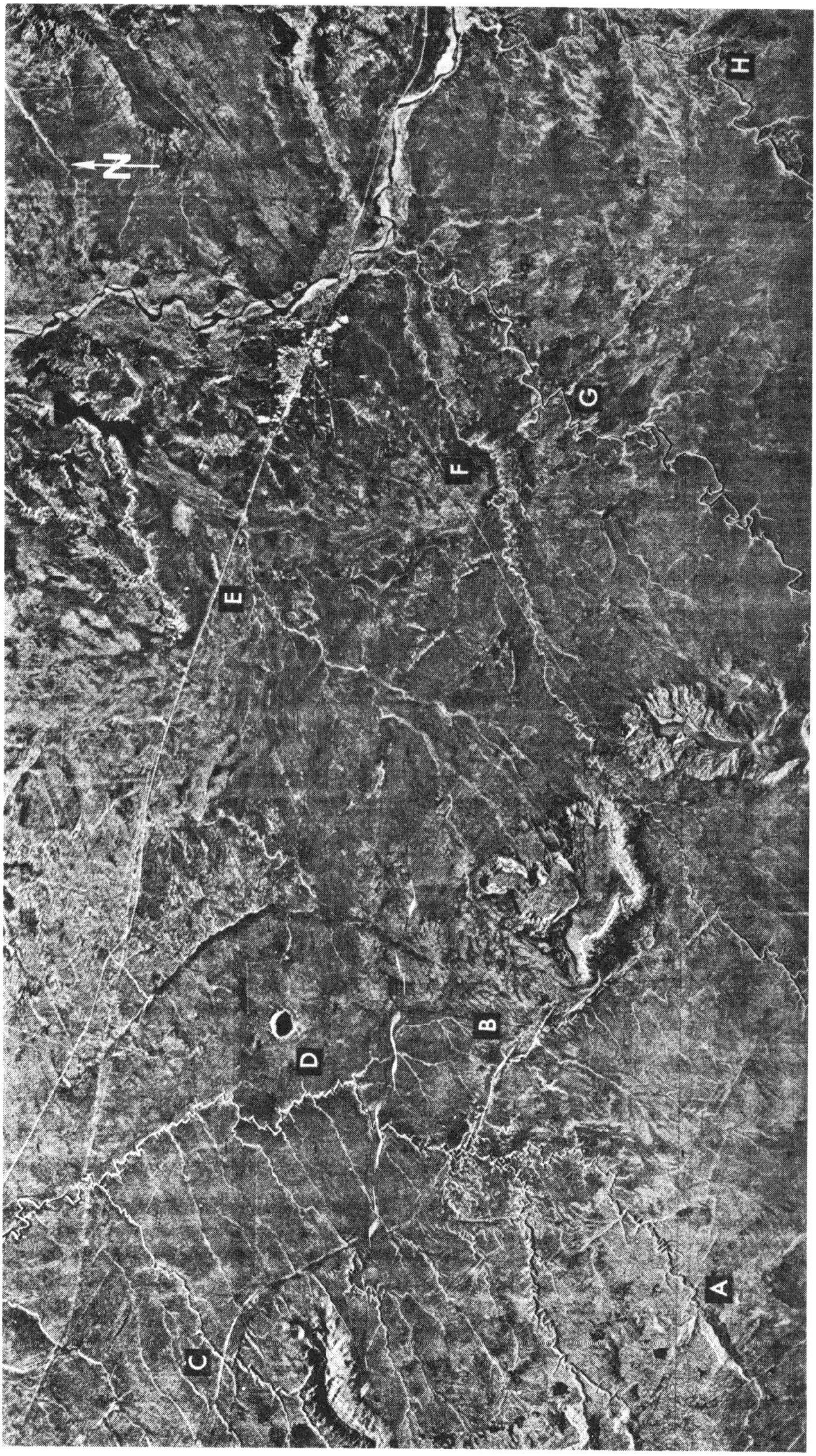

FIGURE 114. Winslow, Arizona.

269

CANYON DIABLO
LIMESTONE CORNER REFLECTORS

GROUND PHOTO

TRIHEDRAL CORNER REFLECTOR

GOODYEAR AEROSPACE

IMAGE

01121-25

FIGURE 115. Canyon Diablo.

270

bly originated after setting of the joint framework. The most prominent faulting shown in the imagery occurs along the line defined by BC in the imagery of Figure 116. This imaged fault zone duplicates all of the nuances of Shoemaker's detailed map, which is reproduced in part in Figure 116 and modified by annotations to match those on the imagery. On the map, heavy dots indicate the downthrown sides of what are classified by Shoemaker as normal faults.

Inspection of the radar signatures of these known faults shows that they form two distinctly different images, depending upon the relation of the radar beam to the scarp. Because the radar beam is directed northerly, the south-facing scarps are imaged as a return (reflections from the rock faces). Examples are the white diagonal lineations at B, which form an en echelon pattern.

Conversely, those faults with a north-facing scarp are imaged as a no-return (the radar shadow cast by the scarp). An example is the diagonal black streak southwest of J.

Comparing these fault scarp signatures with the canyon signatures reveals two important differences. First, the canyon forms an image which is generally compound, a shadow from the near range wall (nearest the radar) and a return from the far range wall, whereas the fault images are simple, either a return or a no-return but not both. Second, images of the fault scarps are softer, having more edge shading and generally weaker returns. Likely reasons for this are that the faults frequently cut the joint system diagonally and probably afford fewer trihedral reflectors for strong energy return, the physical action of faulting produced rather smooth scarp faces, and erosive forces are weak because the top of the scarp is, in effect, a small drainage divide from which runoff in either direction is minimal. The canyons, on the other hand, receive periodic sheet runoff over their rims and flow through their channels, producing more active erosion and exposing more blocky forms.

Other faults mapped by Shoemaker, such as those south and north of K on Figure 116, can be seen on the imagery. At the south fault of this pair, the broad band of lesser return is caused by a relatively smooth surface of sand and silt eroded from the Moenkopi slope south of the scarp. As runoff is impeded by the scarp, its velocity is reduced, and it drops some of its load before entering the northeasterly channels through the scarp.

Northwest of the fault J, the imagery clearly shows a fault (mapped) which changes from a no-return to a return signature (unannotated arrows) as the orientation angle between the scarp and the radar beam changes. This is shown enlarged at M of Figure 117.

About midway between B and C of Figure 116, a variation of the fault signatures can be seen. Here, the very weak returns are interpreted as indicating a small displacement or a fracture trace in which a shallow erosional trough has developed along the zone of weakness.

Oil-Producing Structures Located from Radar Imagery

Remote sensing for oil exploration is classified as a surface geological method; that is, remote sensors describe the surface and each tells the story in its own par-

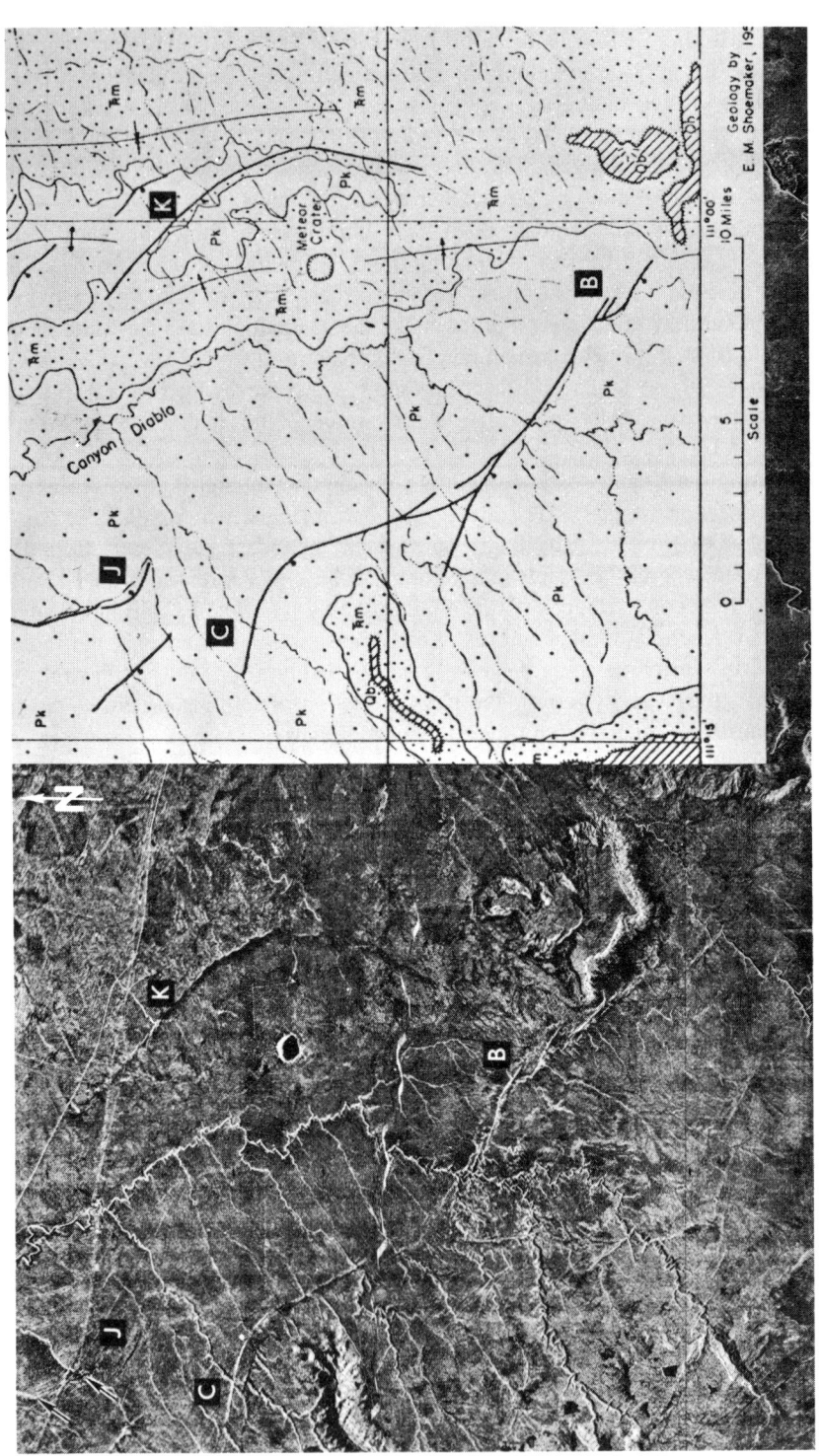

FIGURE 116. Meteor Crater.

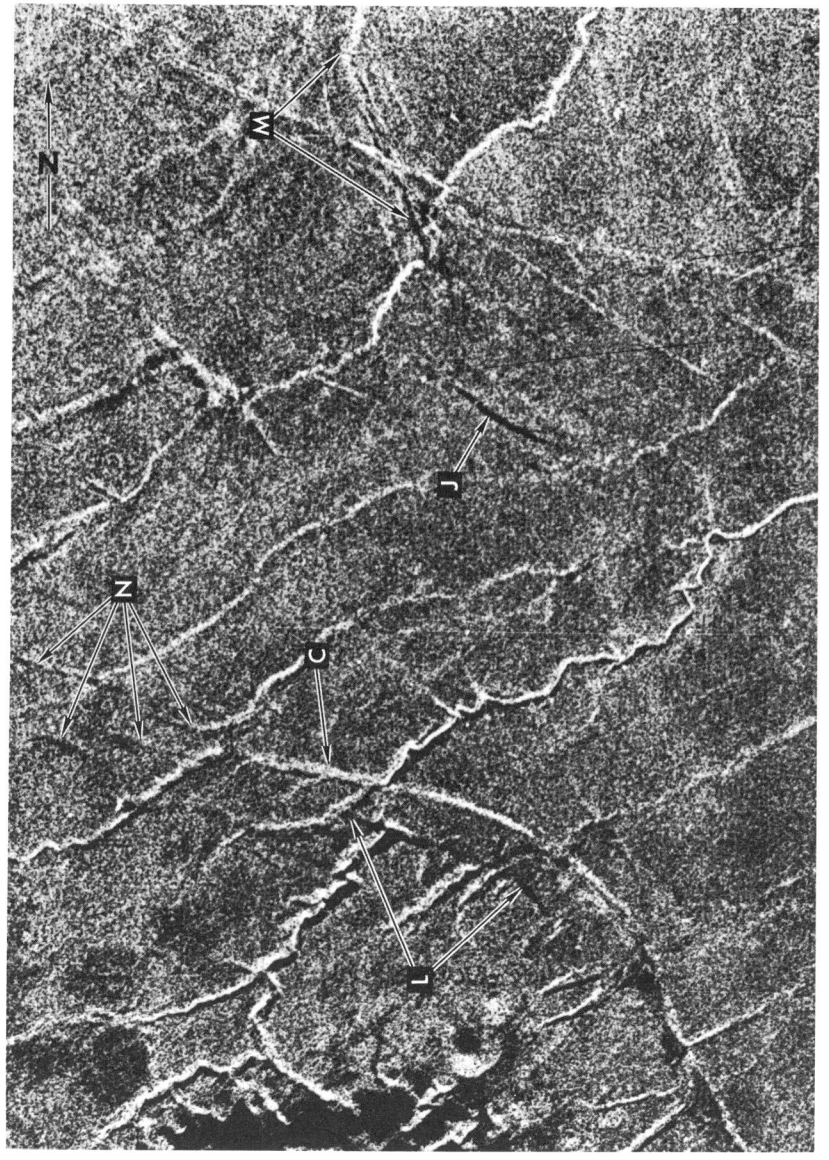

FIGURE 117. Fault detail.

273

ticular manner. The surface geologist relies strongly on sets of indications or trends which are apparent at the surface and may be equatable with the structure or lithology that is pertinent to his particular investigation. A few of the things he notes and places on his base map are the type, thickness, and attitude of formations and fault traces and their disposition. He also makes use of associative clues indicating structure and lithology — stream patterns and vegetation, for example.

The employment of radar imagery to supply geologic data is similar to aerial photographic techniques. Each one provides surficial information from which certain aspects of geology may be determined either directly or by proxy. However, the two sensors do not necessarily supply identical information, nor do they present the same information in the same way. Part of the reason for this is that they respond to different wavelengths of electromagnetic energy.

Aerial photography has been useful to the geologist for a number of years, and new ways of interpreting it or giving added depth to previous interpretations are still being found. But radar is in its infancy, particularly as applied to the environmental sciences, and its potential for geologic exploration is relatively untested. It is the purpose of this study to investigate the application of radar to the exploration for petroleum as a step toward describing ways in which it can provide assistance to the geologist.

Selection of the Imagery

Imagery near Wichita, Kansas, was specifically selected because it was known to be in a petroleum province. This type of area was considered necessary to the study because the radar imagery indications of structure and its relation to producing fields could be verified in the literature. Specific locations of oil fields were unknown to the investigator when the exercise was undertaken.

Plotting the Large Apparent Structures

Inspection of the radar imagery (Fig. 118) showed well-defined and easily detectable drainage alignments with very definite indications in their curving courses of control by fold structure. Because there is no evidence on the imagery of ridge patterns attributable to fold limbs, it was concluded that folding was gentle. The flatness of the area, common in Kansas, is apparent from the absence of topographic shadowing. Although gentle surface undulations are undetectable on this image, such subdued features can be emphasized by a radar flightpath which provides a very small depression angle, forming tonal variations or shadows from very slightly elevated features. Generally, in an area as flat as this one, one that also showing regular erosion and no ridges or hills, the rocks are flat-lying sedimentaries. When this is the case, over a distance such as the 45 mi east to west of this image, these rocks will have a gentle regional dip. This dip appears to be southwesterly, judging by the tributary streams in the right half of the image. Without greater imagery coverage, however, this assumption can only be tentative because the large streams (left) flow southerly and easterly, causing the westward flow of the tributaries.

Certain structures with oil-producing potential may be delineated from surface clues revealed on radar imagery. However, as every geologist is aware, a surface

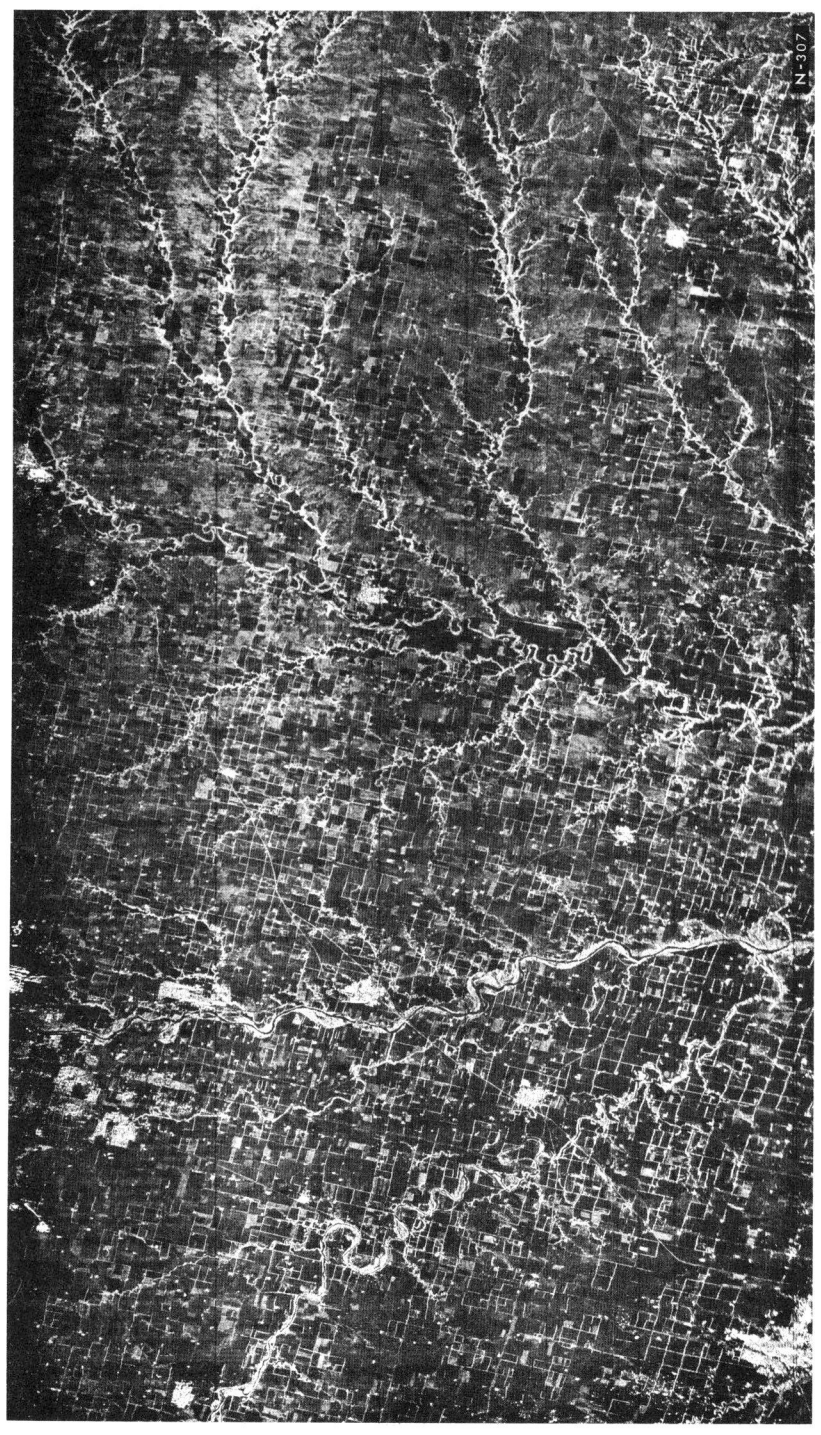

FIGURE 118. Wichita, Kansas.

structure may be entirely different than the subsurface structure in which petroleum is eventually discovered. In other cases, the trap may be stratigraphic or a complex combination of structure (both faulting and folding) and stratigraphy.

Many of these may form unique surfaces which are expressed on the radar imagery and can be interpreted in terms of oil potential.

In connection with the employment of radar in petroleum exploration, it must be understood at the outset that no known surface or subsurface method "discovers" oil directly. The ultimate verification always comes with the drill. And despite all the most favorable evidence from numerous sources, such as photogeology, field work, seismograph, magnetometer, or gravity meter, the result may be a dry hole. Radar will not change this picture by providing some new and miraculous way of directly detecting oil-bearing strata. But it does lend sharp, easily detected definition to certain types of probable structure, as in the imagery of this report, and can be employed in areas where other imaging sensors are seriously constrained by weather, clouds, or long periods of darkness. Therefore, radar is considered to have a definite place for exploration operations in remote tropic and arctic regions as well as for reevaluating existing petroleum provinces because of the synoptic view and differences in emphasis which it provides.

Radar Imagery of Luke Salt Dome, Maricopa County, Arizona

For many years, an anomalous surface has been recognized on remote sensor imagery in the Salt River Valley west of Phoenix, Arizona. The anomaly is characterized by a radial drainage pattern and elevation of the terrain unusual to intermontane valleys of the Basin and Range province. A radar image (Fig. 119) illustrates this appearance, although the elevation difference is small and difficult to discern at the depression angle employed. In the figure, arrows emphasize the directions of the radial drainage and the letter A indicates elevated terrain. For security reasons, an airfield appearing on the imagery has been blocked out.

The appearance of the anomaly suggests a doming such as those associated with salt structures in certain oilfields. The possibility of a salt structure has not been held very likely, and attempts to correlate the occurrence with river terraces or related features have been unsuccessful.

Recent work by the U.S. Geological Survey, however, shows that the feature is a salt dome having a pronounced gravity expression and creating anomalies in ground water flow patterns, hydrogeochemistry, and lithofacies.* In addition, a deep hole drilled in the anomalous area was bottomed, still in salt, after passing through 400 ft of valley fill, 90 ft of anhydrite cap, and over 3000 ft of salt. Means for recovery of the salt are available, and it is considered by some to be an economic deposit. Its exploitation appears to hinge on whether extraction will have widespread effects — perhaps pollution or water level declines — on the vital ground water supply of the valley. Among other effects, one that cannot be dismissed is massive sinking of the land surface with attendant destruction of existing land uses, including the multimillion-dollar airfield nearby.

Despite the outcome of proposed exploitation, verification of the nature of the anomaly points up the fact that few if any such unusual occurrences (which can be

*H. Schumann, U.S. Geological Survey, Phoenix, Arizona.

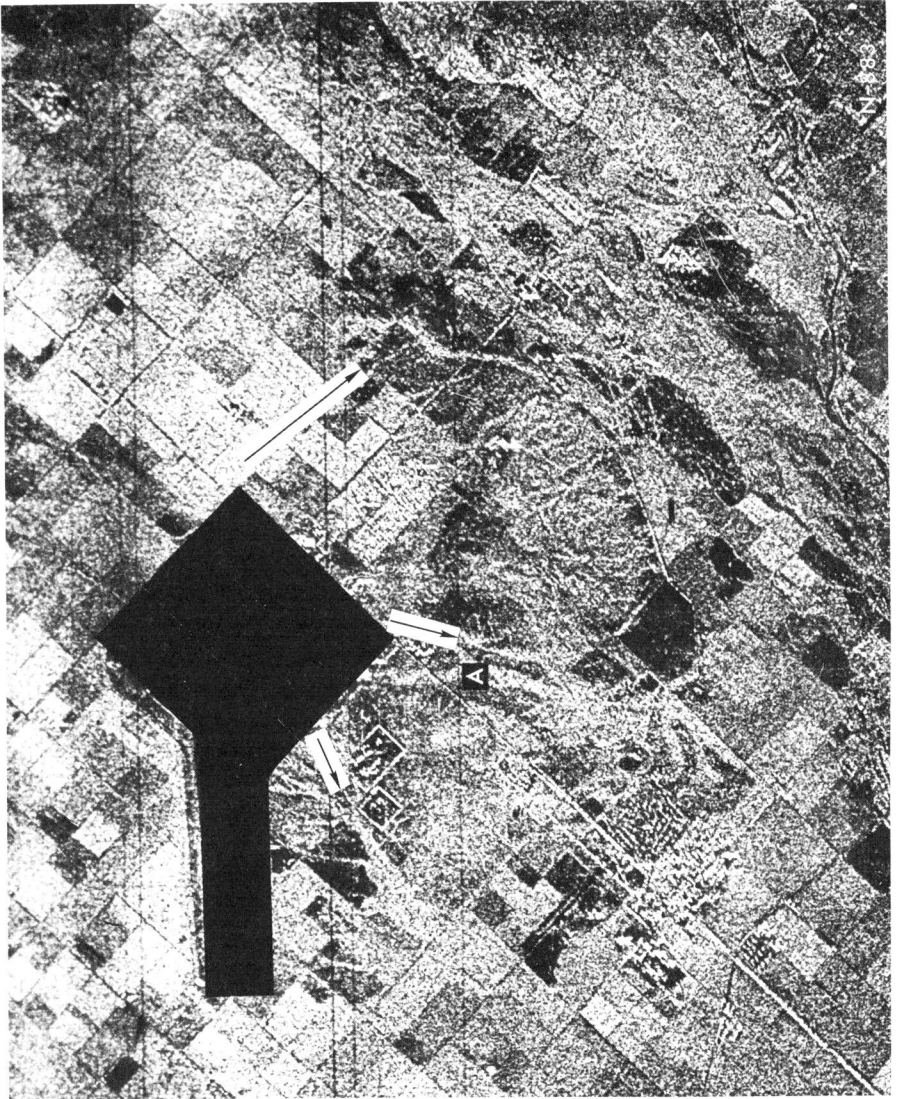

FIGURE 119. Luke Salt Dome, Arizona.

277

readily viewed in regional context on radar imagery) are the result of meaningless happenstance. It could well be that analysis of radar imagery over the entire United States and other areas would be a rapid means of locating important yet undefined anomalies with economic potential.

Radar Imagery of Glacial Landforms

The ice-scoured glacial plain southwest of Cranberry Lake, New York, annotation 1, is shown in Figure 120. The rolling but finely dissected surface in the right and lower portions of the imagery is common to areas where resistant bedrock is overridden by an ice sheet. The ice abrades, quarries, and plucks rock from projecting ridges, lowering and smoothing them, and deposits till thinly and selectively.

In the Cranberry Lake area, the southwesterly advance of the ice apparently accentuated preexisting regional lineaments. Thus, the fine lineations at annotation 2 probably reflect southwesterly trending geologic foliation or fracture, for example, and this same trend can be seen in the prominent topographic alignments, as at annotation 3. The features of 3 have drumlinoid shapes and probably represent drumlinization of rock rather than till drumlins in this area of heavy scouring. The alignment of these features corresponds to the general west-of-south movement of the ice sheet—a movement that seems to have received a more westerly swing as it was influenced by the preglacial Adirondack highland.

Annotation 4 points out the strong, sinuous returns from eskers, ridges composed of poorly sorted sand, gravel, and boulders deposited from glacial meltwater. The probability of a boulder-strewn surface and good stands of a vegetation which requires a well-drained environment can account for the strong return from these eskers. The mere fact that eskers have sloped sides is insufficient to cause these strong returns, and thus diffuse, rough-surface reflection is indicated. (Energy must strike a smooth slope at a 0° incident angle for energy return, and this is impossible in this image because of the 8° depression angle of the beam and the less-than-90° orientation angle of most of the eskers.)

Annotation 5 is interpreted as a possible esker, although its extreme regularity in width and curvature suggests a man-made structure, perhaps an embankment or aqueduct. However, this latter interpretation cannot be substantiated on the imagery from associated features such as the linear no-return of a road or canal.

Annotation 6 is a stream course which bears similarities to the images of the eskers. Note, however, that the two features can be differentiated on the basis of no-return relationships. The no-return (water or shadow or both) of the stream is at near range from the strong returns (bank/vegetation), whereas the no-return (shadow) of the eskers is at far range from the strong returns (ridges).

Analysis of Fold Structures from Radar Imagery and Field Data

On radar imagery of the area south of Winslow, Arizona, strong evidence of folding appears in certain drainage pattern alignments. The form of these alignments suggests stream course deflection such as would be caused by a doubly plunging

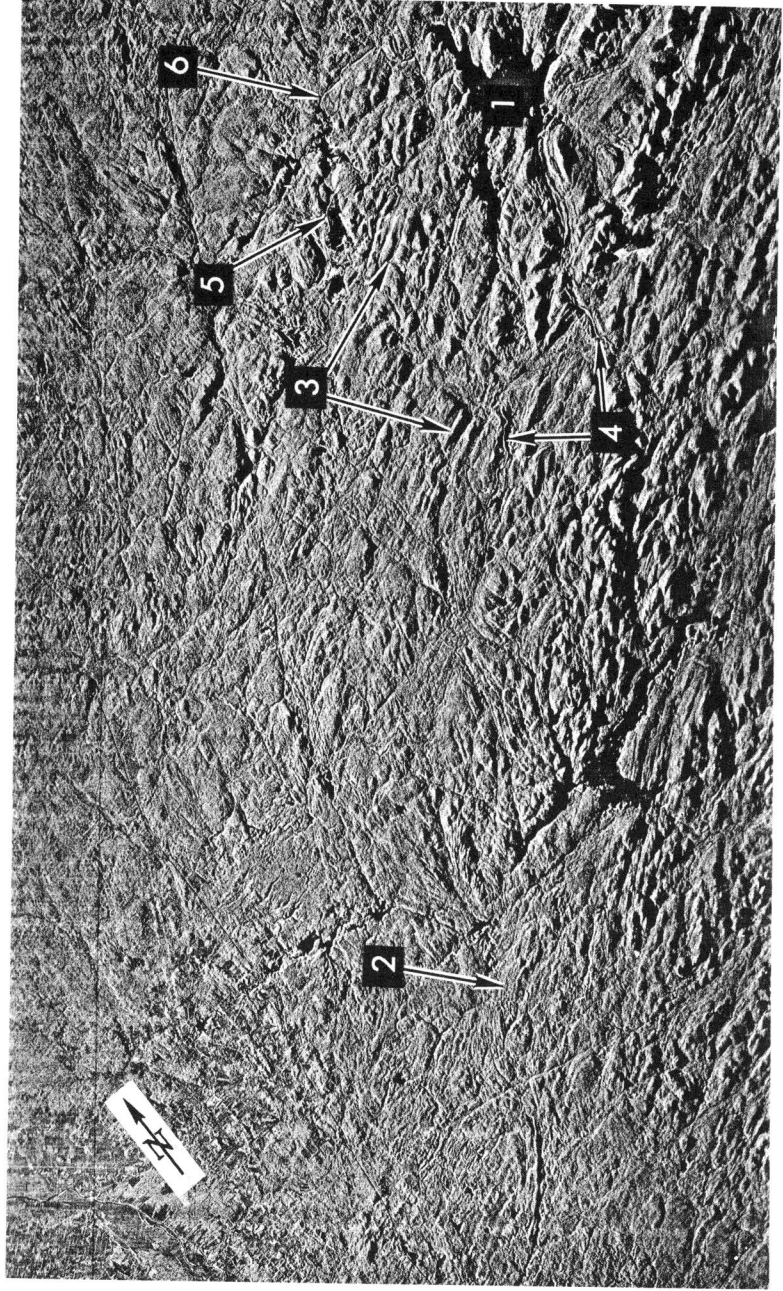

FIGURE 120. Glacial landforms, New York.

anticline. On the radar imagery of Figure 121, the apparent axis of the anticline is labeled A and denoted by a broken line. However, on the Arizona geologic map sheet, an anticline appears several miles to the southwest, located at annotation B rather than at A.

At least two explanations for this lack of correspondence are possible. They are:

The radar imagery provides evidence of an unmapped fold.

The radar imagery and field data indicate an asymmetrical fold which has been eroded.

Rock formations in the area have generally low dips of 5° or less, with a regional northeasterly trend. However, local flexures superimposed on the regional trend occasionally cause reversals of dip and increases in steepness of dip to 10° or more. In an area where low dips prevail, it is difficult to obtain measurements and axial locations. Thus, no two investigators always agree in their mapping of such nebulous structures.

On the other hand, it is necessary to locate the axes of low surface folds as closely as possible because they may be an expression of sharper folding at depth — folding which could perhaps cause closure for a petroleum trap.

During preparation of the geologic map of this area, field evidence of folding may have been inconclusive or unobtainable because of poor exposures. If this were the case, the large view of the radar imagery could have provided the additional evidence required to substantiate the uncertain field data. Alternatively, it may be presumed that the time allocated to the field work did not allow for the detailed study that would have been required to establish the location of this fold.

With this we conclude the discussion of using high-resolution radar in geological exploration. The utility of this less commonly used information source, particularly since conventional aerial or satellite imagery can also be employed, may lead to its wider acceptance in future exploration projects.

3.3 AGRICULTURE

In the middle and late 1970s the USDA, NASA, and NOAA cooperated in the Large Area Crop Inventory Experiment (LACIE). This effort was intended to determine whether the then emerging technologies of multispectral classification of Landsat MSS imagery, weather information from NOAA satellites, and various computer models could substantially improve the crop production estimates and forecasts. Much of the effort was coordinated at the Johnson Space Center in Houston, Texas. The experiment succeeded, although much possible improvement remains to be achieved.

In 1980 the effort transitioned into a permanent cooperative research program termed AgRISTARS. The focus was placed on using aerospace remote sensing technology to provide early warning of changes affecting commodity production and to provide commodity production forecasts.

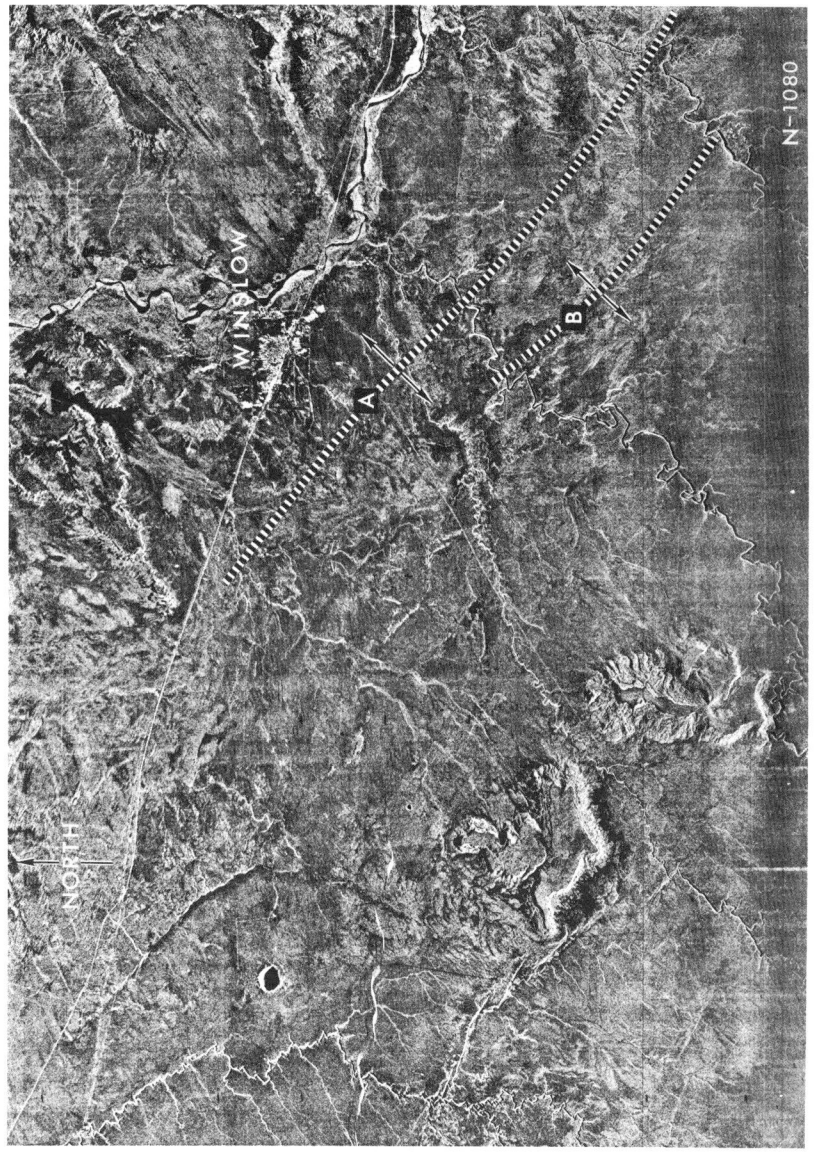

FIGURE 121. Probable anticline.

281

This section addresses the application of digital image processing to agriculture by considering in detail the topic of crop area estimation. The methods of estimating crop area are seen to be considerably more complicated than simply running a maximum likelihood classification program on a Landsat scene.

Yield estimation is not addressed here because it has proven to be an even more complex topic. For example, crop moisture is only one of many factors that affect crop yield, and the estimation of crop moisture is itself a challenge. The Crop Moisture Index (CMI) is currently used in many AgRISTARS projects to assess soil moisture conditions, a major determinant of crop yield. The CMI involves a two-layer soil–water model and potential evapotranspiration calculated by the Thornthwaite method. Some of its requirements restrict its use to regions for which long-term average data, as well as current precipitation and temperature, are available. An evapotranspiration model estimates monthly total evapotranspiration in inches from vapor pressures corresponding to the mean monthly maximum and minimum temperatures. This development of the CMI allowed the use of a simple moisture indicator, termed "sponge." Sponge has a sound physical basis that uses common remotely sensible meteorological variables and is suitable over a broad range of climates. Besides moisture, crop yield is affected by numerous other factors, such as temperature extremes and degree day accumulations, fertilization, soil salinity, plant species, planting practices, insect infestations, and wind speed. Putting all of these elements into a yield estimator employing remotely sensed data is beyond the scope of this section.

3.3.1 Crop Area Estimation

This section describes the results of research done in Kansas related to the estimation of winter wheat by the Economics, Statistics, and Cooperatives Service (ESCS) in 1976. The goal of the project was to utilize data gathered by the Landsat satellite to improve existing winter wheat estimation procedures at the state, multicounty, and individual county levels. Existing ground surveys, especially the June Enumerative Survey (JES), provide crop hectarage estimates with measureable precision at national and state levels. The ESCS approach to utilizing Landsat data is to use it as an auxiliary variable with the JES ground data being the primary variable.[1]

Kansas is the number-one ranking state nationally in winter (and all) wheat planted, harvested, and produced. In 1976, Kansas ranked fourth in area planted to principal crops with 8.8 million hectares. Over half of this hectarage, some 5.2 million hectares (see Figure 122 for distribution) was planted to winter wheat. Final harvested hectarage from the 1976 crop was 4.5 million hectares and produced 9.2 million metric tons, third largest production ever for winter wheat in Kansas.

Other major crops for Kansas include sorghum, corn, rye, barley, soybeans, oats, alfalfa, and other hay. For this study, crops or land uses of interest were only

*Based on M. Craig, R. Sigman, and M. Cardenas, *Area estimates by LANDSAT: Kansas 1976 winter wheat*, U.S. Department of Agriculture, Washington, D.C., August 1978.

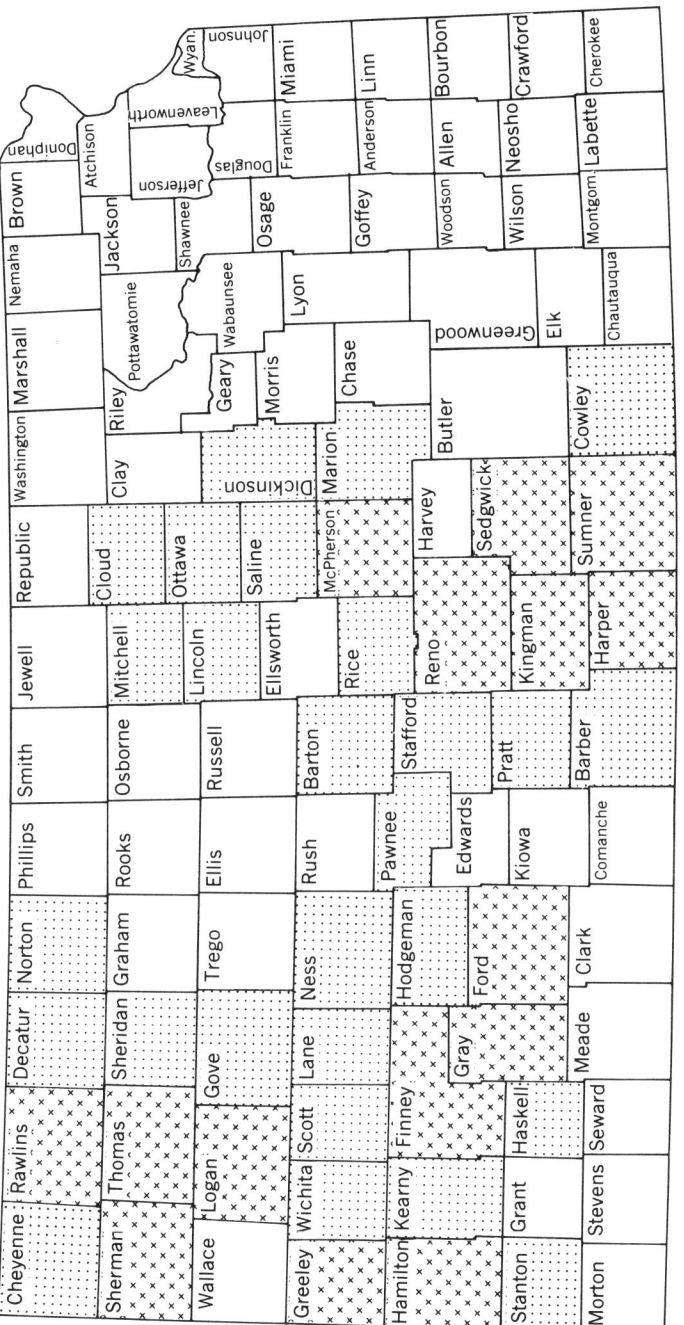

95,200,000–136,000,000 kg ▨   ▨ ≥ 13,000,000 kg

FIGURE 122. Distribution of wheat production.

those that can be spectrally confused with winter wheat. This restriction excluded spring-planted crops (soil still bare in spring imagery). Thus, the possible confusion crops were alfalfa, hay, barley, rye, and pasture (see Figure 123).

Rye and barley hectarages in Kansas (32,000 and 35,000 hectares, respectively) were small compared to winter wheat hectarage. Hence, rye and barley were not significant as confusion crops. This left alfalfa and other hay as the major confusion crops during the period of study.

Ground Truth Management

Data Collection

Published wheat hectarage estimates at the state and national level are based in part on the JES, a sample survey that utilizes area frame sampling. The design of the JES is that of a stratified cluster sample (see Section Appendix C and reference 2). The clusters (referred to as *segments* by ESCS) are land areas consisting of several farms or parts of farms. These segments were used as ground truth information for applying Landsat data estimation procedures.

From a total sample of 435 Kansas JES segments, 87 were subsampled for the Landsat project. Another 87 were available from segments rotated out of the JES after the 1975 survey making a total of 174 segments to be used for Landsat analysis. This number of segments subsampled was determined to reduce the impact of the Landsat research on the 1976 JES data collection effort.

Crops	Usual Planting Dates	Usual Harvesting Dates		
		Begin	Most Active	End
Barley				
Fall	10 Sept.–25 Oct.	10 June	15 June–1 July	5 July
Spring	5 Mar.–30 Apr.	20 June	25 June–1 July	10 July
Corn				
Grain	15 Apr.–10 June	15 Sept.	10 Oct.–5 Nov.	5 Dec.
Silage	20 Apr.–20 June	25 Aug.	1 Sept.–1 Oct.	10 Oct.
Hay				
Alfalfa		10 May		30 Oct.
All		25 May		10 Sept.
Oats	25 Feb.–1 May	25 June	30 June–10 July	20 July
Sorghum				
Grain	10 May–1 July	20 Sept.	10 Oct.–10 Nov.	1 Dec.
Silage	10 May–1 July	5 Sept.	10 Sept.–10 Oct.	15 Oct.
Soybeans	10 May–5 July	20 Sept.	1 Oct.–5 Nov.	20 Nov.
Winter wheat	10 Sept.–25 Oct.	15 June	20 June–5 July	15 July

FIGURE 123. Kansas crop calendar. (From Agricultural Handbook No. 283, USDA, Economics, Statistics, and Cooperatives Service.)

Each set of 87 segments contained two replications each from strata 11, 12, 20 and one replication each from strata 31, 32, and 40 (land use strata definitions are given in Section Appendix A). For the Landsat analysis it was decided to study only the major agricultural strata (11, 12, and 20) because the subsample contained very few segments in the urban and rangeland strata (31, 32, and 40). Thus, the size of the subsample was reduced from 174 to 156 (see Table 38).

The enumerators collected segment data on forms designed by the New Techniques Section with assistance from the Kansas State Statistical Office (SSO). Field boundaries were drawn on the black-and-white Agricultural Stabilization and Conservation Service (ASCS) photos of the segments. Training schools were held on the use of these forms and photos. Enumerated data were collected on two visits to each segment. The first visit, called the April visit, was made during the period from 12 April to 3 May 1976. The second visit, called the June visit, was made during the period from 21 May to 21 June 1976. For fields in subsampled segments, enumerators collected such items as total field and crop area, crop or land use cover, intended uses of crop fields, field appearance, and date of harvest.

To assist with the interpretation of ground truth information, low-level color infrared (IR) aerial photography of the subsampled segments was taken and prepared by the Remote Sensing Institute of the South Dakota State University. These photos were developed at a scale of 5.25 in. to 1 mi. The photo acquisition flights over the segments occurred during the period from 1 May to 8 May 1976.

Data Edit

As soon as both the ASCS and the color IR photos were received, field, tract, and segment boundaries were transferred to the color IR. These boundaries and field numbers were transferred as reported by the enumerators with no attempt made to interpret them. There were 11 segments with unusable or missing color IR photos.

After editing some of the data, the large amount of time required to correct all field boundaries necessitated restructuring the edit to label only the wheat fields, with all other fields called "other."

Field cover type and boundaries were photo-interpreted on the color IR and compared to enumerator data. Inconsistencies between the IR and reported data were rectified. Area and appearance data were then coded and keypunched. Using county maps with JES segments located on them, the segments were located and drawn on USGS quadrangle maps.

A field determination using the color IR photography was made with a computer process called *digitization*. This process related field boundary coordinates

TABLE 38. Kansas Segment Allocation

Stratum	Population (number of segments)	Segment Size (mi²)	JES Sample	Landsat Sample
11	25,058	1.0	170	68
12	21,732	1.0	120	48
20	21,284	1.0	100	40

to a map base (the USGS maps), from which very precise area measurements were available for individual fields. A discussion of the software package used is given in reference 4.

The coded ground data were then merged with the digitized area determinations to make field level records containing (both April and June) reported and digitized area, field appearance codes, strata, segment number, and dates of visits. Checking the ratio of reported to digitized area at the field level was done along with comparing the total digitized segment area to the planimetered area given by the JES master record. Any discrepancies were checked and updates made as needed to get a final data set. This data set was used to create ground data files for analysis.

Landsat Data Acquisition and Management

Characteristics of Landsat Data

The basic element of Landsat data is the set of measurements by the satellite's multispectral scanner (MSS) of a 0.4-hectare area of the earth's surface. The MSS measures the amount of radiant energy reflected and/or emitted from the earth's surface in various regions (bands) of the electromagnetic spectrum. The Landsat satellite used by this study has four bands; one green, one red, and two near infrared bands.

The individual 0.4-hectare MSS resolution areas, referred to as pixels, are arrayed along east–west-running rows within the 185-km wide north-to-south pass of the Landsat satellite. A given point on the earth's surface is imaged once every 18 days by the Landsat satellite. Satellite passes which are adjacent on the earth's surface are at least 1 day apart with respect to their dates of imagery.

Satellite passes are cut into *scenes,* which are just strips of Landsat data covering a length of 185 km (same as the width). Adjacent scenes in the same pass overlap several hundred scan lines. Adjacent scenes east-to-west overlap approximately one-third of the columns.

Scene Selection

In order to cover the state of Kansas with Landsat imagery, six satellite passes were required. Coverage is composed of five passes of three scenes each and one pass consisting of only one scene (to cover the southeastern tip of the state).

It was felt that the separation of other land uses from wheat would be best in early spring imagery. Hence, the first criterion for selection of Landsat imagery was the optimum period for images which was believed to be April or May. The second criterion considered the machine quality of digital data over all four bands. Third was the presence or absence of clouds was considered in the selection.

Cloud cover presented a definite problem.[8] Four passes were available which were nearly cloud-free. For another pass, two counties were lost due to a small cloud-covered area. The remaining pass (over central Kansas) had no cloud-free scenes for the period required. Two partially cloud-covered scenes on one date were used to cover a seven-county area found to be cloud-free in this pass. See

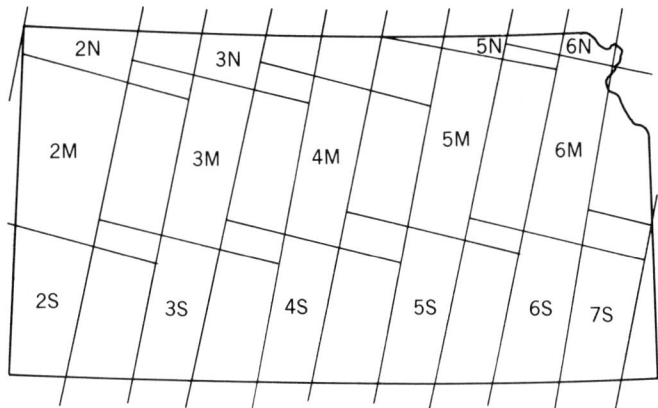

FIGURE 124. Kansas Landsat coverage.

TABLE 39. Landsat II Data, Kansas, 1976

Scene	Date	Landsat ID Number	Comments
2N	4/1/76	2435-16404	Clear
2M	4/1/76	2435-16410	Clear
2S	4/1/76	2435-16413	Clear
3N	5/6/76	2470-16335	Clear
3M	5/6/76	2470-16342	Clear
3S	5/6/76	2470-16344	Clear
4M	4/17/76	2451-16291	Heavy clouds
4S	4/17/76	2451-16293	Heavy clouds
5N	4/16/76	2450-16230	Clear
5M	4/16/76	2450-16232	Clear
5S	4/16/76	2450-16235	Some clouds
6N	5/3/76	2467-16165	Clear
6M	5/3/76	2467-16171	Clear
6S	5/3/76	2467-16174	Clear
7S	5/20/76	2484-16113	Clear

Figure 124 and Table 39 for final scenes used for coverage. Individual scenes were labeled by pass number and position (north, middle, south) in a pass.

An inspection of one pass showed a visible edge separating light pixels in the middle (3M) and dark pixels in the southernmost scene (3S). This edge (or front) ran in a diagonal fashion across the two scenes and was believed to be caused by wet versus dry soil. A possible explanation for this difference was a large rain front over the wet-looking area the day before the imagery. This difference tended to confuse the classification of wheat and other crops between the two areas within the same pass. Healthy wheat fields in the "dry" area looked similar to abandoned wheat or waste fields in the "wet" area.

Registration and Segment Calibration

Registration relates Landsat row–column coordinates to map based latitude–longitude by means of mathematical equations called affine transformations. These equations allow prediction of specific points on maps to corresponding pixels and vice versa. Registration of the 15 scenes picked for Kansas analysis to a map base was done using corresponding points found on 1:500,000 scale Landsat paper products and on USGS quadrangle maps.

Segment calibration is a local movement of the predicted segment area to a more exact location as determined by field patterns in the segment. These field patterns were found in the Landsat data by use of computer generated gray scale printouts. A gray scale in a picture-type line printer product where each printed character represents a pixel. For raw data gray scales, the printed character represents the amount of energy reflected in the specific light band. A categorized grayscale has numbers representing the category into which the pixel was classified.

Segment calibration was done in two ways. First, grayscales from raw data bands were made of the predicted segment area plus a boundary of 20 pixels in width for each segment. Using the digitized segment files, plots of the segment were made at the same scale as the gray scale prints (see Figs. 125 and 126 for an example of segment gray scale plots). Starting from the predicted position of the segment, the plot is overlayed on the gray scale and moved until the field boundaries on the plot best fit the field patterns of light and dark pixels on the gray scale. The new coordinates of the segment, if it needs to be moved, are entered into a local calibration file, which will supplement the precision registration when required.

A second way of checking segment location was to use the same procedure as above, with a categorized gray scale. Each pixel in a categorized gray scale had been given a category number based on a preliminary clustering or classification of the segment area and using all 4 Landsat bands (not just a single band as above). Thus, field patterns were found that were sometimes hidden in the other type of gray scale. This method did take more resources and if distinction was good in the original gray scales may not have been worth the extra time and effort.

Bands 5 and 6 were most helpful in finding wheat fields and their boundaries. The registration accuracies prior to calibration were very good, with many segments needing no movement at all, and the rest were generally moved less than 2 pixels in either direction.

Large field sizes (see Table 40) and rectangular field shapes are common in Kansas. These characteristics greatly facilitated the location of ground truth pixels in the Landsat data.

Segment Grayscale

Segment = 5048   Channel 3, Scene 4M, Band 6;

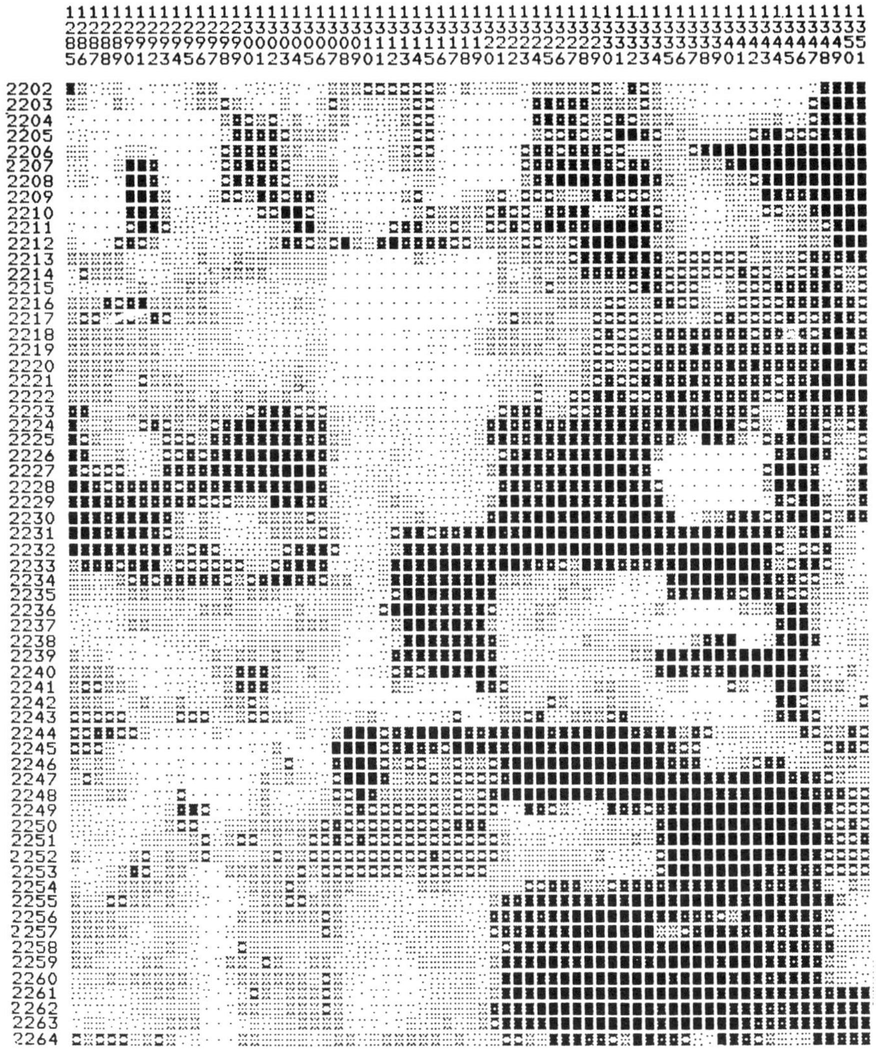

FIGURE 125.   Segment gray scale.

Analysis Procedures and Results

Definition of Analysis Districts

One characteristic of Landsat data is that it does not consider political boundaries when taking imagery. Thus, the state was divided into *analysis districts*, which

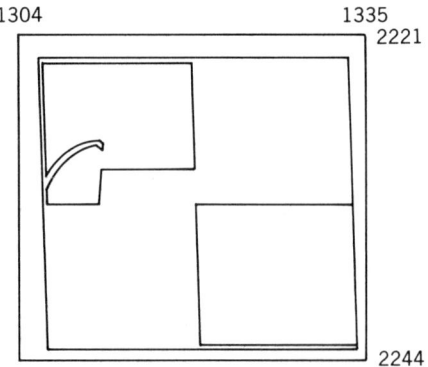

FIGURE 126. Segment plot.

TABLE 40. Kansas Average Wheat Hectares

Agricultural Stratum	Per Segment	Per Field
11	97.1	22.7
12	85.0	16.3
20	46.5	13.7
(11, 12, 20)	80.5	18.0

Source: Reference 3.

were determined by Landsat boundaries and not comparable to ESCS's crop reporting districts. An analysis district is a group of counties or parts of counties that is wholly contained in a Landsat pass. Estimates for these multicounty areas were made and then individual county estimates were derived from them.

County maps with land use strata marked on them were digitized to a latitude–longitude coordinate system (as discussed earlier with segments and fields). The vertices of the outer boundaries of counties were then transformed to the row–column coordinate system of Landsat data using each individual scene's registration. These coordinates were then viewed and counties were assigned to Landsat scenes (see Fig. 127 for final analysis districts for Kansas). Counties were assigned to one and only one pass (and its corresponding analysis district). There were 14 counties with no Landsat data because of clouds and 4 counties lost due to lack of training data. Counties without Landsat data or with other problems that made them unusable were lumped into one analysis district called pass 4C.

Split Counties
In addition to the 18 counties lost due to clouds or no training data, 13 counties were found to be *split* across scene boundaries within 4 of the passes. In earlier experiments (see reference 1) when this situation was encountered, psuedoframes

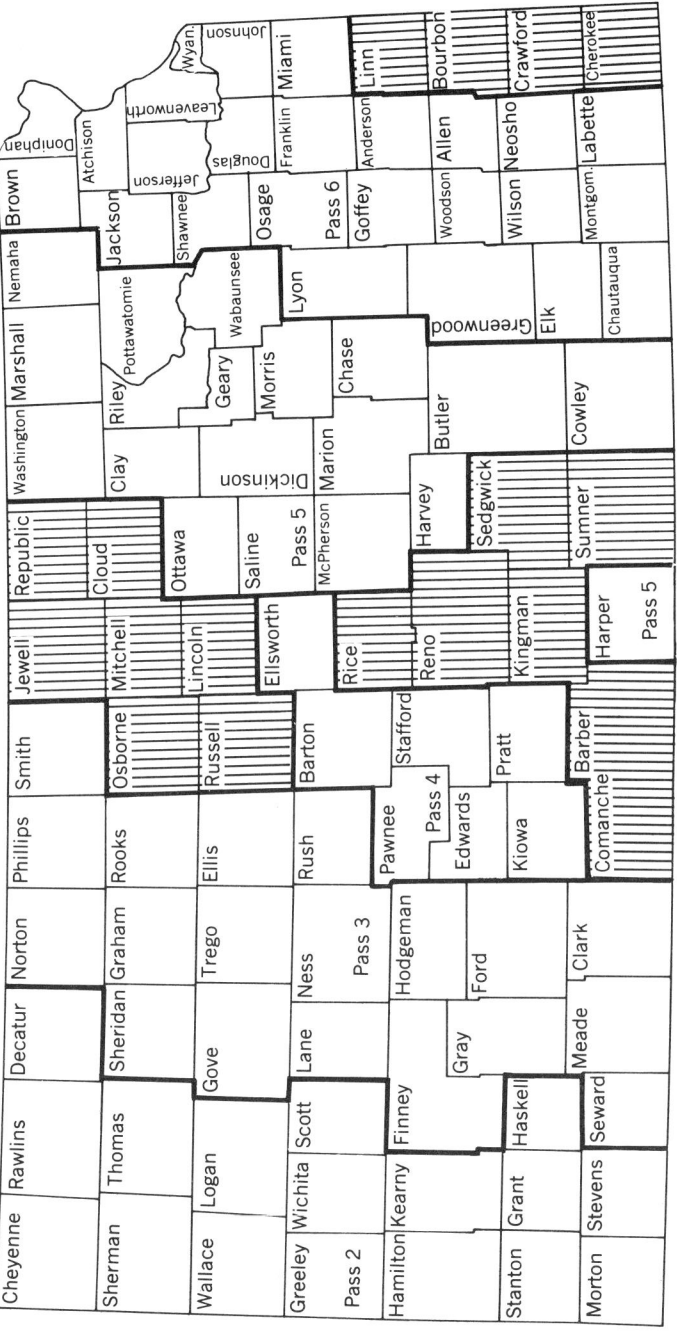

FIGURE 127. Final analysis districts.

were constructed by putting together the bottom part of the northern scene and the top section from the southern scene thus creating a new frame. This method was only valid when the scenes were from the same date and when counties were cut north or south by Landsat lines and not east or west by Landsat column boundaries. Another drawback of the psuedoframe approach was that it requires registration of the new scenes.

A new, quicker method was created to handle these split counties whether they were divided by lines or columns. The new approach, called the psuedocounty approach, was to digitize a figure that divided a county into two (or more) parts, or subcounties, that were each completely within one Landsat scene, utilizing the fact that scenes partially overlap. This figure was then used to cut up the original digitized county file into parts called psuedocounties. Each psuedocounty was distinct from all others and thus was estimated as were the nonsplit counties. In Kansas, only one county was split across analysis districts, and it happened to lie partly in the one-scene pass (pass 7) that did not have enough training data and so it was not used. The other counties were split across scenes within the same pass and thus the only adjustment to the estimation process was to sum the wheat pixels by strata for each county's parts. For estimation when the psuedocounties for a given county are in different analysis districts, each psuedocounty would be considered a separate county all the way through the actual estimation and would require adjustment in the number of area frame units for each analysis district. This did not occur in the Kansas study.

After the analysis districts and split counties in Kansas were all decided upon, the segments were labeled by analysis district also. See Table 41 for segment and county allocation to the various passes (districts).

Pixel Clustering and Classification
Separate analyses were conducted for each analysis district using various clustering and classification procedures. For general information on classification and clustering see Section Appendices B and C. Various factors affecting classifier

TABLE 41. Number of Segments and Counties by Analysis District

	Sample Segments				Frame Units by Stratum			
	By Stratum			11 + 12 + 20				Number of
Analysis District	11	12	20	Total	11	12	20	Counties
Pass 2	21	5	3	29	8067	2681	2189	17
Pass 3	11	12	12	35	4678	5295	5264	19
Pass 4	9	2	0	11	2750	1047	739	7
Pass 5	9	13	9	31	3579	3497	3961	19
Pass 6	2	5	9	16	1518	4520	5478	25
Pass 4C	28	29	15	72[a]	4470	4696	3652	18

[a]Total JES sample used for direct expansion.

performance are discussed in the Illinois report.[1] For this Kansas project, the pixel classifier for each pass was based on training data from that specific pass only. Initially each cover type was clustered into distinct groups or categories and calculations made of the signature* means and covariance matrix for the training set of labeled pixels defined by the digitized segments. Signature statistics for several categories or cover types were grouped together in statistics files used with classification software. Different clustering attempts for each cover type were made.

One approach tried was to set the minimum number of categories to a large number (say 12–15 per cover type) and let the clustering algorithm find the best set of that number of categories. This set of categories (now signatures rather than groups of pixels) was then used in a grouping algorithm[6] which merged signature means and variances one at a time from the total number of categories down to one overall cover type category. The merging criterion in the algorithm was a minimum pattern transmission loss function (an example is shown in Fig. 128). Using the loss function graph, the final number of categories was determined considering the natural breaks and an 80-90 percent pattern retention. In the example shown, 5 or 6 final categories could be a likely choice.

Another approach used was to set a small minimum number of categories and let the clustering algorithm find the best set of these, deleting any that seemed to be outliers. The two approaches came up with much the same categories, although clustering only without grouping seemed to give slightly tighter clusters. Grouping helped to give an idea of the number of categories desired, whereas clustering by itself required predetermination of numbers of categories.

Several methods were used to compare the different clustering and grouping approaches. One method, called a scattergram, shows pixels values labeled by category in a two-dimensional graph (two of the four bands must be chosen). Scattergrams show each pixel's value and may become messy when categories overlap. Another method used for signature comparisons was to plot concentration ellipses (again picking two bands). These ellipses were actually just two-dimensional 90%, confidence intervals computed from statistics files containing the signature means and variances. These two methods allowed visual comparisons of the amount of confusion (seen as overlap) between categories. Two other methods of comparing the signatures involve classification and estimation and are discussed later in this section.

Table 42 gives the final number of categories per cover type used in this study for large-scale classification. On three passes (2, 3, and 5) the number of wheat categories needed came out five, and so in later passes (4 and 5) the wheat training data were clustered directly into five categories. The number of categories needed for "other" was more variable by pass, ranging from four to seven.

The final statistics file was used to create a set of discriminant functions for classification of Landsat pixels. Usually, each analysis district had one and only one statistics file no matter how heterogeneous the data may have been across a

---

*Signature refers to the mean vector and covariance matrix for a specific cover type or category and ideally is separable in the four-dimensional Landsat scanner space from other categories.

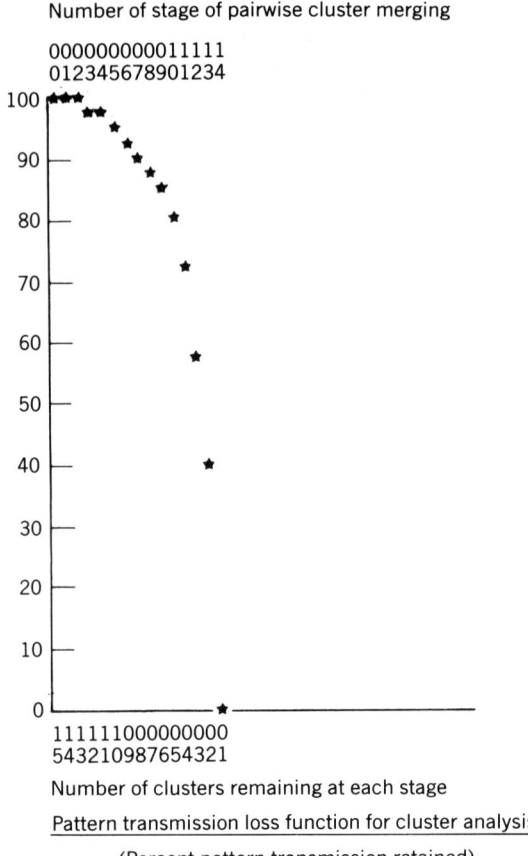

Number of stage of pairwise cluster merging

Pattern transmission loss function for cluster analysis

(Percent pattern transmission retained)

FIGURE 128. Example of pairwise clustering merging tree.

TABLE 42. Number of Categories by Cover Type

Analysis District	Wheat	Other	Total
Pass 2	5	4	9
Pass 3	5	7	12
Pass 4	5	7	12
Pass 5	5	5	10
Pass 6	5	5	10

county, scene, or pass. This classification was done at two levels: One was to classify only those pixels interior to the sampled segments, the other was to classify entire landsat scenes using the same statistics file. The segment level classifications were used to test the performance of the classifier and for estimation of regression parameters. The large-scale (entire-scene) classifications were used for the actual acreage regression estimates at district and county levels.

After examining the visible differences in pass 3 (discussed earlier), it was decided that another level of classification was needed to allow more than one statistics file per scene. This classifier would take into account completely different signatures for the various covers as a function of location in the scene. New software was written* to apply this classifier as well as to allow its results to be used in estimation. Two statistics files were then created (although both have the same number of categories per cover, the application of the new classification method to estimation did not require this) for pass 3 by splitting up the set of segments into two parts, and then following the normal clustering procedure. Table 43 shows the actual pixels classified by the analysis districts.

Another factor involved in the classification of pixels is the use of "different prior probabilities" (different weighting factors for the likelihood functions in the discriminant functions). The priors used were unequal priors proportional to expanded digitized acres (called PED) and equal priors (called EP). For a given analysis district the PED prior probability for a specific cover was defined as the ratio of the current year direct expansion estimate to the total land area in the region.

Some idea of the performance of the classifier may be obtained from the percent correct, that is, the percentage of the digitized segment information that was classified correctly. Since the classifier was trained and tested on the same data (called resubstitution), the numbers may be somewhat optimistic. One drawback of the percent correct as a measure of classifier performance is that it takes into account only the errors of omitting pixels of a specific cover and does not measure the error of calling a pixel that specific cover type when it is actually another cover type. Table 44 contains percents correct for the final statistics files used in this project, with both unequal priors (PED) and equal priors (EP). Notice that the

TABLE 43. Pixel Classification Table by Stratum and Analysis District

Analysis District	Cover Type	Stratum				Total over All Strata
		11	12	20	31–62	
Pass 2	Wheat	1542556	479453	294008	90795	2406812
	Other	3015429	1028932	951584	803402	5799347
Pass 3	Wheat	832104	823850	574361	169791	2400106
	Other	1828384	2177453	2378133	1110488	7494458
Pass 4	Wheat	691804	214542	97788	29275	1033409
	Other	864889	377724	324671	356944	1924328
Pass 5	Wheat	883916	676351	457624	171630	2189521
	Other	1156347	1297593	1799603	2310827	6564420
Pass 6	Wheat	121051	269503	217076	91486	699116
	Other	745203	2303272	2905911	1645465	7599851

*This software was supplied by Martin Ozga and Walter Donovan of the Center for Advanced Computation at the University of Illinois, Champaign–Urbana.

TABLE 44. Percent Correct Pixel Classification for Segments

Analysis District	EP Priors			PED Priors		
	Wheat	Other	Overall	Wheat	Other	Overall
Pass 2	86.83	89.38	88.73	80.19	94.61	90.97
Pass 3	71.73	87.07	82.81	[a]	[a]	[a]
Pass 4	77.28	73.43	75.10	84.74	78.99	81.49
Pass 5	70.07	84.70	80.10	64.93	88.78	81.29
Pass 6	62.45	79.12	77.76	43.48	97.17	92.79

[a]Not calculated.

unequal priors statistics file was used for large-scale classification in pass 6 only, even though PED tends to increase the percent correct. The final criterion for picking a statistics file will be discussed in the following subsection on estimation.

Estimation Procedures

Regression Approach

As mentioned earlier, ESCS uses Landsat data as an auxiliary variable in a regression procedure. In past estimation projects where more segments per analysis district were available, a separate regression equation was estimated for each land use stratum. The statistical methods involved with the separate regression have been described in the paper by Sigman et al.[5] and are excerpted in Section Appendix C. The technique of pooling strata, used in the Illinois project[1] to alleviate the problem of small sample sizes within strata, was rejected because pooling strata tends to overestimate the variance of the estimate.

A "combined" regression estimator was developed (see Section Appendix D) for use with this project. This method assumes that the regression coefficient of the estimator is the same for all strata but that the intercepts are obtained from the stratum means. Using the small-scale classifications of the sampled segments the regression coefficients for each stratum in each analysis district (shown in Table 45) were apparently estimates of a common value within each analysis district.

Selection of a Classifier for Estimation

Besides estimating the regression coefficients, the small scale classification and estimation provides a measure of the performance of the classifier with respect to the variances of the estimates. The various types of regressions are "separate," "pooling," and the combined strata.

In the "separate" regression, the sample coefficients of determination (r-squared) between digitized wheat acres and classified wheat pixels are determined for each stratum. As shown in Appendix C, maximizing the r-square values minimizes the variance of the regression estimates resulting from a classification. Thus, one criterion used to compare classifier performances on the same strata

TABLE 45. Estimated Regression Coefficients for Each Stratum

Stratum	Analysis District				
	Pass 2	Pass 3	Pass 4	Pass 5	Pass 6
11	1.1738	1.1785	1.0435	0.9155	1.2140
12	1.1973	1.1132	—	0.6962	1.6004
20	0.9618	1.0929	—	0.3788	1.6604
Combined:	1.0648	1.1206	1.0117[a]	0.7909	1.6206

[a]Only two data values existed in stratum 12 and none in stratum 20.

was the respective r-squares. These values were calculated in all Kansas analysis except for strata 12 and 20 in pass 4 and stratum 11 in pass 6. In some analyses districts, however, the small sample sizes per stratum make this figure somewhat unreliable. The pooling of data to derive the classifier relationships in this case assumed that only a single stratum was sampled. All segment data from strata 11, 12, and 20 were pooled together and regression was calculated as for an unstratified population. The various r-squared values for the Kansas analysis districts are shown in Table 46 for both EP and PED prior probabilities.

Since the major objective of this project is estimation of winter wheat acreages with reduced variances, maximization of percent correct or reduction of the classification error was not considered in the choice of classifiers. Maximization of the r-squared values was the final criterion used for selection of a statistics file to use for large-scale classification in a given analysis district. The equal priors (EP) file was selected in all analysis districts except pass 6. In this analysis district, the classifier tends to classify a large portion of "other" pixels into the wheat categories. Wheat in this area was not a very large crop percentagewise, and thus the application of PED priors with small probabilities for wheat tended to give a more reasonable classification and thus better r-squares.

Although in most analysis districts the unequal priors classifier was not chosen for full frame classification, the r-squares found using the PED priors are very close to the corresponding equal priors (EP) values (except in pass 6). Thus, if the

TABLE 46. r-Square Values by Analysis District and Priors

Type and Stratum	Pass 2		Pass 4[a]		Pass 5		Pass 6		Pass 3
	EP	PED	EP	PED	EP	PED	EP	PED	EP
Separate — 11	0.8516	0.7762	0.6161	0.6398	0.8522	0.8361	[b]	[b]	0.6719
Separate — 12	0.9953	0.9920	[b]	[b]	0.4785	0.3883	0.1454	0.9836	0.9430
Separate — 20	0.9965	0.9950	[b]	[b]	0.3962	0.5098	0.0832	0.7429	0.7100
Pooling	0.8818	0.8215	0.5975	0.5614	0.7450	0.7181	0.1911	0.7659	0.8073

[a]Pass 4 pooling includes strata 11 and 12 only.
[b]Not calculated due to lack of data.

objective of the study was yield computation or some type of stratification based on classified pixels, and not estimation of acreage, the better classifier would be the unequal priors classifier.

Large-Scale Estimation

Multicounty regression estimates for winter wheat area planted were calculated for the various analysis districts. The regression estimates were compared to estimates calculated by direct expansion of the subsample segments and direct expansion of the total 435 JES segments and to estimates obtained from the summation of final 1976 county estimates published by the Kansas SSO. The final SSO estimates in Kansas are predominately based on the Kansas State Farm Census. Note that the SSO estimates do not have a calculable variance associated with them because they are based on several nonprobability indications, not just the JES direct expansion.

For the multistratum and multicounty analysis districts, performance of the combined regression estimator was compared to the direct expansion estimator in terms of the relative efficiencies (REs) of the resulting estimates. RE measures the gain, in terms of increased precision, of the combined regression estimate over the respective JES or subsample direct expansion estimate. The equation for calculating the RE follows:

$$RE = \frac{var(\text{direct expansion})}{var(\text{combined regression})}$$

Table 47 gives the estimated wheat area, coefficients of variation (CVs), and relative efficiencies for all passes with Landsat classifications available. Note that the direct expansion estimates shown are based on the subsample chosen for the Landsat project.

TABLE 47. Planted Area Estimates of Winter Wheat for Strata 11, 12, and 20

Analysis District	Number of Segments	Number of Counties	Estimator	Estimate (hectares)	CV	RE
Pass 2	29	17	Regression	886,500	4.9	13.1
			Direct expansion	876,300	18.1	—
Pass 3	35	19	Regression	946,900	6.7	4.8
			Direct expansion	1,114,400	12.5	—
Pass 4[a]	11	7	Regression	382,800	7.8	1.3
			Direct expansion	459,300	7.3	—
Pass 5	31	19	Regression	876,700	5.5	3.2
			Direct expansion	889,800	9.8	—
Pass 6	16	25	Regression	358,900	4.8	10.6
			Direct expansion	258,500	21.7	—
Overall[b]	122	87	Regression	3,488,600	2.8	—

[a]Strata 11 and 12 only
[b]Stratum 20 estimate was prorated from state estimate in Pass 4.

When the relative efficiency was computed for the subsampled segments regression with respect to the whole JES sample, smaller REs were found. Even with this restriction, the regression estimates showed a significant reduction in variance as measured by the relative efficiency, ranging from 1.6 to 4.8 with a RE of 2.7 computed over the 87-county area.

Since some strata were deleted from the classification analysis, "Swiss cheese" estimates were computed in order to compare regression estimates with the summations of SSO published county estimates. A Swiss cheese estimate consists of obtaining regression estimates on the strata included in the classification analysis and prorating the direct expansion estimates of the whole state with respect to area frame units on the strata excluded from the classification analysis. Table 48 gives the pass-level "swiss-cheesed" estimates for both regression and direct expansion along with the summation of SSO county estimates. The prorated estimate for strata 31, 32, 33, 40 and 50 range from 2.9% of the total for pass 2 to 11.3% for pass 6. The state level estimate uses a direct expansion for pass 4C.

County Estimates

Single-county estimates were made for 87 of the 105 Kansas counties. As mentioned before, the unavailability of Landsat data (due mostly to cloud cover) prevented the estimation of the remaining 18 counties. Estimates were computed with the "Swiss cheese" technique as just discussed. The combined regression estimator as discussed in Appendix D was employed in calculating the estimates for

TABLE 48. Planted Area Estimates of Winter Wheat for All Strata

Analysis District	Estimator	Estimate (hectares)	CV	RE
Pass 2	Regression	912,900	4.8	13.0
	Direct expansion	902,700	17.6	
	SSO sum	1,035,600	—	
Pass 3	Regression	984,200	6.5	4.8
	Direct expansion	1,151,700	12.1	
	SSO sum	1,106,400	—	
Pass 4	Regression	431,300	6.9	1.3
	Direct expansion	507,700	6.7	
	SSO sum	494,500	—	
Pass 5	Regression	947,500	5.3	3.1
	Direct expansion	960,600	9.1	
	SSO sum	945,800	—	
Pass 6	Regression	404,700	4.7	9.0
	Direct expansion	304,300	18.6	
	SSO sum	382,400	—	
State	Regression	5,141,900	2.7	
	SSO sum	5,220,400	—	

Note: Regression and direct expansion estimators are based on the "Swiss cheese" technique and use only the subsample segment data for strata 11, 12, and 20.

strata 11, 12, and 20. The estimates, their standard errors, and coefficients of variation are listed in Appendix E.

Recently a family of county estimators was developed.[7] County estimates of Kansas winter wheat were computed for strata 11, 12, and 20 using the "ratio" estimator from this family. An empirical comparison of the "ratio" and regression estimates was made. Since the actual county wheat acreage totals were not available, the SSO estimates were used as the "true" totals in the comparison. The county estimates computed by the "ratio" estimator for the total wheat acreage in all strata are also shown in Appendix E along with their standard errors.

Each county in Figure 129 has two numbers associated with it. The first (top) was computed by the following formula:

$$\frac{\text{Regression estimate} - \text{SSO estimate}}{\text{SSO estimate}} \times 100$$

and the second (bottom) was computed in a like manner with the "ratio" estimate replacing the regression estimate in the formula. The figure actually sheds little if any light on which estimator might be superior; however, it does show that the bigger relative differences between both the regression and "ratio" estimates from the SSO estimates occur in counties which are on the border of the state. Also of interest was the fact that both the "ratio" and regression estimates were generally larger than the SSO estimates in the northern counties. The coverse was true for the southern counties. These occurrences were due to the fact that both the "ratio" and regression estimates were highly correlated with the number of pixels classified as wheat which also exhibited these properties. But the reason why such a pattern should be present in the number of pixels classified as wheat was not known and might well serve as a topic for future research.

Since the "ratio" and regression estimates did not exhibit a pronounced difference, we shifted our attention to their coefficients of variation. Figures 130 and 131 show respectively the distribution of the coefficients of variation for the regression and "ratio" estimates of strata 11, 12, and 20. Except for pass 6 (i.e., the eastern part of Kansas), the ratio estimates generally had smaller coefficients of variation. In pass 6, however, the converse was true. In fact, the smallest coefficient of variation achieved by the "ratio" estimator in pass 6 was larger than all but two of the CVs attained by this estimator outside pass 6. This was due, at least in part, to the fact that the pooled within-county variance for the whole state was used to estimate the within-county variance in pass 6 since there were not enough data values available in pass 6 alone.

Actually, 67% of the "ratio" estimates had coefficients of variation less than 20%, whereas 48.3% of the regression estimates had CVs less than 20%.

Intervals of one and two standard errors were computed for both the "ratio" and the regression Swiss cheese estimates in each county. We then checked to see if the SSO estimates fell within these intervals. Table 49 shows the results.

The percentage of county SSO estimates falling within a given standard deviation interval was lower for the regression estimates in pass 6 than in the other

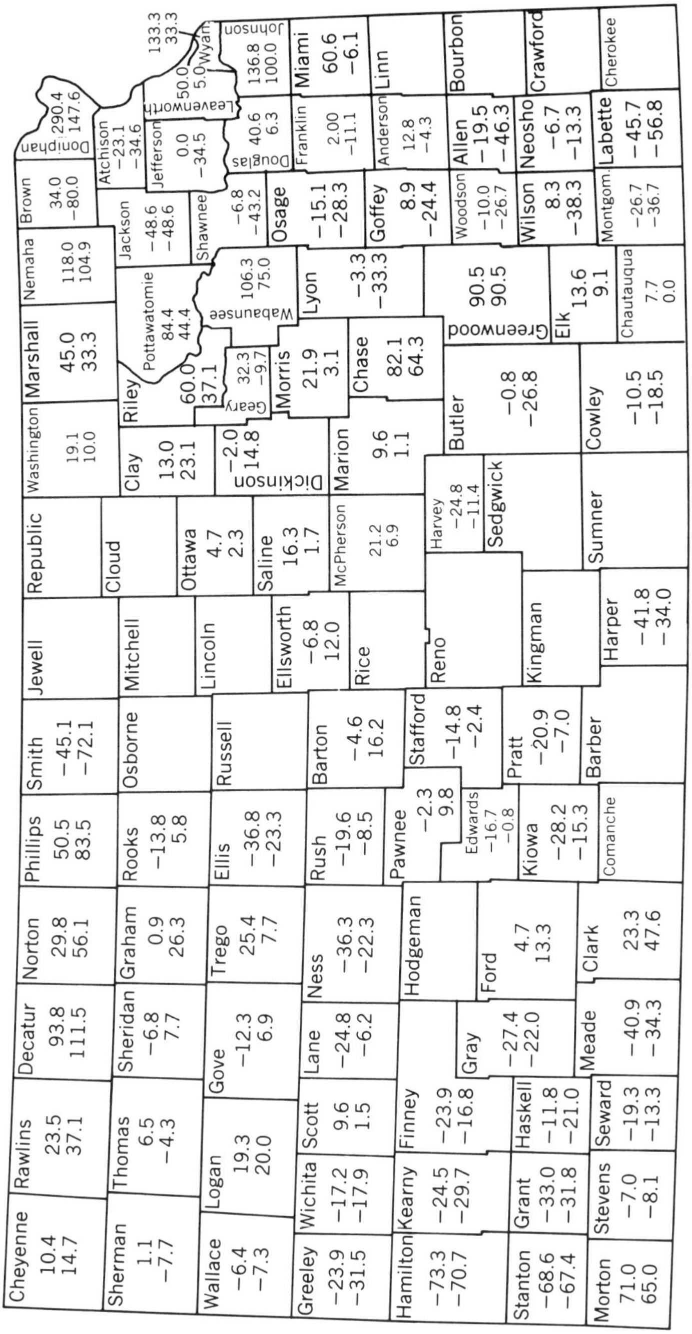

FIGURE 129. Percent changes from SSO estimates.

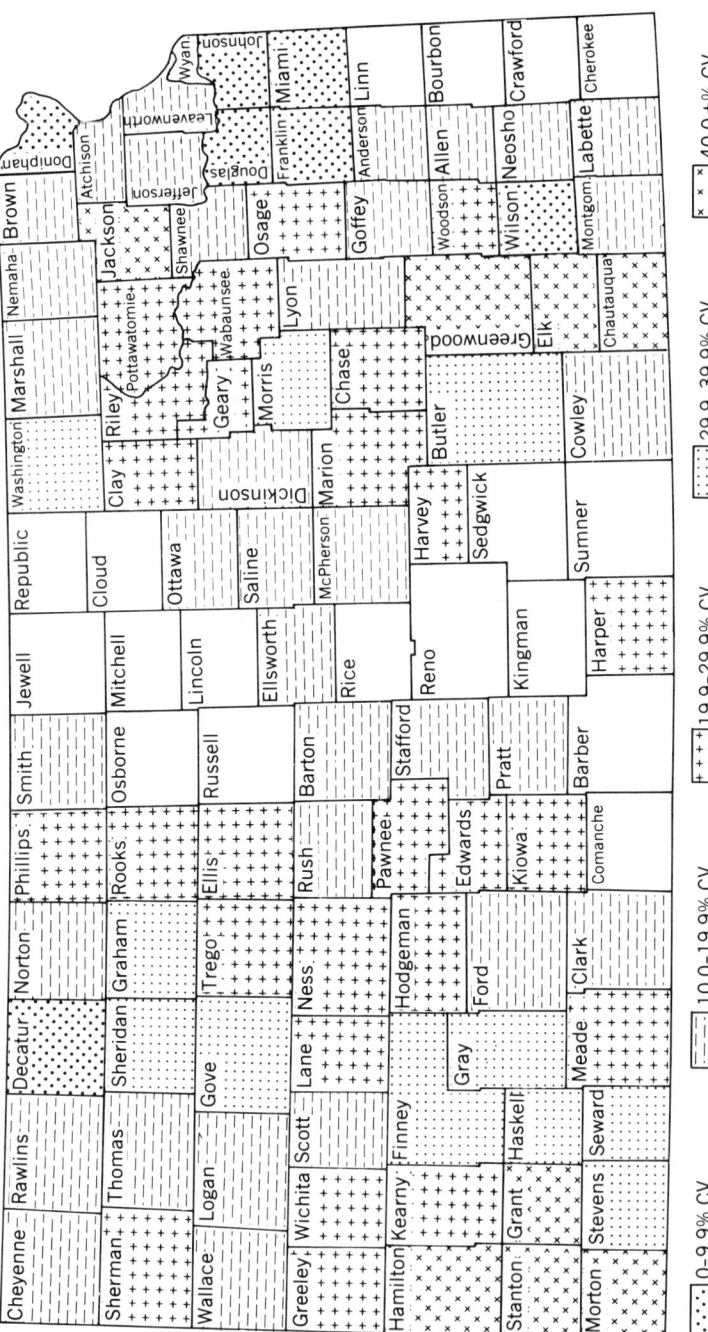

FIGURE 130. County level CVs — regression.

0-9.9% CV　　10.0-19.9% CV　　19.9-29.9% CV　　29.9-39.9% CV　　40.0+% CV

Otherwise: no estimates made due to cloud cover

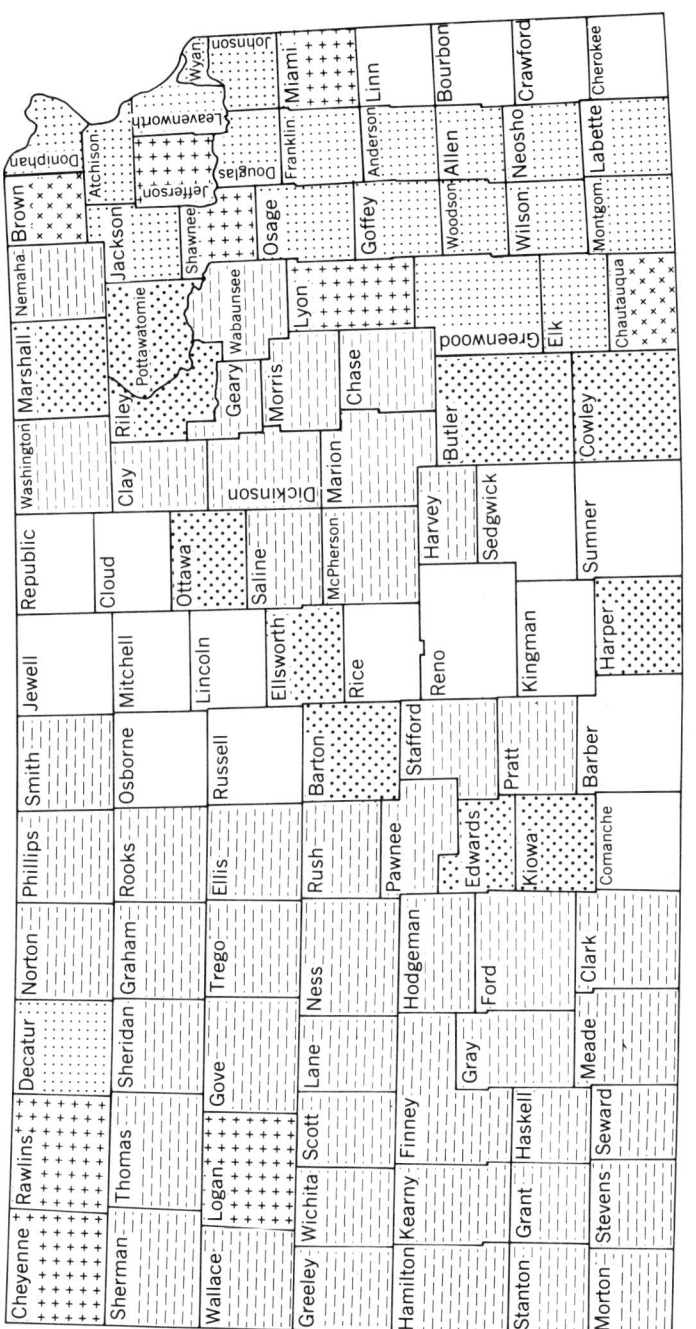

FIGURE 131. County level CVs—"ratio."

Otherwise: no estimates made due to cloud cover

0-9.9% CV

10.0-19.9% CV

19.9-29.9% CV

29.9-39.9% CV

40.0+% CV

TABLE 49. Number of Counties for Which the SSO Estimate Falls within a One- or Two-Standard-Error Tolerance Interval for the "Ratio" and Regression Estimates by Pass

Pass	Number of Counties in Pass	Number of Counties within One-Standard-Error Tolerance Interval		Number of Counties within Two-Standard-Error Tolerance Interval	
		"Ratio" (%)	Regression (%)	"Ratio" (%)	Regression (%)
2	17	8 (47.1)	8 (47.1)	11 (64.7)	13 (76.5)
3	19	8 (42.1)	9 (47.4)	12 (63.2)	16 (84.2)
4	7	4 (57.1)	5 (71.4)	5 (71.4)	6 (85.7)
5	19	7 (36.8)	8 (42.1)	10 (52.6)	14 (73.7)
6	25	9 (36.8)	9 (36.0)	17 (68.0)	10 (40.0)
Total:	87	36 (41.4)	39 (44.8)	55 (63.2)	59 (67.8)

passes. Otherwise the percentages were fairly consistent by passes for each estimator (if we exclude pass 4, which consists of only seven counties, and note that a change of one county SSO estimate falling with a tolerence interval brings about a change of over 14 percentage points). The percentages of SSO estimates falling within a one-standard-error tolerence interval for both estimators were 41.4–44.8%. For a two-standard-error tolerence interval the percentage was a little higher for the regression estimator (67.8%) than for the "ratio" estimator (63.2%).

The coefficients of variation were plotted versus the estimate produced by the regression (Fig. 132) and "ratio" (Fig. 133) estimates. At first glance it appears that for the ratio estimates small CVs are associated with the large estimates and vice versa. However, upon closer inspection, we see that the size of the estimates and the satellite pass are confounded. The small estimates with the large CVs all belonged to pass 2. In all the other passes the CVs seemed to be independent of the size of the estimate, with the possible exception of pass 2, which apparently exhibited the opposite trend. On the other hand, for the regression estimator the small CVs generally corresponded to large estimates and conversely. The degree to which this relationship is true varied from pass to pass. Further research is needed to determine all relationships.

Conclusion

The goal of this project was to utilize Landsat data to improve existing winter wheat estimation procedures. Winter-wheat-planted area estimates were made at multicounty and county levels for 87 of the 105 Kansas counties. Eighteen counties were not estimated due to cloud cover or lack of training data.

Attainment of the project goals was measured by the reduction in variance of the area estimate computed using Landsat data as compared to the direct expansion estimate over the same area. The use of Landsat data as an auxiliary variable was seen to reduce the variation of the multicounty (17–25 counties each) areas

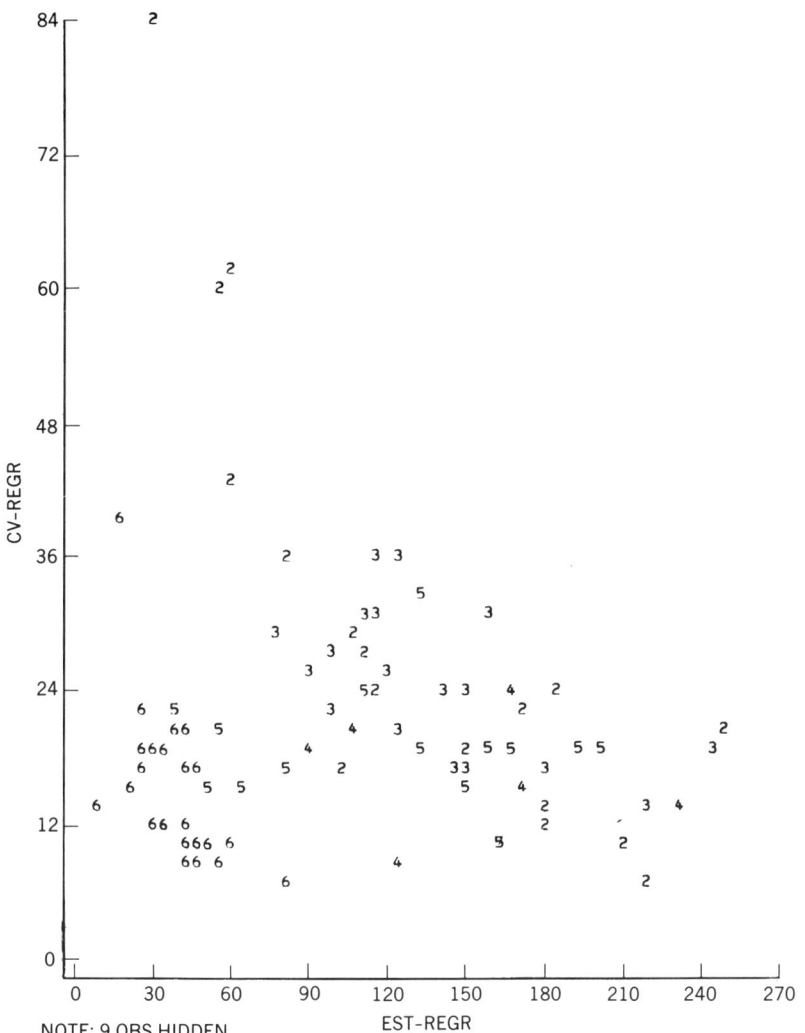

FIGURE 132. Regression estimates vs. CVs.

NOTE: 9 OBS HIDDEN

from 68% to 92%. One analysis area which contained only 7 counties and 11 segments showed a 23% reduction in variance due to use of Landsat data. For the 87-county area as a whole, a reduction of 64% was seen for the variation of the planted area estimate.

Several new procedures for analysis of Landsat data in general were also explored in support of this project. The split-county approach will be especially useful in the future. Combined regression techniques were seen to be applicable to Landsat-based estimation. More study is needed to compare the suggested county estimators.

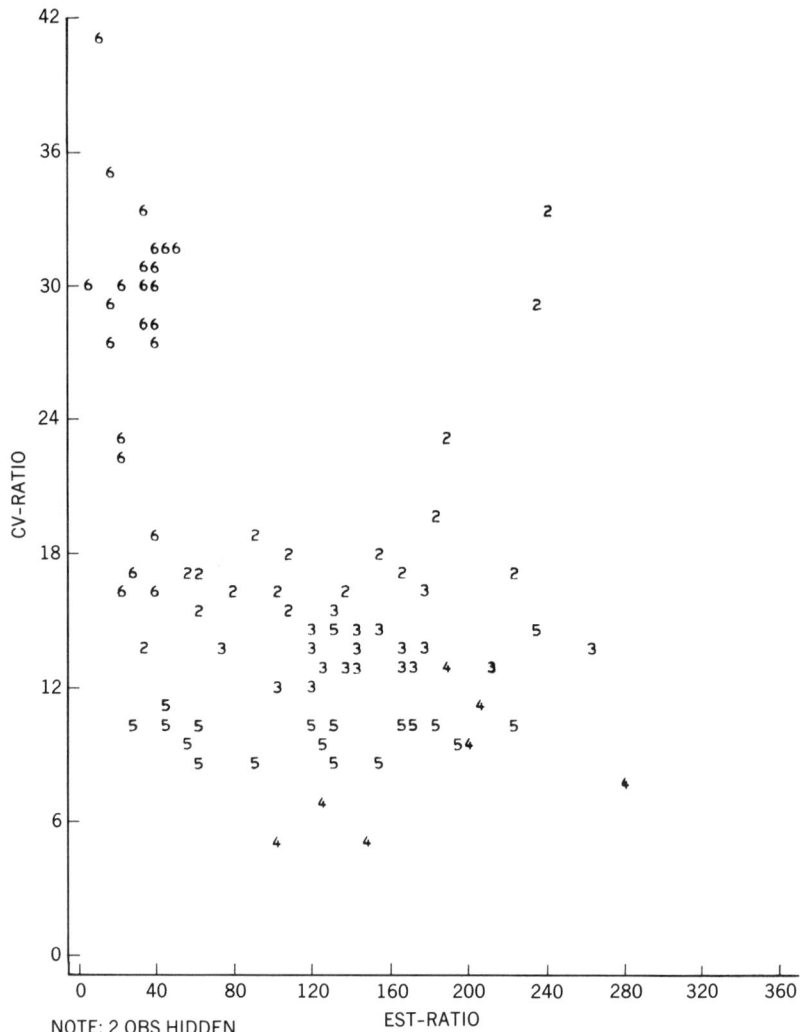

FIGURE 133. Ratio estimates vs. CVs.

REFERENCES

1. Gleason, C., R. R. Starbuck, R. S. Sigman, G. A. Hanuschak, M. E. Craig, P. W. Cook, and R. D. Allen, The auxiliary use of LANDSAT data in estimating crop acreages: results of the 1975 Illinois Crop-Acreage Experiment, Statistical Reporting Service, U.S. Department of Agriculture, Washington, D.C., October 1977.

2. Pratt, W. L., *The use of interpenetrating sampling in area frames,* Statistical Reporting Service, U.S. Department of Agriculture, Washington, D.C., May 1974.

3. Craig, M. E., and M. Cardenas, *Kansas wheat non-sampling error analysis,* Economics, Statistics, and Cooperatives Service, U.S. Department of Agriculture, Washington, D.C., April 1978.

4. Ozga, M., W. E. Donovan, and C. P. Gleason, An interactive system for agricultural acreage estimates using LANDSAT data, in *Proc. 1977 Symp. Machine Processing of Remotely Sensed Data,* Purdue University, West Lafayette, Ind., 1977.

5. Sigman, R. R., C. P. Gleason, G. A. Hanuschak, and R. A. Starbuck, Stratified acreage estimates in the Illinois Crop-Acreage Experiment, in *Proc. 1977 Symp. Machine Processing of Remotely Sensed Data,* Purdue University, West Lafayette, Ind., 1977.

6. Ray, R. M. III, *An entropy maximization approach to the description of urban special organization,* CAC Document 237, University of Illinois, Champaign–Urbana, September 1977.

7. Cardenas, M., M. E. Craig, and M. M. Blanchard, Small area estimators: county crop acreage estimates using LANDSAT data, paper presented at the 1978 annual American Statistical Association meeting, San Diego, Calif.

8. Hanuschak, G. A., The effect of the LANDSAT cloud cover domain on winter wheat acreage estimation in Kansas during 1976, in *Proc. 1977 Symp. Machine Processing of Remotely Sensed Data,* Purdue University, West Lafayette, Ind., 1977.

Appendix A: Kansas Strata Definitions

Stratum	Description	Population Size	Average Segment Size (mi^2)
11	80+% cultivated	25,028	1.00
12	50–80% cultivated	21,704	1.00
20	15–49% cultivated	21,286	1.00
31	Agricultural–urban	2,774	0.25
32	Urban	2,941	0.10
33	Resort	247	0.25
40	Range	3,147	4.00
50	Nonagricultural	294	1.00
61	Water	29	0.50

Appendix B: Categorization or Classification Procedures

Introduction*

The total radiance from an object is composed of two components, reflected radiance and emitted radiance. In general, the reflected radiance forms a dominant portion of the total radiance from an object at shorter wavelengths of the electromagnetic spectrum, while the emissive radiance becomes greater at the longer wavelengths. The combination of these two sources of energy would represent the total spectral response of the object. This, then, is the "spectral signature" of an object, and it is the differences between such signatures which allows the classification of objects using multivariate statistical techniques. Every picture ele-

*Excerpted from W. Wigton, The technology of LANDSAT imagery and its value in crop estimation for the U.S. Department of Agriculture, Statistical Reporting Service, U.S. Department of Agriculture, Washington, D.C., March 1976.

ment (pixel) is recorded with four variables corresponding to one of the following four bands.

Sensor spectral band relationships

Sensor	Spectral Band Number	Wavelengths (μm)	Color	Band Code
MSS	1	0.5–0.6	Green	4
MSS	2	0.6–0.7	Red	5
MSS	3	0.7–0.8	Near infrared	6
MSS	4	0.8–1.1	Infrared	7

Discriminant Analysis

This background is intended to be general and enable the reader to understand the detailed computations and results. Kendall and Stuart formulate discriminant analysis and classification by stating:

> We shall be concerned with problems of differentiating between two or more populations on the basis of multivariate measurements. . . . We are given the existence of two or more populations and a sample of individuals from each. The problem is to set up a rule, based on measurements from these individuals, which will enable us to allot some new individual to the correct population when we do not know from which it emanates.

For example, suppose the land population of interest were a portion of San Joaquin Valley in California. Cotton, wheat, and barley are the major crop populations of interest. From every acre in the San Joaquin Valley we have light intensity readings for green light, red light, and two infrared wavelengths. These light intensities are multivariate measurements that will be used to allot or classify each data point into a crop type such as cotton, wheat, or barley.

A sample of fields from each crop type is selected and their respective light intensities obtained. These sample points are plotted on a two-dimensional graph showing relative positions of each crop in the measurement space (MS). The problem is to partition the measurement space in some optimal fashion so that points are allotted as nearly correct as possible.

There are many ways to partition a measurement space. We have done a simple nonstatistical partition, merely by drawing lines, in Figure 134. Visually partitioning the measurement space may work when it is one- or two-dimensional, but for more-than-two-dimensional measurement spaces, a visual partition is not possible. For most Landsat and aerial photography classification studies a four-dimensional measurement space has been used.

The method used in this example was that of constructing contour "surfaces" in the MS. These dividing surfaces were constructed so that points falling on the dividing surface have equal probabilities of being in either group on each side. Those points not on the dividing surface always have a greater probability of being classified into the crop for which the point is interior to the contour surface. If prior knowledge of the population density function indicates that the density is

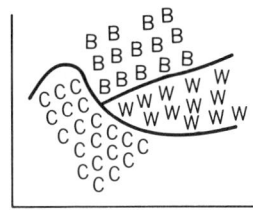

FIGURE 134. Two-dimensional measurement space.

multivariate normal, then a multivariate normal density distribution will be estimated for each crop. It is hoped that the data are approximately multivariate
normal, since only the mean vector and covariance matrix are required to estimate
a discriminant function. Usually small departures from normality will not invali-
date the procedure, but certain types of departures (for example, bimodal data)
may be very detrimental to the statistical technique. However, the error rate and
estimator properties are dependent on the assumptions of the distributions and
prior information.

For example, in this example a multivariate normal density was assumed, so it
becomes quite simple to estimate the density functions and the discriminant
scores, which in turn determine boundaries.

The discriminant score for ith population is

$$P_i(2\pi)^{-q/2}|\Sigma_i|^{-1/2}e^{-1/2}(\chi - \mu_i)'\Sigma_i^{-1}(\chi - \mu_i)$$

where P_i = prior probability for the ith crop
Σ_i = covariance matrix ($q \times q$) for the ith crop
μ_i = mean vector (q length) for the ith crop
χ = set of measurements of an individual from the ith population or its
equivalent discriminant score the $\log_e$ of

$$D_i = \log_e(P_i) - \tfrac{1}{2}\log_e|\Sigma_i| - \tfrac{1}{2}(\chi - \mu_i)'\Sigma_i^{-1}(\chi - \mu_i)$$

The boundary between two populations is quadratic (curved), and the point χ
that falls on the boundary has an equal probability of being in either population.

When an unknown land point is classified, its measurement vector is compared
to the mean vector for each crop represented. The point is assigned to the crop
whose mean point is "nearest" in a statistical sense.

The procedure used for finding the "nearest" mean uses the Mahalanobis measure of distance, not the Euclidean. This is illustrated in Figure 135. The point is
actually closest (Euclidean distance) to the mean vector (center point) of B. However, when one takes into account the variance and covariances, χ is found to be
closest to group A based on a probability concept and an outlier of group B.
Therefore, the point would be classified into group A, because the probability that
the point (χ) is a member of group A is much greater than for group B.

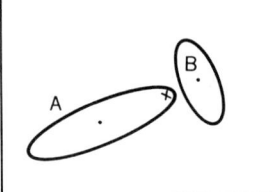

FIGURE 135. Measurement space showing two crop density functions.

Thus the partitioning of the MS is done by computing the means for each crop type and using the Mahalanobis distances from this mean. This distance depends on the covariance matrix and is a measure of probability. The discriminant functions without prior probabilities are:

1. $(X - \overline{X}_i)'S_i^{-1}(X - \overline{X}_i)$, which is a sample estimate of $(X - \mu_i)'\Sigma_i^{-1} \times (X - \mu_i)$ if linear discriminant functions are used.
2. $-\frac{1}{2}\log_e|S_i| - \frac{1}{2}(X - \overline{X}_i)'S_i^{-1}(X - \overline{X}_i)$ if quadratic discriminant functions are used. These functions involve the exponent of the density formula of the multivarate normal distribution

$$C_{\exp} = -\tfrac{1}{2}(X - \mu_i)'\Sigma_i^{-1}(X - \mu_i)$$

of the ith crop. If $\Sigma_i = \Sigma_j$ for all $i \neq j$, linear discriminant functions are used.

It is worth pointing out that if linear discriminant functions are used, one assumes that (1) $\Sigma_i = \Sigma_j$, (2) for all crops in the MS the major and minor axes are equal, and (3) the sample data for each crop have the same slope. Such an event in two-space is shown in Figure 136. This space can be partitioned effectively with straight lines. Thus, we can use linear discriminant functions.

Figure 137 shows a MS where covariance matrices are not equal and therefore linear discriminant functions are not appropriate. In either case, the Mahalanobis distance is used.

In Figure 136, even though a common center point is not present, a common covariance (ellipse) matrix would be computed. In Figure 137 a different covari-

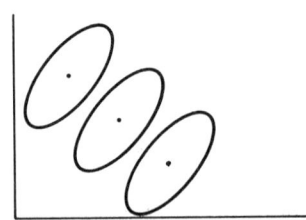

FIGURE 136. Measurement space showing same covariance.

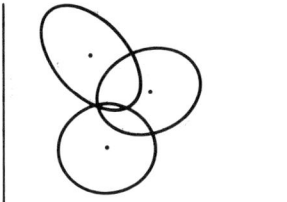

FIGURE 137.   Measurement space showing different covariance.

ance matrix will be needed for each crop type. When the off-diagonal elements in the covariance matrix are unequal, the slopes of the data are different and linear discriminant functions are *not* appropriate.

The above techniques follow from our first assumption that the data is normally distributed in the MS. In practice, however, one does not decide what the distribution of the population density is in the MS and program the correct procedure. One uses the available procedures for analyzing data. Most available programs assume multivariate normal data because the program and the calculations are greatly simplified.

In order to explain better how a parametric procedure can reduce the workload, consider that the first step in the discriminant analysis (DA) is to estimate the population density function in the MS, with a sample of points from each crop. Once these population density functions have been estimated, then partitioning the space is extremely simple.

To estimate a multivariate population density in MS for cotton where we have no prior information except sample data on cotton is extremely difficult. If a sample of 1000 points were available, each of these 1000 data points would need to be stored in the computer. On the other hand, if we are working with a multidimensional normal distribution, theory tells us that the sufficient statistics are computed (mean vector and covariance matrix) and stored in the computer.

The individual data points could be discarded because no additional information about the population distribution in the MS is available in these points. (There would be information about how well the data fit the normal distribution in these 1000 data points.)

Another consideration is that all the techniques we have described require independent random samples from each crop in order to estimate the population density in the MS (training data). This point is mentioned because most remote sensing analysts do not work with randomly selected points. In this study, we have tried to work with randomly selected fields. However, the points within these fields are not a random sample of all possible points in a given crop, but the data are nested within fields. Consequently, the random selection is restricted to the selection of fields within the randomly selected segments.

One type of prior information that can be used in the classification procedure is the relative frequency or occurrence (prior probabilities) for each of the K populations in the total land population. For example, if one-third of all land is cotton and one-fourth is barley, this information would be used and it would effect the partitioning of the measurement space accordingly. If a crop has a high chance of selection, then the area in the MS would be increased. Conversely, if a certain crop has a very low change of occurrence, then the area in MS would be adjusted downward.

Clustering*

Clustering is a data analysis technique by which one attempts to determine the natural or "inherent" relationships in a set of observations or data points. To get an intuitive idea of what is meant by natural or inherent relationships in a set of data, consider the examples in Figure 138. If one were to plot height versus weight for a random sample of students, without regard to sex, on a college campus, it is likely that two relatively distinct clusters of observations would result, one corresponding to the men in the sample (heavier and taller) and another corresponding to the women (lighter and shorter). Similarly, if the spectral reflectance of vegetation in a visible waveband were plotted against reflectance in an infrared waveband, dry vegetation and green vegetation could be expected to form discernible clusters.

If the data of interest never involved more than two attributes (measurements or dimensions), cluster analysis might always be performed by visual evaluation of two-dimensional plots such as those in Figure 138. But beyond two or possibly three dimensions, visual analysis is impossible. For such cases it is desirable to have a computer perform the cluster analysis and report the results in a useful fashion.

With regard to the application of clustering to remote sensing research, the greatest use of cluster analysis has been for the purpose of assuring that the data used to characterize the crop or land use classes do not seriously violate the assumption of Gaussian statistics. In general, it may be expected that each distinct cluster center will correspond to a mode in the distribution of the data. Therefore, with the objective of defining a crop or land use subclass for each cluster center, the possibility of multimodal (and hence definitely non-Gaussian) crop or land use distributions is essentially eliminated.

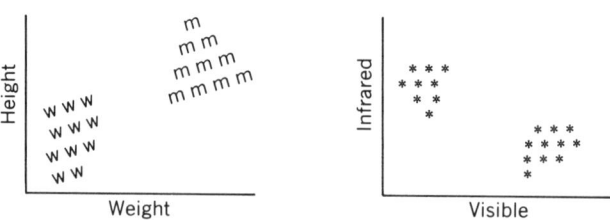

FIGURE 138. Clustering patterns.

*Excerpted from P. H. Swain, *Pattern recognition: a basis for remote sensing data analysis*, LARS Information Note 111572.

A more detailed report on the technical development of several clustering algorithms is provided by Swain.

Appendix C: Crop Acreage Estimation Procedures and Classifier Design Methods

Stratified Acreage Estimation

Direct Expansion Estimation (Ground Data Only)

Aerial photography obtained from the Agricultural Stabilization and Conservation Service is photo-interpreted using the percent of cultivated land to define broad land use strata. Within each stratum, the total area is divided into N_h area frame units. This collection of area frame units[†] for all strata is called an area sampling frame. A simple random sample of n_h units is drawn within each stratum. The Statistical Reporting Service then conducts a survey in late May, known as the June Enumerative Survey (JES). In this general-purpose survey, acres devoted to each crop or land use are recorded for each field in the sampled area frame units. Intensive training of field statisticians and interviewers is conducted providing rigid controls to minimize nonsampling errors.

The scope of information collected on this survey is much broader than crop acreage alone. Items estimated from this survey include crop acres by intended utilization, grain shortage on farms, livestock inventory by various weight categories, and agricultural labor and farm economic data.

Let $h = 1, 2, \ldots, L$ be the land use strata. For a specific crop (corn, for example) the estimate of total crop acreage for all purposes and the estimated variance of the total are as follows. Let

Y = total corn acres for a state (Illinois, for example)

$\hat{Y}$ = estimated total of corn acres for a state

y_{hj} = total acres in the jth sample unit in the hth stratum

Then

$$\hat{Y} = \sum_{h=1}^{L} N_h \left(\sum_{j=1}^{n_h} y_{hj} \right) \Big/ n_h$$

The estimated variance of the total is

$$v(\hat{Y}) = \sum_{h=1}^{L} \frac{N_h^2}{n_h(n_h - 1)} \cdot \frac{N_h - n_h}{N_h} \cdot \sum_{j=1}^{n_h} (y_{hj} - \bar{y}_h)^2$$

*Excerpted from R. R. Sigman, C. P. Gleason, G. A. Hanuschak, and R. S. Starbuck, Stratified acreage estimation in the Illinois Crop-Acreage Experiment, in *Proc. 1977 Symp. Machine Processing of Remotely Sensed Data*, Purdue University, West Lafayette, Ind., 1977.

[†]In this context, all area frame units mean all the segments in the population and is not the same concept of area frame unit (count unit) used in the body of this discussion.

Note that we have not yet made use of an auxiliary variable such as classified Landsat pixels. The estimator is commonly called a direct expansion estimate, and we will denote this by $\hat{Y}_{DE}$. As an example, for the state of Illinois in 1975, the direct expansion estimates were

Corn $\hat{Y}_{DE}$ = 11,408,070 acres
Relative sampling error = 2.4% = $\sqrt{v(\hat{Y})}/\hat{Y}$
Soybeans $\hat{Y}_{DE}$ = 8,569,209
Relative sampling error = 2.9% = $\sqrt{v(\hat{Y})}/\hat{Y}$

Regression Estimation (Ground Data and Classified LANDSAT Data)

The regression estimator utilizes both ground data and classified Landsat pixels. The estimate of the total Y using this estimator is

$$\hat{Y}_R = \sum_{h=1}^{L} N_h \cdot \bar{y}_{h(reg)}$$

where $\bar{y}_{h(reg)} = \bar{y}_h + \hat{b}_h(\bar{X}_h - \bar{x}_h)$

$\bar{y}_h$ = average corn acres per sample unit from the ground survey for the hth land use stratum

$$= \sum_{j=1}^{n_h} y_{hj}/n_h$$

$\hat{b}_h$ = estimated regression coefficient for the hth land use stratum when regressing ground-reported acres on classified pixels for the n_h sample units

$$= \frac{\sum_{j=1}^{n_h}(x_{hj} - \bar{x}_h)(y_{hj} - \bar{y}_h)}{\sum_{j=1}^{n_h}(x_{hj} - \bar{x}_h)^2}$$

$\bar{X}_h$ = average number of pixels of corn per frame unit for *all* frame units in the hth land use stratum (thus *whole* Landsat frames must be classified to calculate $\bar{X}_h$; note that this is the mean for the population and not the sample)

$$= \sum_{i=1}^{N_h} X_{hi}/N_h$$

X_{hi} = number of pixels classified as corn in the ith area frame unit of the hth stratum

$\bar{x}_h$ = average number of pixels of corn per sample unit in the hth land use stratum

$$= \sum_{j=1}^{n_h} x_{hj}/n_h$$

x_{hj} = number of pixels classified as corn in the jth sample unit in the hth stratum.

The estimated (large sample) variance for the regression estimator is

$$V(\hat{Y}_R) = \sum_{h=1}^{L} \frac{N_h^2}{n_h} \frac{N_h - n_h}{N_h} \cdot \sum_{j=1}^{n_h} (y_{hj} - y_h)^2 \cdot \frac{1 - r_h^2}{n_h - 2}$$

where r_h^2 = sample coefficient of determination between reported corn acres and classified corn pixels in the hth land use stratum.

$$r_h^2 = \frac{[\sum_{j=1}^{n_h} (y_{hj} - \bar{y}_j)(x_{hj} - \bar{x}_h)]^2}{[\sum_{j=1}^{n_h} (y_{hj} - \bar{y}_h)^2][\sum_{j=1}^{n_h} (x_{hj} - \bar{x}_h)^2]}$$

Note that

$$v(\hat{Y}_R) = \sum_{h=1}^{L} \frac{n_h - 1}{n_h - 2}(1 - r_h^2)v(\hat{Y})$$

and so $\lim v(\hat{Y}_R) = 0$ as $r_h^2 \to 1$ for fixed n_h. Thus a gain in lower variance properties is substantial if the coefficient of determination is large for most strata.

The relative efficiency of the regression estimator compared to the direct expansion estimator will be defined as the ratio of the respective variance:

$$RE = v(\hat{Y}_{DE})/v(\hat{Y}_R)$$

Ratio Estimation and Classifier Design*

Ratio Estimation

A ratio estimate of the total Y for a particular cover type is

$$\hat{Y}_{ratio} = \sum_{h=1}^{L} (\bar{y}_h/\bar{x}_h)X_h$$

$$= \sum_{h=1}^{L} r_h X_h$$

where $r_h = \bar{y}_h/\bar{x}_h$.

*Excerpted from M. Ozga, W. E. Donova, and C. P. Gleason, An interactive system for agricultural acreage estimates using LANDSAT data, in *Proc. 1977 Symposium on Machine Processing of Remotely Sensed Data,* Purdue University, West Lafayette, Ind., 1977.

The variance of the ratio estimate is

$$v(\hat{Y}_{\text{ratio}}) = \sum_{h=1}^{L} \frac{(N_h - n_h)N_h}{n_h}(S_{h,y}^2 + r_h^2 S_{h,x}^2 - 2r_h\rho_h S_{h,y} S_{h,x})$$

where ρ_h = sample correlation coefficient between x and y for the hth stratum
$\quad\quad S_{h,y}^2$ = sample variance for the hth stratum for the y variate
and $S_{h,x}^2$ is similarly defined.

Designing a Classifier

The pixel classifier is a set of discriminant functions corresponding one-to-one
with a set of classification categories. Each discriminant function consists of the
category's likelihood probability multiplied by the category's prior probability. If
the prior probabilities used are correct for the population of pixels being classified,
then the resulting Bayes classifier minimizes the posterior probability of misclassi-
fying a pixel for a 0–1 loss function.

In crop acreage estimation, however, the objective is to minimize the variance
of resulting acreage estimates. Since minimizing the posterior probability of
misclassification does not necessarily achieve this objective, optimum acreage estimation may require the use of prior probabilities different than the optimum
Bayes set.

For the case of multivariate normal signatures, the category likelihood functions are completely specified by the population means and covariances of the
category signatures. Thus, the calculation of category discriminant functions involves the estimation of signature means and covariances and category prior probabilities.

Designing the classifier for this experiment consisted of the following steps:

1. Identification of classification categories
2. Calculation of signature means and covariances and category prior probabilities from a training set of labeled pixels (called "training the classifier")
3. Measurement of classifier performance on a test set of labeled pixels (called
   "testing the classifier")
4. Heuristic optimization of the classifier by repeating steps 1 through 3 for
   different numbers of categories and/or different prior probabilities, and then
   proceeding to step 5 for the "optimized" classifier
5. Estimation of classifier performance in classifying the entire pixel population

Because of the availability of ground data, which supplied the location and
cover type of agricultural fields, supervised identification of classification categories was possible. A classification category was created for each cover type
in which the number of training pixels exceeded a specified threshold, usually
100 pixels. In addition, a classification category for surface water was created
using pixels from rivers, lakes, and ponds.

Appendix D: The Combined Regression Estimator

Estimation of the Total

The combined regression estimator utilizes both ground data and classified Landsat pixels. Since the entire state of Kansas was not covered by Landsat passes in one date, it was necessary to work with analysis districts' poststrata which were wholly contained within a Landsat pass. In this study, the analysis districts were collections of counties wholly contained in a Landsat pass. The estimate for the state total, Y, is

$$\hat{Y}_R = \sum_{i=1}^{P} \hat{Y}_{Ri}$$

where P = number of passes required to cover the state

$\hat{Y}_{Ri}$ = combined regression estimate for the ith poststratum

$$= \sum_{h=1}^{L_i} N_{ih} \bar{y}_{ih}(\text{reg})$$

$\bar{y}_{ih}(\text{reg}) = \bar{y}_{ih} + \hat{b}_i(\bar{X}_{ih} - \bar{x}_{ih})$

$\bar{y}_{ih}$ = average wheat acres per sample unit from the ground survey for the hth land use stratum within the ith poststratum.

$\hat{b}_i$ = estimated regression coefficient when regressing ground-reported acres on classified pixels for the combined data contained in the n_i sample units within the ith poststratum

$\bar{X}_{ih}$ = average number of pixels classified as wheat per population segment in the hth land use stratum within the ith poststratum

$$= \frac{\displaystyle\sum_{h=1}^{L_i} \frac{W_{ih}(1 - f_{ih})}{n_{ih}(n_{ih} - 1)} \sum_{j=1}^{n_{ih}} (y_{ihj} - \bar{y}_{ih})(x_{ihj} - \bar{x}_{ih})}{\displaystyle\sum_{h=1}^{L_i} \frac{W_{ih}^2(1 - f_{ih})}{n_{ih}(n_{ih} - 1)} \sum_{j=1}^{n_{ih}} (x_{ijh} - \bar{x}_{ih})^2}$$

y_{ihj} = total wheat acres in the jth sample segment of the hth stratum within the ith poststratum

x_{ihj} = number of pixels classified as wheat in the jth sample segment of the hth stratum within the ith poststratum

x_{ih} = average number of pixels of wheat per sample segment in the hth landuse stratum within the ith poststratum

L_i = number of strata in the ith poststratum

n_{ih} = number of sampled segments in the hth land-use stratum in the ith post-stratum

$$W_{ih}^2 = \frac{N_{ih}}{N_i} = \text{number of population segments in the } h\text{th land-use stra-}$$

tum in the ith post-stratum divided by the number of population segments in the ith post-stratum

$$f_{ih} = n_{ih}/N_{ih}$$

The estimated variance of $\hat{Y}_R$ is given by

$$V(Y_R) = \sum_{i=1}^{P} V(Y_{Ri})$$

$$= \sum_{i=1}^{P} \sum_{h=1}^{L_i} N_{ih}^2 V(\bar{y}_{ih}(\text{reg}))$$

where

$$V(\bar{y}_{ih}(\text{reg})) = \sum_{h=1}^{L_i} \frac{W_{ih}^2(1 - f_{ih})}{N_{hi}} (s_{iyh} + 2b_i s_{iyxh} + b_i^2 s_{ixh}^2)$$

with

$$s_{ihy}^2 = \frac{\sum_2 (y_{ihj} - y_{ih})^2}{j - 1}$$

$$s_{iyxh} = \frac{\sum_i (y_{ihj} - \bar{y}_{ih})(x_{ihj} - \bar{x}_{ih})}{j - 1}$$

$$s_{ixh}^2 = \frac{\sum_j (x_{ihj} - \bar{x}_{ih})^2}{j - 1}$$

County Estimates

Let

$N_{ih,c}$ = total number of segments in the hth stratum of the cth county within the ith poststratum

$\bar{X}_{ih,c}$ = total number of pixels in the cth county classified as wheat for the hth stratum in the ith poststratum divided by N_{ih}

Then the estimate based on the combined regression estimator is the total wheat acreage for the cth county is

$$\hat{Y}_{\text{reg},c} = \sum_{h=1}^{L_{i,c}} N_{ih}[y_{ih} + \hat{b}_i(\bar{X}_{ih,c} - \bar{x}_{ih})]$$

where $L_{i,c}$ = number of strata in county c within the ith poststratum.

The estimated variance of $\hat{Y}_{\text{reg},c}$ is

$$V(\hat{Y}_{\text{reg},c}) = \sum_{h=1}^{L_{i,c}} N_{ih}^2 \left\{ W_{ih,c}^2 + \frac{W_{ih}^2}{n_{ih}} + \left[\sum_{k=1}^{L_i} \frac{(d_{ih}^2 s_{ixh}^2/n_{ih} - 1)}{(\sum_{k=1}^{L_i} d_{ih} s_{ixh}^2)^2} \right] (W_{ih,c} \bar{X}_{ih,c} - W_{ih} \bar{x}_{ih})^2 \right\}$$

where $d_{ih} = \dfrac{W_{ih}^2}{n_{ih}}(1 - f_{ih})$

$w_{ih,c} = \dfrac{W_{ih,c}}{N_{i,c}}$

$N_{i,c}$ = number of segments in county c in the ith poststratum

$x_{ih,c}$ = mean of classified pixels classified as wheat per segment in the hth stratum in county c of the ith poststratum

The variance formula given above for the county estimator is derived by treating the part of county c in stratum h as a single segment.

Appendix E: Kansas 1976 Winter Wheat County Estimates*

County	Pass	SSO Estimate	Combined Regression Estimated	SD	CV	"Ratio" Estimated	SD	CV
Cheyenne	2	65.96	72.72	8.91	12.25	76.76	17.36	22.92
Decatur	2	45.73	88.75	6.03	6.79	96.92	32.05	33.07
Grant	2	35.61	23.88	10.12	42.37	24.32	4.13	16.97
Greeley	2	89.84	48.51	15.74	22.98	61.39	11.09	18.06
Hamilton	2	91.05	24.40	15.15	62.07	26.59	4.13	15.52
Haskell	2	48.16	42.53	12.63	29.69	38.04	7.29	19.15
Kearny	2	62.73	47.35	11.39	24.07	44.03	6.84	15.53
Logan	2	60.70	72.32	10.15	14.03	73.01	14.23	19.49
Morton	2	40.47	11.53	10.01	86.79	14.12	1.93	13.70
Rawlins	2	68.80	85.11	9.11	10.70	94.17	27.20	28.88
Scott	2	54.63	60.06	11.74	19.54	55.36	9.19	16.60
Sherman	2	73.65	74.58	17.56	23.55	68.11	11.45	16.82
Stanton	2	70.82	22.38	13.35	59.67	23.07	4.05	17.54
Stevens	2	34.80	32.42	11.66	35.96	32.01	5.34	16.69
Thomas	2	93.89	99.88	19.83	19.85	90.00	15.22	16.91
Wallace	2	44.52	41.52	6.83	16.44	41.36	6.91	16.70
Wichita	2	54.23	44.96	12.67	28.17	44.39	8.13	18.32
Clark	3	41.68	58.84	10.16	17.27	61.59	9.08	14.74
Ellis	3	53.82	37.27	9.61	25.78	41.16	5.01	12.17
Finney	3	79.72	64.51	20.37	31.58	66.25	8.97	13.55
Ford	3	94.29	99.43	19.34	19.46	106.92	14.37	13.44
Gove	3	52.61	46.30	16.67	36.01	56.25	7.20	12.81
Graham	3	46.13	47.63	14.93	31.35	58.27	8.62	14.79
Gray	3	67.99	50.18	17.85	35.57	52.85	8.22	15.55
Hodgman	3	57.06	50.26	10.28	20.46	57.63	7.29	12.65
Lane	3	52.20	40.19	8.70	21.66	48.81	7.12	14.60
Meade	3	73.25	47.71	12.61	26.43	48.24	5.88	12.19
Ness	3	87.01	57.06	14.09	24.69	67.70	8.87	13.10
Norton	3	46.13	88.22	12.71	14.40	71.83	9.75	13.58
Phillips	3	39.25	61.16	14.29	23.37	71.87	11.74	16.34

County	Pass	SSO Estimate	Combined Regression			"Ratio"		
			Estimated	SD	CV	Estimated	SD	CV
Rooks	3	55.85	50.67	10.49	20.71	58.92	7.94	13.48
Rush	3	76.49	61.39	10.52	17.14	69.97	8.94	12.78
Seward	3	33.59	30.59	8.72	28.51	29.22	4.05	13.87
Sheridan	3	47.35	45.08	13.68	30.34	50.87	6.56	12.89
Smith	3	49.37	72.20	12.51	17.32	84.82	11.21	13.22
Trego	3	52.61	40.71	11.09	27.25	48.44	6.48	13.38
Barton	4	97.53	92.92	12.07	12.99	113.19	9.15	8.09
Edwards	4	50.99	42.53	8.59	20.20	50.79	3.55	6.99
Ellsworth	4	53.82	50.30	4.08	8.12	60.50	2.99	4.95
Kowa	4	50.18	36.14	6.89	19.06	42.37	2.25	5.32
Pawnee	4	70.01	68.23	16.15	23.67	76.93	9.80	12.73
Partt	4	87.01	68.72	10.25	14.91	80.90	7.46	9.22
Stafford	4	84.98	72.44	12.88	17.77	83.04	9.48	11.41
Butler	5	49.78	49.21	12.75	25.91	36.50	3.01	8.25
Chase	5	11.33	20.76	3.20	15.41	18.74	2.10	11.22
Clay	5	43.71	49.49	10.16	20.53	53.86	5.75	10.67
Cowley	5	65.56	58.52	10.28	17.57	53.42	4.44	8.32
Dickinson	5	78.32	77.62	14.85	19.13	91.22	9.39	10.29
Geary	5	12.55	16.39	3.79	23.10	11.37	1.17	10.25
Harper	5	120.19	70.01	15.10	21.56	79.52	7.53	9.47
Harvey	5	60.30	45.16	10.64	23.57	53.58	7.69	14.35
Marion	5	71.63	63.54	12.43	19.59	69.57	7.41	10.66
Marshall	5	44.92	68.15	12.88	18.90	62.77	5.38	8.57
McPherson	5	104.81	80.82	15.30	18.93	95.95	13.80	14.38
Morris	5	25.90	31.65	9.37	29.59	25.09	2.49	9.94
Nemaha	5	24.69	53.94	10.52	19.51	50.59	4.86	9.60
Ottawa	5	69.20	66.09	7.36	11.14	67.70	6.76	9.98
Pottawatomie	5	18.21	33.63	6.01	17.87	26.35	2.23	8.48
Riley	5	14.16	22.54	4.68	20.78	19.59	1.98	10.09
Saline	5	72.03	60.34	9.48	15.72	73.17	7.55	10.31
Wabaunsee	5	12.95	26.79	4.12	15.37	22.62	2.18	9.62
Washington	5	44.52	52.93	17.56	33.18	48.76	4.86	9.96
Allen	6	16.59	13.23	1.58	11.93	9.11	2.75	30.23
Anderson	6	19.02	21.41	2.15	10.03	18.25	5.75	31.49
Atchison	6	10.52	8.22	1.30	15.77	6.88	2.43	35.30
Brown	6	20.23	13.44	2.63	19.58	3.97	1.62	40.83
Chautauqua	6	10.52	11.21	1.97	17.62	10.52	1.79	17.00
Coffey	6	18.21	19.67	3.48	17.70	13.72	4.25	30.98
Doniphon	6	8.50	33.14	2.02	6.11	21.04	6.64	31.54
Douglas	6	12.95	18.21	1.74	9.56	14.00	4.37	31.22
Elk	6	8.90	10.12	2.28	22.57	9.87	1.63	16.52
Franklin	6	16.19	19.55	1.82	9.32	14.41	4.78	33.15
Greenwood	6	8.50	16.11	3.44	21.37	16.19	2.57	15.88
Jackson	6	14.97	7.73	3.00	38.76	7.69	2.27	29.49
Jefferson	6	11.74	11.90	2.19	18.38	7.85	2.19	27.85
Johnson	6	7.69	18.13	1.47	8.11	15.22	4.74	31.15
Labette	6	32.78	17.85	2.27	12.71	14.33	4.29	29.95

County	Pass	SSO Estimate	Combined Regression			"Ratio"		
			Estimated	SD	CV	Estimated	SD	CV
Leavenworth	6	8.09	12.10	1.42	11.75	8.62	2.55	29.61
Lyon	6	24.28	23.55	2.43	10.30	16.07	3.02	18.81
Miami	6	13.35	21.41	2.02	9.45	14.20	4.05	28.49
Montgomery	6	24.28	18.01	3.12	17.33	15.58	4.41	28.33
Neosho	6	18.21	19.34	2.15	11.10	15.94	5.06	31.73
Osage	6	21.45	18.05	3.70	20.48	15.38	4.67	30.34
Shawnee	6	17.81	16.75	1.90	11.32	9.96	2.34	23.46
Wilson	6	24.28	22.14	2.00	9.02	15.05	4.13	27.47
Woodson	6	12.14	10.81	2.12	19.62	8.90	1.96	22.05
Wyandotte	6	1.21	2.63	0.36	13.76	1.54	0.46	29.77

*In thousands of hectares, CV in percent.

3.4 MARITIME

Maritime applications are not as mature as applications in other areas of endeavor. They have lagged but now seem to be emerging. Seasat's early failure and the cancellation of NOSS contributed to this delay. This section begins by identifying the observation requirements. Then the operational products are identified and the experimental applications of the Coastal Zone Color Scanner are described.

3.4.1 Civil Oceanic Remote Sensing Needs*

Accurate analyses and predictions of oceanic and coastal conditions are important for ever-increasing maritime activities. The requirements and needs of the civil maritime community for satellite-derived information have been evolving for over a dozen years. The first formal documentation occurred in 1973.[1] These needs are defined in detail in the original plan for the National Oceanic Satellite System (NOSS). In spite of the demise of this system, the civil and military needs which gave rise to NOSS remain largely unfulfilled and have focused attention of the civil marine community on alternative approaches.

The scope of the civil marine data requirements include the following 14 ocean parameters:[2]

Precipitable water	Surface currents
Precipitation rate	Tides — astronomical and storm
Surface winds	Salinity at the surface
Surface air temperature	Chlorophyll concentration
Waves — directional energy spectra	Water turbidity
Sea ice conditions	Shallow water bathymetry
Sea surface temperature	Oil spills

*Based on J. W. Sherman III, *Civil oceanic remote sensing needs*, International Society of Optical Engineers, vol. 481, 1984.

In addition to these requirements, NOAA is developing another set of remote sensing requirements for estuaries.

Within the 14 ocean parameters, the highest-priority requirements, in decreasing order, are surface winds, sea surface temperature, and waves (directional energy spectra). With regard to these three requirements, the military and civil priorities are the same. The next three civil priorities are for sea ice, ocean color related (chlorophyll and turbidity), and currents and circulation. There is no consensus within the civil marine community on the priorities of the second grouping of marine parameters; most civil users agree those parameters which contribute to the loss of life and property at sea demand the highest ranking.

REFERENCES

1. Apel, J. R., and J. W. Sherman III, *Monitoring the seas from space: NOAA's require-ments for oceanographic satellite data*, Report AOML-LORS 6.73.1, NOAA/DOC, Miami, Fla., June 1973.

2. Sherman, J. W. III, ed., *NOAA Workshop on Oceanic Remote Sensing*, vol. I, *Action summary report*, vol. II, *Workshop support documentation*, NOAA/DOC, Washington, D.C., vol. I—December 1979, vol. II—March 1980.

3.4.2 Civil Oceanic Remote Sensing Operational Products*

NOAA, through the National Environmental Satellite, Data, and Information Service (NESDIS), operates two classes of satellites: two in polar orbits and two in geostationary orbits. The important operational instrument for the oceanographic community is the Advanced Very High Resolution Radiometer (AVHRR) on the polar-orbiting NOAA series for which NOAA-7 and -8 are presently used. The AVHRR operates with detectors in the visible and infrared portions of the spectrum. A Data Collection System (DCS) provides the relay of in situ data. The Geostationary Operational Environmental Satellite (GOES) system has proven invaluable for the monitoring of severe weather conditions used to provide marine warnings.

The AVHRR system provides information on sea surface temperature, currents and gradients, and sea and lake ice. Because of the constraints of clouds on visible and infrared radiation, the AVHRR is limited to oceanic data collection in cloud-free areas. The AVHRR collects data at a basic resolution of 1.1 km and an altitude of 835–870 km. AVHRR images along with computer-generated global sea surface temperature products using a Multi-Channel Sea Surface Temperature (MCSST) algorithm are routine oceanic remote sensing products provided by NOAA. The MCSST products take two basic forms: globally mapped data on 100-

*Based on J. W. Sherman III, *Civil Oceanic Remote Sensing Needs*, International Society of Optical Engineers, vol. 481 (1984).

km grids derived from global area coverage (GAC) collected daily at 4-km resolution, and local area coverage (LAC) data derived from the 1.1-km resolution data collected on a regional basis.

3.4.3 CZCS Applications*

The practical applications to the maritime community of digital image processing of remotely sensed data are perhaps best illustrated by the Coastal Zone Color Scanner. The CZCS, launched in October 1978 on the Nimbus-7 spacecraft, was orbited as a research instrument with an expected lifetime of 1 year. Since the Nimbus-7 passed several years in operation with the CZCS functioning, attempts have been made to use the instrument in a quasi-operational mode to evaluate its potential for full operational utilization from one of the NOAA series of satellites. Although hampered by the lack of a specific data-processing system for rapid turnaround, several demonstrations have been carried out in which data was processed within several days of acquisition and used in a near-operational mode to guide scientific vessels to areas of particular interest and to guide fishermen to areas of optimum probability for catching fish. The unexpectedly long life of the CZCS has also allowed for the analysis of the degradation of such an instrument in orbit over a long period of time and for the development of techniques to recalibrate the instrument in flight so that it may be used continuously throughout its lifetime. These lessons provide valuable insight into how a future operational CZCS should be built, including a full-system on-board calibration for optimum use of such an instrument over a long lifetime in space.

The CZCS is a multichannel scanning instrument using a conventional rotating 45° mirror for scan perpendicular to the spacecraft track, a Cassegrain telescope to image the ocean surface onto a field stop, and a small spectrometer to disperse the energy from the field stop into the desired spectral bands and onto the appropriate detectors. The unique feature of the CZCS is that the scan mirror can be tilted up to 20° forward or behind the spacecraft to avoid sun glint. The dynamic range of the channels of the CZCS is also adjusted for use over water, making maximum use of the 8-bit telemetry for study of ocean color. This unfortunately leads to a saturation condition in several of the bands over land, especially during the summer months.

Because Nimbus is a nonoperational satellite and has a large complement of experiments vying for power, the CZCS can run only on request and only at a maximum of 2 hours per day of operation. Operation can be programmed to anywhere in the world and data recorded by on-board tape recorders. Such things as the mirror tilt and the gain of the various amplifiers can also be programmed for optimum data collection over the area of interest. Both mirror tilt and gain are determined by calculating the solar elevation angle for the scene area and selecting the best values for that circumstance.

---

*Based on Warren A. Hovis, *Practical applications of Nimbus-7 Coastal Zone Color Scanner data,* International Society of Optical Engineers, vol. 481, 1984.

The principal problem encountered in analyzing the data from the CZCS was the calculation of the atmospheric interference. Both the Rayleigh component of the atmosphere and the aerosol scattering must be removed to a high degree of accuracy before such things as the pigment concentration and the diffuse attenuation coefficient of the water can be determined.

The most notable practical use of the CZCS has been the West Coast Fisheries Demonstration Project, wherein the data are collected by the Scripps Institution of Oceanography in La Jolla, California, utilizing their antenna facility. It is then processed overnight into a contour drawing that can be sent by telefax to Ocean Data Services Incorporated in Monterey, California. The data are then made available to fishermen on shore or is radioed to fishermen at sea who have telefax systems on-board. The NOAA Southwest Fisheries Research Laboratory has undertaken a study that compares fish catch of tuna with the location of water mass boundaries. The study has shown quite clearly that the fishermen who fish along the edge of the rich water but remain in the reasonably clear water catch the largest number of fish per day. Prior to this study it was assumed that the tuna stayed in the clearer water because it was warmer by approximately 2°C than the rich water in the upwelling areas. It was assumed that the tuna had a physiological aversion to cooler temperatures and hence stayed out of the rich water. Studies conducted by Dr. Michael Laurs of the NOAA Southwest Fisheries Service in which tuna were captured and fitted with transponders have shown that the tuna throughout the period of a day sound up and down up to 100 m in depth with temperature changes up to 12°C. Obviously, with such large temperature changes being counted many times throughout the day, the tuna do not seem to care much about a 2° change. One possible explanation for this behavior is that the tuna hunt by sight and cannot see the prey in turbid water. By staying on the edge of the turbid water, tuna can look upward from the lower depths and then swim upward to catch the smaller as they leave the turbid water where the bait fish, on which the smaller fish feed, are located.

Another study, conducted in the same area, was used to determine the location of anchovy eggs during the spawning season off the West Coast. Anchovy eggs were found almost entirely in the rich upwelling waters and in waters with surface temperature measured by the NOAA AVHRR exceeding 14°C. Since the anchovy provide the bait fish upon which most of the larger fish depend for their food, the determination of the area of water hospitable to these eggs is quite important in estimating the recruitment rate of the larger fish along the west coast of the Americas. The satellite provides large area coverage and coverage of areas that cannot be penetrated by U.S. Fisheries vessels for political reasons. This allows the Fisheries personnel to estimate the total amount of hospitable area and hence the total amount of anchovy hatch each year.

Another application, with somewhat negative results, was an investigation conducted by the NOAA Atlantic Oceanographic and Meteorological Laboratory (AOML), specifically Dr. Peter Ortner of that facility. An expedition had been planned to the west coast of South America off Peru to search for upwellings and the ship's time had been allocated. When the ship went to South America in 1983, the satellite was reporting no upwellings in the area due to the well-known El Niño that was occurring at the time. The ship upon arrival off the coast of South America verified that there were no upwellings, confirming the satellite measurements. When operational data become available from instruments such as the CZCS, it is clear that the planning of ship location for research can be optimized to utilize the ship time in the most profitable manner by locating the best area for research.

Personnel of the Southeastern Fisheries Laboratory of the National Marine Fisheries Service, located in Bay St. Louis and headed by Dr. Andrew Kemmerer, have used the CZCS data to determine the area of hypoxic water that has been found off the Gulf Coast of the United States to the west of the Mississippi mouth. Shrimp fishermen in the Gulf of Mexico were experiencing a severe degradation in 1982 in their fish catch. Ships from the Southeastern Fisheries Laboratory went into the area in question and found that there was an area of hypoxic water with 2 ppm or less of dissolved oxygen through which the shrimp could not pass on the way from the inshore breeding ground out to the Gulf of Mexico. The CZCS cannot measure dissolved oxygen, but it could easily determine the location and size of the phytoplankton bloom that had absorbed all the oxygen and caused the hypoxic area. Ship measurements in the area of hypoxia agreed in every case with the satellite measurements showing a large phytoplankton bloom. Since inshore fishing is controlled out to the 3-mile limit by the states, we have a limited period of time for shrimp fishing. Such information can be used by the states when making decisions as to whether to extend the time of shrimp fishing because of such an event. In 1982 the total shrimp catch was down 30% because of this phenomenon. It should also be noted that when the dissolved oxygen drops below 2 ppm, the shrimp caught in the area die. Since the area in question was approximately 5000 mi^2, considerable damage was done to the shrimp catch in the Gulf of Mexico. The personnel of the Southeastern Fisheries Service were prepared in 1983 to carry out the same type of measurement and make recommendations to the states concerning extension of the shrimp fishing season. However, in 1983 fortunately very little hypoxia developed. This again was confirmed by the spacecraft observations as well as the ship measurements.

Another practical use of the CZCS data in the Gulf of Mexico was carried out in conjunction with the AOML of NOAA located in Miami and personnel located aboard a NOAA research vessel off the Mississippi mouth. Data were acquired by NASA's Goddard Space Flight Center in their tracking station and provided to NOAA/NESDIS on the same day. The data were processed in the NESDIS interactive imaging system and sent by overnight courier to oceanographers in Miami, who interpreted the data and radioed directions to the ship as to where to locate the various mass boundaries. The pattern of water mass boundaries off the Mississippi is quite complex, quite changeable, and very difficult to observe from the height of a ship. This near-real-time direction provided to the ships considerably increased the efficiency of the operation and allowed personnel to find the types of water they desired without steaming around over large areas, wasting both fuel and time.

Data from the CZCS have been used in the United States' submission to the World Court concerning the dispute over the location of the economic zone boundary in the Gulf of Maine. Canada claimed a boundary line roughly halfway between Cape Cod and Nova Scotia that would give Canada jurisdiction over approximately one-third of Georges Bank. The United States argued that the proper line is the clear water mass distinction between the waters of Georges Bank and that of the Scotian Shelf. The CZCS data have been prepared for each month of the year showing that the water mass boundary remains valid throughout the year and that there is indeed a clear discrimination between the waters adjacent to the United States and those adjacent to Canada, with a clear demarcation line to the east of Georges Bank.

An instrument such as the Coastal Zone Color Scanner, designed for research, can still have many practical applications. It would be far more usable for such applications if it were operational in nature with near-real-time data processing. Attempts to make an operational sensor out of a research sensor are inherently inefficient, expensive, and limited in scope because of the nature of the spacecraft and the limitations due to power and transmission capability. Practical applications of the CZCS data will continue both in the retrospective and in the near-real-time mode. The CZCS has carried out its original mission of demonstrating that measurement of ocean color can be carried out from space.

Appendix:

Landsat Pixel Printout

With the advent of the personal computer the interest in actually trying some image-processing algorithms on real image data has grown considerably. Unfortunately, digital satellite imagery is not available on floppy disks or other media readable by personal computers. This appendix seeks to remedy that by listing the pixel values of part of a Landsat frame. A window from band 5 of Landsat frame 5137-17505 acquired 3 September 1975 is presented. The scene is Woodside, California and has coordinates 37°27′N, 121°55′W. The presentation is arranged as 180 rows of 180 pixels each. The northwest corner pixel is pixel 1312 on line 1255 of the full Landsat MSS frame. The southeast pixel is pixel 1491 of line 1434. The USGS map of this area is reproduced at reduced scale on the following page.

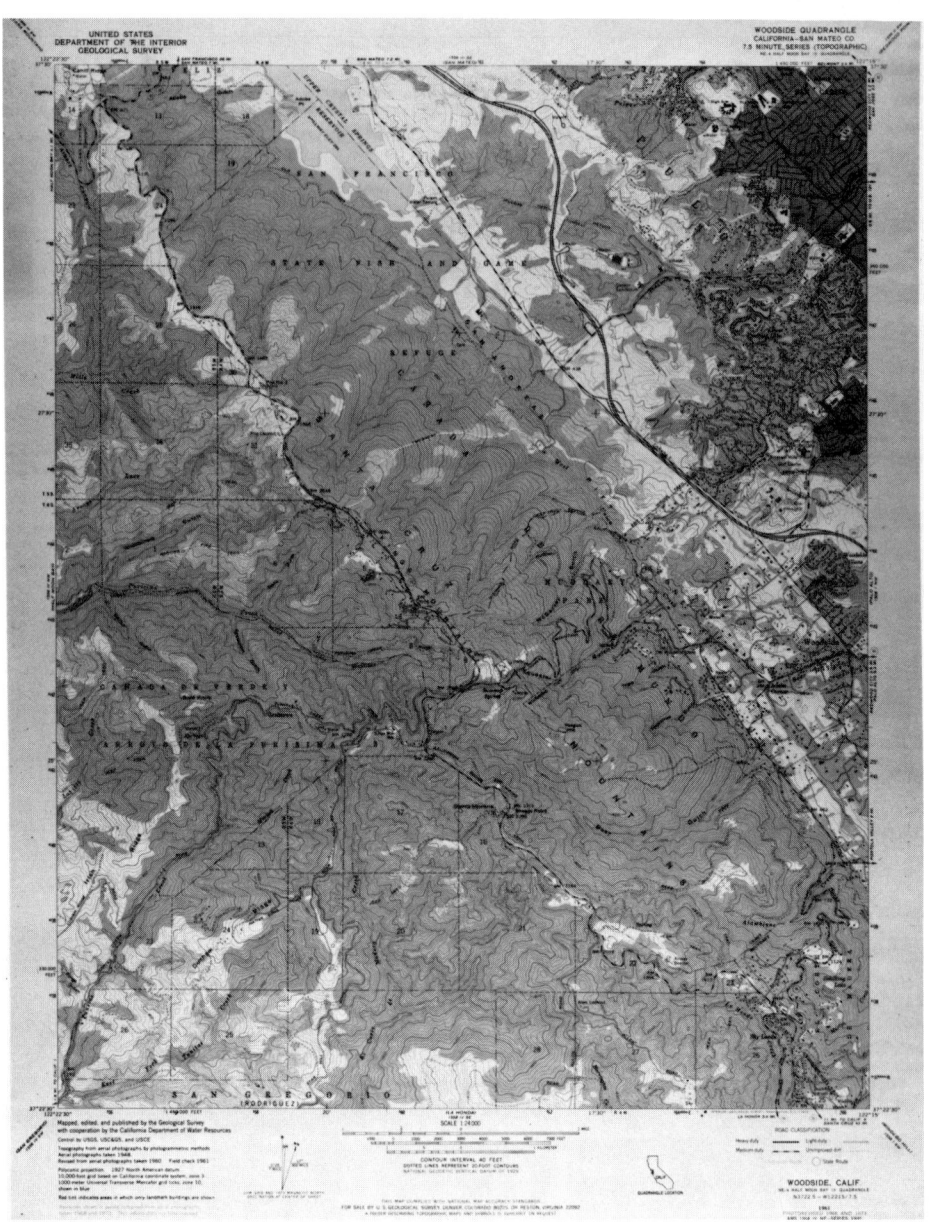

328

```
19 15 14 15 18 24 24 25 28 25 19 15 15 13 14 14 14 14 13 14
15 16 16 17 17 16 17 15 16 15 14 14 15 21 22 24 20 19 19 19
20 16 16 19 19 18 10 10 10 10 10  8 10 10 10  9  9 21 34 35
31 23 22 21 27 36 36 30 30 31 31 33 36 31 31 32 37 37 34 32
32 32 24 25 25 25 28 24 23 28 28 24 24 24 21 23 27 27 29 35
34 34 27 43 47 45 33 21 21 15 21 19 18 16 16 16 18 19 18 18
19 18 18 19 29 25 19 18 18 15 15 24 19 17 19 23 23 29 31 29
31 37 37.31 31 30 27 28 27 20 21 22 21 19 23 23 24 23 20 22
25 22 22 22 18 22 19 24 28 29 36 37 28 25 29 28 25 28 31 29

20 16 13 16 19 24 24 23 24 23 18 14 15 15 15 15 15 15 14 14
14 16 14 16 17 20 16 16 17 13 13 17 20 19 21 23 21 23 19 19
20 24 24 23 19 20 24 10  8  3 10  8  8  8 10 10 10 11 11 23
31 27 27 23 23 31 40 35 28 26 27 33 33 33 31 33 33 31 33 31
31 32 32 34 32 31 31 29 24 24 23 21 19 15 18 21 24 31 31 28
29 29 26 29 34 37 27 27 26 15 16 16 16 14 16 20 16 16 16 15
16 18 19 19 26 26 18 16 16 16 18 16 15 16 18 24 29 33 33 31
33 33 28 29 29 26 26 27 22 21 21 21 22 17 20 20 20 21 24 21
23 24 21 19 19 20 26 25 26 34 44 31 29 31 28 25 31 29 36 33

19 25 23 18 18 22 25 26 26 24 22 16 14 14 14 15 14 15 15 16
16 16 15 15 15 12 20 20 21 17 12 12 20 20 22 22 22 22 22 22
21 21 21 21 23 23 21 22 16 13 11 10 10 10 10  8  8 10 10 11
24 24 25 31 29 25 19 17 34 31 27 22 23 23 34 33 33 30 30 30
27 33 31 33 31 31 31 33 26 26 19 15 16 18 20 21 25 28 25 25
31 31 25 23 24 28 31 31 36 20 17 14 16 16 16 17 20 17 15 14
15 14 16 21 25 28 23 18 16 18 18 16 16 18 26 37 31 28 31 28
24 25 28 28 25 23 25 23 21 23 21 22 21 17 21 23 23 21 22 27
26 23 26 21 19 23 25 25 29 29 36 37 34 31 26 29 32 31 31 31

15 25 27 25 23 23 23 24 25 25 23 22 13 15 13 13 15 16 16 14
15 15 16 15 14 11 12 15 16 14 11 11 15 21 17 17 21 21 21 22
21 21 22 22 22 22 21 23 23 16 15 11  8 10 10  9  9 11 10 10
 9 37 36 31 31 24 23 24 19 24 29 28 24 16 31 34 27 21 17 14
16 21 21 22 26 28 28 28 30 33 33 23 19 24 24 25 24 25 25 24
24 25 25 21 21 25 25 25 25 31 23 28 33 36 33 28 21 17 17 16
13 16 16 16 16 17 27 23 21 27 19 16 16 21 29 29 28 29 29 32
25 26 25 25 24 22 22 18 16 18 18 19 19 17 19 25 28 22 22 23
26 27 26 27 21 22 23 27 33 30 28 25 23 28 25 28 29 28 29 32

13 22 28 23 25 23 23 23 25 23 22 16 15 18 18 18 16 18 15 15
15 16 15 13 12 11 14 14 11 11 10 12 21 21 21 23 24 23 19 22
21 20 21 21 20 20 17 20 21 17 15 13 10  8  8 10 10  8  8 10
11 11 12 18 22 22 24 25 24 18 21 25 25 16 17 17 19 17 17 17
17 16 16 20 21 22 23 34 43 33 30 28 26 23 24 24 25 24 24 24
23 25 20 22 26 25 25 24 31 45 45 42 50 45 40 29 24 17 15 14
15 21 17 16 21 34 37 29 22 28 28 28 26 28 30 27 33 31 29 29
23 23 24 25 20 16 18 22 29 32 23 25 25 22 22 28 24 23 31 31
28 24 24 26 23 23 23 23 26 27 28 26 26 26 26 26 27 28 29 29

15 19 29 31 22 26 27 26 22 23 20 20 20 17 16 15 15 16 18 18
18 18 19 16 12 12 11 12 13 11 11 13 18 19 19 18 18 16 16 16
15 16 18 14 14 14 20 13 13 17 20 16 16 10 10  9 10 10 10 10
10 10 10 10 10 10 10 11 18 19 24 19 20 25 16 15 18 18 16 15 18
17 19 17 16 16 17 21 23 26 34 31 34 31 31 27 26 26 27 26 23
22 21 18 23 25 21 21 21 24 25 29 37 41 41 41 42 33 16 16 18
24 42 23 29 45 40 25 19 19 20 20 21 22 34 35 31 28 27 26 27
26 21 22 23 16 19 29 33 36 33 25 25 24 24 29 26 31 31 33 24
29 29 29 25 28 21 21 23 25 29 29 27 26 27 29 27 23 28 28 30

18 19 19 23 31 31 29 25 24 25 24 20 17 17 20 14 14 16 16 20
22 23 20 16 14 12 15 16 15 12 11 12 15 13 15 15 13 13 13 16
15 14 12 12 12 11 14 12 14 18 21 19 16 14 11  9  9  9 11 11
```

```
10 10  9 10 10 10 11 19 23 23 24 24 23 22 19 19 20 22 24 22
23 23 23 16 16 19 23 23 25 29 25 28 29 29 31 29 27 27 29 27
26 21 21 26 27 21 21 20 25 25 25 25 25 29 36 34 32 34 37 39
44 34 45 40 33 19 19 19 18 19 21 28 25 29 24 23 23 22 21 21
22 20 21 23 21 27 33 38 34 26 29 29 31 28 29 31 41 47 31 29
32 32 31 28 24 28 28 23 23 25 31 29 31 31 29 28 25 26 31 29

18 16 16 28 29 25 28 25 24 24 21 24 29 25 21 21 19 14 16 17
16 16 16 16 14 14 13 11 11 13 11 12 11 12 11 12 12 14 13 13
13 13 15 16 16 12 12 14 16 19 23 19 15 18 16 11 10 10 11  9
 9  9  9  9  9  9 23 23 17 23 26 26 21 18 19 19 23 25 21 22
24 22 22 24 24 24 23 21 28 31 29 28 24 31 31 31 31 33 29 31
26 26 23 17 16 15 14 20 20 23 20 20 21 22 37 33 36 42 48 37
21 24 19 19 16 19 20 18 18 19 23 25 29 24 21 23 21 19 21 19
17 24 22 23 23 26 27 27 27 27 30 30 23 30 28 29 28 24 31 42
40 40 26 32 32 26 31 29 31 29 33 31 31 31 28 28 31 31 29 29

15 14 14 25 29 25 24 24 22 20 24 31 21 14 15 16 16 19 21 19
16 15 15 12 12 12 11 12 12 12 11 11 13 11 11 14 13 13 14 14
14 15 16 15 15 16 15 16 19 24 22 20 15 15 15 15 14  8  8 10
10 10 10 10 10 11  9 12 22 22 22 28 26 17 23 23 21 22 23 21
23 24 23 23 23 24 20 19 20 22 24 22 24 23 29 29 29 31 31 29
33 28 24 24 25 25 29 13 17 20 20 17 17 22 30 38 47 49 47 34
22 18 18 16 15 16 14 15 18 18 20 18 18 18 19 21 24 23 16 13
19 23 23 21 21 21 21 23 24 27 27 34 29 31 37 28 30 27 26 28
30 33 37 42 36 33 31 29 31 31 31 31 32 31 29 26 28 29 28 29

15 13 15 16 23 24 22 19 22 23 25 31 24 15 15 15 13 15 15 14
12 12 14 12 10 11 12 14 14 14 11 11 12 12 11 12 12 12 14 15
15 16 17 17 16 18 16 19 18 16 15 14 16 12 13 15 15 13 10  8
10 10 10  8 10 10 11 11 15 16 21 29 31 31 26 21 20 21 22 22
22 23 23 23 26 23 21 18 19 21 23 23 23 26 29 29 26 25 25 25
43 21 15 16 17 16 15 13 15 16 16 16 21 21 16 18 18 20 16 19
25 19 21 19 19 18 16 21 29 23 23 23 23 31 37 43 35 34 31 27
27 31 34 41 43 34 30 30 33 33 31 33 33 31 29 31 31 26 26 32

13 17 17 20 20 20 17 20 21 19 22 29 28 16 14 14 12 12 13 13
13 13 15 10 12 15 14 14 12 11 11 11 10 11 12 12 14 13 14 14
13 14 14 17 16 15 15 15 14 14 15 15 15 18 18 18 22 12  9 10
10  9 10 10 10  8 10 12 21 19 19 28 24 28 25 25 24 23 22 21
20 21 22 26 23 21 21 21 23 26 23 21 23 19 19 25 29 24 21 26
29 25 25 24 24 29 33 31 31 33 31 29 29 25 28 25 28 36 42 45
31 20 20 20 20 14 14 16 20 16 17 21 16 14 18 19 18 19 25 25
18 18 20 20 19 18 24 31 31 29 24 29 31 33 37 40 36 36 36 31
29 31 31 34 29 29 27 34 36 33 33 30 28 23 26 28 29 31 24 25

15 16 17 16 17 20 17 16 16 21 22 23 23 23 22 17 14 14 12 14
15 15 15 13 13 16 16 15 12 11 10 10 11 11 14 14 14 15 14 14
15 14 18 21 16 16 16 14 14 14 13 14 16 17 20 23 23 11  8  8
 8 10 10  8  8  9 10 10  9 13 11 19 23 28 33 28 21 19 23 23
19 23 25 24 24 23 22 21 21 22 22 23 22 21 22 23 30 26 21 24
21 21 19 19 19 18 29 26 31 31 32 32 31 29 24 21 24 29 40 48
42 42 42 37 28 25 31 16 17 16 17 21 21 14 20 21 20 22 23 22
20 19 24 21 18 19 24 23 29 25 26 29 29 31 33 33 31 29 25 28
28 31 29 29 25 25 24 28 34 31 34 31 27 27 26 30 33 30 27 26

16 15 15 13 10 16 16 16 15 21 23 21 21 22 23 23 14 13 13 15
16 16 14 14 10 15 15 15 10 11 11 10 11 13 15 15 13 12 15 15
15 14 12 14 10 16 14 16 18 18 16 14 14 20 22 23 21 13 10 10
10  9 10  9  9 10 10 10  8  8 10 10 16 26 26 25 31 34 21 19
18 23 24 23 21 23 21 23 23 23 23 19 21 21 21 20 26 27 22 22
17 17 20 17 17 21 19 19 25 31 33 31 33 29 22 24 24 26 34 31
```

```
50 45 42 42 45 45 45 37 25 21 21 17 16 14 16 21 20 20 23 22
17 16 22 27 27 28 23 22 25 28 31 31 31 29 31 34 32 31 29 29
29 29 29 29 29 28 28 28 28 31 25 25 25 25 28 22 26 27 26 29

15 14 14 15 10 16 14 14 16 19 21 21 21 21 23 22 21 20 16 14
13 12 15 15 15 14 13 13 14 11 11 12 15 15 12 14 12 13 15 13
12 13 13 14 12 12 15 16 18 16 15 15 19 23 18 18 19 11  9 11
 9  9 11  9 10  9  9 10 11 10 10 12 25 28 23 28 24 25 20 16
16 16 19 24 21 21 23 24 21 19 19 19 19 19 19 24 25 21 21 20
17 16 16 16 14 22 20 20 27 34 34 34 31 24 21 24 25 25 23 31
41 47 49 52 47 44 48 33 16 16 16 16 14 19 24 21 19 24 23 21
21 20 23 26 26 26 29 21 33 41 36 38 38 33 33 31 36 33 33 29
26 26 29 31 29 29 32 31 29 24 21 28 29 29 28 24 25 28 28 28

15 13 16 18 18 15 13 15 15 18 16 14 19 23 18 21 21 21 15 13
15 16 16 14 12 13 14 14 14 15 15 15 11 13 13 13 14 15 14 12
12 14 14 13 13 13 15 15 15 13 15 15 15 15 16 23 23 16 11 10
10 11 10 10  9  9  9  9  9  9 10 10 15 17 20 25 23 22 25 24
23 22 24 28 25 24 22 24 24 20 18 18 18 18 16 18 23 19 21 19
18 16 16 16 15 16 17 14 16 17 16 21 33 33 28 22 27 22 20 23
24 24 28 36 40 45 45 32 18 13 15 18 13 16 24 21 18 23 21 18
19 16 15 16 21 29 33 35 34 34 37 34 23 34 34 34 36 34 33 30
29 25 31 33 33 33 29 29 26 26 26 25 29 25 28 24 25 28 31 29

13 16 15 16 15 14 15 14 15 18 16 19 25 22 19 19 18 15 12 14
15 15 16 13 13 13 16 16 16 16 14 14 13 14 14 12 14 15 14 13
13 14 14 14 14 14 12 12 14 14 13 13 13 15 16 16 22 23 14 10
10  8 12  8 10 10 12 10 10 11  9 11  9 11 13 20 26 25 25 28
33 27 25 23 28 25 24 24 19 18 18 16 18 18 16 16 19 20 16 18
14 14 14 14 15 15 16 15 16 18 18 18 31 27 28 31 31 23 20 22
21 21 22 23 20 27 37 33 31 37 23 15 16 15 24 19 20 24 19 16
15 16 15 15 15 21 31 33 33 36 36 31 33 26 34 31 29 27 29 27
27 26 27 31 30 28 23 28 23 25 29 31 31 25 22 26 26 29 31 31

15 16 16 14 14 14 15 16 18 16 24 28 24 24 20 18 18 19 19 16
15 14 12 14 15 15 15 12 13 13 13 15 15 12 13 13 14 14 12 13
14 14 14 14 15 14 15 14 14 14 12 14 12 12 14 15 22 18 11 11
11 13 18  8 10 11 10 10 10 10 10 13 15 25 25 23 26 26 23 26
26 28 34 21 22 21 22 20 15 15 18 19 18 22 18 19 19 18 13 12
13 12 12 13 18 14 15 16 16 16 16 19 16 18 23 29 31 28 21 20 16
16 22 22 23 23 30 34 36 41 33 23 33 18 25 28 25 23 15 16 19
16 16 18 18 22 32 28 29 29 24 28 29 36 36 36 36 33 33 31 23
26 26 26 29 29 23 26 27 27 28 27 30 27 25 28 29 28 25 28 28

17 16 16 20 20 13 13 20 20 20 23 27 23 25 23 24 23 22 24 19
14 12 13 13 15 15 15 12 10 11 10 10 11 11 12 15 13 12 13 13
13 15 14 14 14 14 13 13 14 14 13 13 13 14 16 16 22 22 15 12
19 23 18 15 11 10  9  9 10  8  8  8 10 16 23 25 24 25 28 23
23 24 25 27 22 20 22 20 17 16 16 16 17 20 16 17 17 16 14 12
12 14 14 16 16 16 16 16 16 19 19 16 18 23 28 29 28 18 18 18
19 23 21 21 23 23 31 35 35 37 35 40 33 27 28 21 17 17 20 18
15 19 29 40 37 28 34 34 34 31 26 25 31 31 28 28 31 33 31 25
29 29 28 23 23 24 23 23 22 26 29 29 27 28 27 28 28 27 28 27

23 15 16 16 10 16 15 13 20 21 26 23 23 26 23 22 23 25 20 14
13 11 12 15 15 18 14 12 12 12 11 11 10 11 12 14 12 14 14 12
12 14 15 13 15 13 13 13 13 14 14 13 12 13 14 16 21 21 17 16
22 27 21 19 24 19 14  8 13 10 10  9 10 10 11 19 25 28 28 24
23 25 25 28 25 23 21 16 15 21 16 21 26 20 14 20 20 15 13 14
15 16 16 16 16 10 16 15 14 16 18 18 16 16 19 24 26 31 25 12
15 16 16 19 20 29 31 28 24 29 33 31 43 45 43 37 35 35 34 17
22 34 30 23 26 27 31 33 31 31 31 33 29 31 31 32 31 26 31 31
29 29 29 21 21 25 25 31 29 31 31 31 29 27 27 27 29 29 27 23
```

```
15 16 15 14 16 15 16 17 21 23 23 24 21 22 22 23 23 23 20 14
13 14 15 15 14 13 13 12 11 10 10 11 11 14 13 15 13 13 13 15
15 14 14 14 12 12 12 11 13 12 12 13 12 12 15 16 21 22 21 22
23 23 20 22 22 23 16 10 10 10  8 10 10 10 11 26 26 26 26 29
31 32 24 24 28 24 18 18 18 29 37 31 25 19 18 23 20 13 12 14
14 14 14 16 15 16 16 16 15 17 16 16 18 18 19 29 31 18 16 15
16 22 25 24 24 24 23 23 29 28 25 40 40 40 37 31 24 23 34 34
34 22 17 20 20 22 27 28 28 34 30 29 28 29 25 28 29 28 29 25
18 22 26 29 31 31 29 31 29 28 29 25 28 29 31 31 31 28 23 27

18 15 15 15 15 16 14 15 18 16 15 15 16 17 17 19 21 19 21 17
12 13 14 16 17 13 12 10 10 11 11 11 13 14 14 14 14 15 14 12
14 15 15 13 13 11 10 10 10 12 12 14 12 12 14 15 13 13 15 15
16 13 20 16 20 22 21 20 10 10 10 10 10 11 13 18 24 18 19 25
28 25 22 25 29 32 26 20 19 23 28 31 31 21 15 19 25 15 12 15
16 16 18 17 14 13 14 16 16 17 17 17 20 15 16 28 33 24 22 23
24 22 19 22 20 20 24 22 19 19 29 40 40 42 37 28 19 18 24 29
29 19 19 19 18 31 28 23 27 27 27 27 33 27 23 22 22 27 21 19
21 23 28 31 29 34 31 31 29 26 26 26 25 28 31 31 28 29 28 24

15 14 15 16 16 16 15 15 15 16 16 19 18 14 16 19 19 21 21 19
15 16 16 15 13 13 15 11 11 11 11 11 11 11 14 14 15 14 14 14
15 14 12 11 10 10 11 12 11 10 11 12 12 13 16 15 16 16 15 14
14 14 15 15 15 17 24 24 20 13 11  9  9 11 12 14 13 13 14 13
16 27 24 25 28 29 31 29 24 18 19 20 20 24 18 13 18 14 15 16
19 19 18 21 19 24 19 16 16 18 17 16 17 16 13 16 31 25 23 25
22 20 17 17 22 22 23 19 19 29 31 39 39 39 34 29 20 18 21 19
25 33 37 29 25 18 28 31 28 28 28 28 22 23 22 21 20 22 28 23
22 23 26 23 27 31 29 31 29 29 28 28 29 29 29 29 29 31 31 25

15 14 14 14 13 18 18 19 16 18 23 23 19 22 22 20 22 20 18 16
16 16 14 12 14 14 12 10 11 11 11 12 13 13 14 14 14 14 14 13
15 13 11 10 10 13 11 11 12 12 12 12 15 14 15 15 13 12 12 13
12 15 15 14 16 23 21 23 23 21 12 11 12 12 13 13 13 13 13 13
17 28 27 25 27 28 30 33 23 19 22 19 22 16 14 15 16 18 18 16
16 18 18 15 16 18 15 16 16 19 19 18 15 15 16 28 31 20 17 22
21 17 17 20 20 17 17 30 36 27 25 36 37 29 24 23 22 18 19 24
26 31 29 24 29 29 33 25 16 18 21 16 19 19 19 19 23 23 26 27
28 31 28 26 23 26 27 28 30 30 26 29 31 31 28 28 29 28 29 29

17 14 13 12 13 14 16 14 15 15 15 15 25 24 22 18 16 18 18 16
16 16 19 18 12 11 11 11 11 12 12 14 14 15 15 13 15 13 13 13
16 14 11 11 12 11 12 13 13 13 13 13 11 13 12 12 12 12 12 14
14 12 12 12 13 16 20 18 23 24 18 12 15 19 15 16 17 16 28 29
23 23 23 23 23 27 23 21 22 17 16 17 17 20 21 16 16 18 16 15
15 15 18 16 15 16 18 16 16 16 18 16 12 15 16 29 33 21 18 23
24 18 18 17 17 20 17 16 23 35 25 34 36 36 36 38 43 19 19 22
25 28 29 33 26 31 26 18 18 19 18 18 18 21 25 28 31 29 29 25
24 28 29 31 26 27 27 27 31 34 27 28 30 33 33 30 28 27 25 25

15 15 14 14 14 12 14 14 14 13 12 12 16 15 13 14 15 14 15 18
18 23 19 11 10 12 11 13 15 13 13 13 12 12 14 15 15 14 15 14
13 11 11 12 11 12 13 12 13 13 12 12 12 13 13 15 14 13 13 11
13 12 12 12 14 15 15 20 16 13 15 15 20 24 22 23 21 23 24 23
21 21 21 25 28 21 13 13 26 26 23 17 20 28 28 22 20 16 16 15
15 17 16 19 19 19 18 18 16 18 18 15 13 15 19 26 33 23 18 23
23 15 16 21 21 18 18 18 21 21 37 34 35 37 37 40 40 34 33 30
33 30 25 21 33 28 16 16 22 24 19 19 18 16 20 24 24 31 31 24
23 23 24 24 33 29 29 28 25 28 28 26 31 27 28 31 28 28 26 26

14 12 15 16 15 15 16 16 14 14 14 13 12 12 12 12 12 13 17 16
15 14 10 15 17 14 15 12 12 14 14 14 13 13 13 13 13 13 11 11
```

```
 8 10 11 11 11 12 13 13 12 13 15 16 16 12 13 14 13 12 12 12
14 13 13 13 14 13 15 14 12 14 14 16 16 20 22 22 20 20 19 18
16 14 15 18 15 15 21 19 21 29 29 25 24 29 20 16 13 13 16 16
17 17 20 16 17 20 20 16 16 15 14 14 12 15 25 32 26 31 32 25
22 22 18 18 16 18 21 16 18 29 37 33 33 28 31 36 40 37 27 31
27 20 20 17 20 22 27 27 27 22 16 16 16 24 23 23 24 26 26 26
20 25 26 28 25 23 24 28 28 24 25 25 24 31 31 25 24 28 22 21

19 20 20 20 16 15 15 14 14 12 12 15 14 12 14 12 14 16 13 14
14 20 22 20 21 15 14 13 13 11 13 14 14 14 15 15 15 12 10 10
11 11 11 12 12 11 11 11 12 12 12 12 14 11 15 12 12 12 12 12
12 13 13 13 14 13 13 14 16 16 13 14 14 15 16 22 22 19 18 18
16 16 18 19 20 25 28 23 24 24 31 31 29 54 37 37 23 21 19 19
17 16 16 16 16 16 16 17 15 14 14 14 14 17 30 36 31 28 18 18
18 18 20 20 20 18 18 18 19 24 31 31 28 25 25 24 31 31 33 33
28 31 37 17 22 27 34 23 26 25 20 17 15 28 34 27 28 29 28 19
22 28 29 29 26 26 29 31 31 31 24 29 28 28 24 24 29 29 25 24

22 19 22 18 18 15 13 12 15 13 12 12 12 11 15 15 16 18 15 15
19 24 24 23 17 17 12 13 12 12 13 13 15 14 14 13 13 10 11 11
11 12 11 11 11 11 11 12 12 13 13 13 11 11 12 12 12 12 12 12
13 15 13 15 13 15 14 16 16 16 16 16 13 15 15 17 20 22 21 20
16 16 16 18 25 25 28 28 31 28 28 32 39 69 91 91 57 25 18 18
15 17 16 15 15 13 13 13 16 13 13 13 14 21 34 27 20 22 20
17 21 19 19 19 16 18 16 15 20 20 29 34 34 31 31 28 37 37 33
29 37 45 36 33 29 31 31 28 28 31 28 28 28 31 28 21 21 20 21
25 25 22 29 29 33 40 37 36 31 29 31 29 29 25 24 26 25 28 23

21 17 16 15 15 12 14 15 15 14 12 13 15 16 16 16 16 18 18 21
18 21 23 19 17 14 14 14 15 15 14 14 14 12 13 12 12 12 11 13
14 13 11 10 11 11 11 11 12 12 11 12 12 13 15 15 13 13 12 12
14 15 14 16 18 16 16 16 15 13 15 15 13 13 13 16 21 21 21 21
20 16 15 20 23 15 23 19 23 29 31 29 33 92 92 73 45 24 24 15
15 14 14 15 15 14 12 13 13 13 15 15 15 17 34 34 28 23 17 17
20 20 16 15 14 14 16 17 16 18 23 25 24 23 25 37 41 39 34 34
37 37 42 42 33 31 37 31 28 29 28 29 25 21 21 28 23 21 22 23
26 21 28 33 30 36 43 36 28 31 36 28 23 23 23 28 26 26 25 24

21 21 20 16 14 11 13 16 16 14 14 15 14 15 12 12 12 15 16 15
18 20 22 19 15 14 15 14 14 14 12 14 12 12 14 15 14 12 12 12
12 12 13 12 11 13 13 13 13 13 13 14 14 15 15 15 14 14 12 13
15 13 13 12 11 15 14 12 12 11 14 14 13 13 12 13 13 15 15 20
17 14 14 16 20 20 14 15 22 25 27 25 23 54 45 31 37 22 22 16
15 15 16 19 20 20 18 12 14 14 14 14 14 18 29 33 30 21 16 17
17 20 16 14 14 14 16 17 16 16 16 20 20 21 23 29 37 37 40 37
37 37 39 39 34 34 32 32 31 29 29 29 23 23 23 18 18 21 25 29
28 31 22 27 28 31 35 31 28 28 28 27 20 20 30 28 25 23 23 24

19 19 17 16 13 12 14 14 13 13 13 13 14 14 13 14 14 14 16 22
22 23 22 16 12 13 13 13 13 13 11 12 11 12 14 12 11 11 12 14
14 14 11 11  9 11 13 13 12 12 13 15 15 15 15 13 11 13 14 15
12 12 14 12 12 12 13 13 13 12 12 12 12 14 12 12 14 14 15 17
17 17 21 21 17 16 16 13 17 26 26 23 28 28 25 23 21 21 25 23
24 23 19 19 18 18 16 16 16 16 16 15 19 28 31 29 21 16 15 14
15 13 16 16 16 17 16 17 16 17 20 17 17 22 23 22 27 33 33 36
36 28 31 36 36 33 28 25 29 34 26 24 31 39 24 24 24 29 33 29
29 36 28 28 23 21 21 28 27 27 27 22 22 21 20 23 28 25 23 30

21 19 17 17 15 11 12 16 15 12 12 13 13 13 13 12 15 16 17 20
22 22 22 19 12 14 15 14 15 15 12 13 15 13 15 12 11 11 12 12
12 11 10 10 11 11 12 14 14 12 15 14 16 14 13 12 11 13 15 14
13 13 14 15 14 15 15 14 14 14 14 12 15 18 15 13 12 12 15 21
21 18 18 15 14 15 15 13 13 16 21 24 26 26 26 26 23 23 26 25
```

```
20 16 20 21 20 16 15 15 15 18 16 15 16 28 32 22 15 15 15 15
14 16 18 18 18 16 16 15 16 19 21 21 24 25 23 20 23 27 34 35
37 36 28 25 36 30 22 23 23 28 19 24 24 24 31 25 22 22 22 26
24 24 21 21 23 24 21 23 36 36 28 19 21 24 24 21 23 27 27 34

15 18 19 18 12 11 11 14 15 14 15 16 14 14 12 14 13 12 14 20
21 22 26 22 16 13 14 14 14 14 12 12 15 15 14 14 11 11 11 10
12 12 12 11 11 10 11 12 12 12 12 16 14 12 14 12 14 15 14 12
12 13 13 13 14 14 15 14 13 14 15 14 14 14 14 15 15 12 13 13
15 16 18 19 15 15 18 16 15 14 15 16 25 23 19 24 30 30 29 26
22 23 26 23 17 17 17 17 17 17 17 20 21 31 36 28 28 29 31 28
15 16 16 16 15 15 16 16 18 19 18 16 18 18 23 19 23 23 28 36
36 37 35 27 26 35 31 17 21 23 23 28 25 23 24 25 24 23 19 19
29 24 26 25 26 26 19 25 29 28 19 19 29 36 28 25 24 25 31 37

18 18 16 13 11 12 15 16 16 21 18 16 18 15 16 17 17 17 19 21
23 19 22 17 17 16 13 13 14 13 13 14 15 14 13 11 11 10 11 11
12 12 10 10 10 13 12 12 13 13 12 11 11 14 14 15 12 14 12 12
14 12 14 14 10 14 14 14 13 16 14 14 14 14 15 15 14 12 14 14
14 15 14 14 15 15 18 16 15 13 15 15 21 19 21 29 31 31 33 25 25
24 23 19 21 23 22 20 17 20 22 23 23 34 36 30 33 33 28 28 24
28 28 25 24 22 23 16 15 16 16 18 19 20 21 21 19 18 23 31 31
36 30 21 23 33 37 26 22 21 22 22 17 21 27 25 28 28 16 22 28
28 28 28 22 18 24 29 37 32 24 25 32 37 32 29 28 29 29 31 33

21 18 15 15 12 14 18 23 16 16 16 16 16 16 16 15 16 15 14 16
16 18 21 21 19 17 16 16 16 13 13 16 14 16 14 12 10 10 11 14
14 11 10 10 11 12 14 14 12 11 12 11 12 15 13 12 11 12 11 12
11 12 15 15 16 16 15 12 14 15 16 14 13 13 13 13 13 14 14 13
13 14 14 14 14 14 14 14 15 14 15 16 16 24 28 28 32 31 25
24 21 19 16 19 19 21 24 25 24 24 25 22 27 34 34 31 23 22 28
22 22 25 28 27 28 24 23 25 28 28 22 19 19 16 18 20 20 24 26
37 28 21 33 36 24 30 24 24 23 17 17 28 28 27 27 31 17 17 27
27 23 21 25 33 33 33 28 29 28 24 31 37 33 37 39 37 37 31 29

23 20 16 15 15 15 17 17 22 16 18 18 16 15 16 15 13 13 13 13
15 15 19 18 16 16 18 15 15 15 16 16 15 15 12 11 11 11 11 13
13 12  9 11 11 13 14 14 13 13 11 12 14 14 12 11 12 12 13 12
12 15 13 15 16 14 12 12 14 14 15 15 14 14 15 15 14 13 13 12
13 13 13 13 13 14 14 14 13 14 14 16 16 22 25 25 33 34 32
28 22 22 22 18 23 24 23 19 19 21 18 24 24 30 33 33 28 19 20
26 31 27 26 28 31 28 33 38 38 30 30 28 22 22 16 18 19 24 23
34 32 32 44 44 39 28 25 24 19 18 29 31 30 28 28 23 19 28 25
27 27 27 31 34 27 31 30 27 23 22 28 34 38 40 51 42 36 36 31

26 22 20 17 14 14 16 16 15 16 20 17 16 16 18 14 12 12 14 15
16 19 19 19 19 18 19 16 15 15 15 14 15 12 12 11 12 14 14 14
11 10 11 11 12 13 13 13 13 11 13 15 14 11 11 13 15 14 11 14
14 15 13 13 13 13 13 15 14 14 12 14 15 14 12 14 14 14 14
14 14 15 13 13 13 13 13 14 14 14 15 27 30 25 22 22 25 33 31 29
25 24 24 23 24 26 26 24 18 19 24 16 21 28 36 36 28 24 19 23
23 25 24 25 25 28 23 22 31 31 26 27 23 22 22 20 21 22 30 19
29 37 37 37 40 28 24 20 19 25 32 32 24 28 28 19 23 24 19 25
29 24 24 25 28 29 34 31 27 22 22 31 40 45 45 34 30 36 28 23

21 23 21 16 15 16 16 17 16 14 16 17 16 16 15 13 15 14 15 15
14 15 16 18 19 18 19 18 13 13 13 15 16 13 11 12 12 14 12 10
10 10 11 12 14 14 14 14 15 13 14 14 13 12 12 12 14 13 11 14
15 14 15 15 14 14 14 14 15 16 16 15 15 12 13 13 12 14 12 12
12 12 12 14 15 14 14 14 15 23 20 28 34 34 26 26 23 28 28 33
33 33 21 21 24 25 28 24 25 25 25 24 25 20 26 34 34 24 21 18
16 16 16 16 15 24 25 21 24 19 19 21 22 20 21 21 21 21 20 27
23 25 33 28 33 36 25 22 22 29 29 25 28 26 25 19 31 34 32 28
```

```
37 29 31 29 26 28 30 29 29 24 23 30 37 40 35 34 31 34 34 26

19 18 18 16 15 15 16 16 17 17 17 15 13 12 13 12 13 13 13 13
13 14 14 14 15 16 16 17 15 14 14 15 15 12 11 12 12 13 12 12
11 11 11 14 14 12 14 14 12 14 15 15 12 12 12 12 13 14 12 13
13 14 15 15 13 13 14 14 15 15 15 14 14 16 16 16 15 12 12 12
13 12 12 12 12 14 12 14 23 36 33 36 31 28 29 29 36 23 28 28
31 34 34 31 27 25 27 27 28 27 28 28 28 24 28 33 36 33 19 16
18 20 18 16 16 16 18 18 18 18 18 18 21 17 19 21 19 17 19 22
16 20 27 28 26 34 33 30 27 27 27 23 31 31 28 31 31 29 29 31
31 28 26 24 25 26 31 28 21 21 18 33 36 30 24 25 30 36 30 25

18 20 18 15 16 18 15 16 15 12 11 11 11 12 13 15 13 13 13 12
13 12 13 13 13 13 14 15 14 14 14 14 13 13 12 14 14 15 14 15
15 13 15 12 15 15 13 12 14 14 11 12 11 12 12 14 14 11 12 14
15 14 13 14 13 13 13 14 14 15 14 15 16 15 13 14 12 12 14 15
16 15 13 15 15 16 16 26 31 42 37 25 31 33 33 25 28 19 25 29
29 29 28 27 27 26 26 28 27 26 27 28 27 23 23 28 34 29 25 24
19 18 22 24 19 15 15 16 16 16 15 21 19 23 25 18 18 16 19 16
16 16 16 17 27 27 26 21 20 16 17 22 22 17 20 17 20 21 28 25
23 22 18 19 22 26 24 19 19 24 31 28 28 21 23 25 28 25 24 28

21 19 21 18 16 15 16 16 12 11 11 12 11 11 14 14 14 14 14 12
12 13 13 13 13 13 12 17 20 14 14 12 13 13 14 14 14 14 16 16
20 15 12 12 15 15 14 14 13 12 11 11 11 11 12 12 14 14 11 12 12
14 15 15 14 12 12 14 15 14 14 14 14 16 14 13 13 13 13 14 15
15 15 15 15 19 24 16 18 23 34 34 28 31 28 28 32 28 19 16 16
23 28 25 28 26 25 25 25 24 25 26 27 23 21 20 17 27 33 30 28
27 23 22 20 18 14 15 16 16 16 18 16 18 20 20 18 16 13 14 18
16 16 18 21 17 19 19 19 19 17 17 20 20 17 17 20 17 17 20 20
20 23 22 20 21 18 19 23 25 28 25 28 24 26 31 28 25 26 28 25

21 20 16 16 14 14 16 14 11 14 12 12 14 12 13 13 13 15 13 12
12 14 14 14 12 15 16 13 17 17 12 12 13 13 11 12 13 14 16 17
16 15 14 14 15 14 15 14 12 11 12 12 14 12 14 12 12 11 12 13
15 12 14 12 11 10 11 12 15 16 16 17 16 14 12 13 13 13 14 16
14 13 15 15 16 22 21 27 22 29 31 37 33 31 31 31 25 13 18 26
26 20 19 23 25 29 25 24 24 23 24 23 19 17 16 16 19 27 31 34
35 28 27 23 15 14 16 17 20 21 20 23 23 18 16 16 15 15 16 19
22 24 24 20 23 19 18 21 21 21 25 19 19 19 17 21 17 16 17 16
16 17 17 16 17 20 17 20 25 25 27 34 25 29 33 29 28 29 29 25

21 20 16 16 16 14 14 13 13 14 14 13 12 12 14 15 14 14 15 13
13 15 16 13 13 13 14 15 18 16 12 14 15 12 13 15 16 17 16 15
16 14 13 14 14 14 13 13 11 13 14 14 14 12 14 15 12 14 15 15
15 12 11 10 11 12 13 15 18 16 14 11 11 12 14 14 15 16 15 14
14 12 13 14 14 13 23 31 17 21 30 34 30 30 28 28 25 18 25 31
29 28 20 24 25 31 26 25 20 18 12 15 18 16 15 21 24 23 28 33
36 31 28 20 21 23 26 26 21 21 22 17 15 15 16 15 16 19 23 23
22 16 15 20 16 16 18 19 22 26 31 25 18 19 21 18 16 16 19 23
19 19 19 19 16 20 23 26 26 21 21 41 43 36 33 28 33 34 36 29

21 21 17 19 21 16 16 14 13 13 13 12 11 13 14 15 16 14 14 14
14 14 12 12 12 12 15 18 18 16 19 18 15 11 12 15 16 15 15 15
15 15 15 15 15 15 12 12 11 12 13 16 14 13 14 13 14 13 14 13
12 12 14 14 12 14 15 16 16 13 12 11 11 15 14 15 14 14 14 14
14 15 16 16 16 15 16 21 23 22 13 16 27 27 31 27 20 17 17 17
17 22 22 22 24 29 31 28 22 16 18 20 20 18 20 18 21 28 31 36
36 36 31 29 29 28 28 29 24 21 21 22 16 17 21 23 20 25 25 16
15 15 16 16 16 18 18 19 18 18 24 22 28 28 25 22 25 18 23 29
28 24 21 25 19 28 25 23 24 24 23 27 34 43 34 28 31 34 28 27

19 19 23 23 22 17 16 15 12 13 12 13 13 16 16 13 13 13 14 14
```

```
14 14 13 11 13 15 15 18 18 16 16 19 19 15 15 13 15 15 15 15
15 15 14 15 15 11 12 13 11 13 15 13 15 13 14 14 13 13 14 13
11 11 13 14 14 14 14 14 14 11 11 12 14 12 13 12 13 13 13 15
15 15 16 15 16 16 16 12 29 25 17 15 16 16 23 28 27 22 20 13
12 13 28 27 17 26 27 26 26 16 22 22 16 19 16 22 16 19 26 28
26 32 31 29 29 24 24 31 25 24 17 17 17 17 19 24 28 20 27 21
16 17 17 17 17 17 17 20 16 17 25 31 37 31 29 29 28 26 31 28
28 26 26 31 31 29 28 29 24 28 21 23 28 28 28 28 30 36 31 28

20 25 25 24 18 16 14 12 14 12 14 15 15 12 12 13 13 15 14 13
13 14 14 12 14 15 13 15 20 22 20 17 16 15 16 21 23 18 15 15
15 15 15 12 12 13 14 12 12 14 15 15 15 15 15 13 13 15 13 12
12 12 12 14 14 16 14 13 10 11 13 14 14 12 12 14 14 14 15 15
16 18 16 16 15 13 13 12 14 15 15 14 14 16 28 24 21 16 14 14
15 13 13 14 17 21 20 21 16 16 17 20 16 17 20 19 18 19 22 19
33 36 32 31 26 34 37 26 25 19 18 18 15 14 16 23 25 25 24 19
21 24 17 17 16 20 21 21 21 23 21 20 22 28 28 28 28 29 31 29
36 31 33 32 28 31 28 25 28 26 25 25 19 24 28 31 31 30 29 29

16 21 21 16 19 16 16 15 13 12 11 12 11 11 12 15 15 15 16 13
13 16 16 15 16 17 14 20 22 20 17 17 21 17 16 17 16 16 20 15
16 15 12 12 14 15 15 12 12 15 13 13 15 14 12 12 14 12 11 12
13 13 13 13 13 13 13 12 11 12 12 13 14 13 13 14 14 15 15 14
16 15 15 15 15 14 14 12 12 13 15 15 13 18 15 14 15 15 19 23
23 15 14 14 15 23 23 21 16 21 21 17 17 20 16 20 20 21 22 22
28 33 29 33 40 37 28 25 29 24 25 19 18 22 24 25 25 28 25 28
31 29 21 17 17 21 17 15 15 17 17 17 21 26 23 27 28 33 30 34
36 36 30 33 33 29 25 28 31 29 22 26 28 34 32 31 28 24 21 28

16 20 21 22 15 15 16 15 14 12 12 12 11 13 16 16 12 12 12 14
14 14 16 16 16 16 16 17 17 16 16 16 20 21 17 20 20 23 16 17
15 14 14 14 14 15 15 14 12 12 14 14 14 15 13 13 12 11 11 12
12 16 14 12 12 11 12 12 11 12 13 15 15 13 13 16 16 14 12 12
14 13 13 14 13 14 14 12 12 14 16 18 18 15 15 13 13 13 16 18
15 15 14 16 16 14 16 21 21 23 24 21 23 23 21 14 16 16 16 17
20 27 30 34 49 41 30 28 27 29 31 31 29 29 28 31 26 28 28 28
28 31 23 18 19 18 14 15 18 17 16 16 16 19 29 33 27 28 34 28
34 35 34 30 33 27 25 27 28 33 28 28 29 31 31 31 28 24 26 32

21 21 23 21 15 14 15 14 11 13 12 12 15 16 12 15 19 16 18 13
13 13 15 20 15 16 15 14 15 16 16 15 16 16 15 13 13 15 14 16
17 14 14 14 14 16 14 13 14 15 15 14 14 12 12 12 14 12 12 12
13 13 13 12 11 10 11 10 11 14 15 16 15 16 15 12 13 15 13 12
14 14 13 14 14 14 13 14 14 14 14 14 16 14 12 14 14 14 14 14
22 24 18 19 16 13 15 25 23 16 23 28 24 21 21 16 17 16 17 17
17 26 27 35 31 35 31 27 27 28 27 21 17 23 28 33 33 31 29 28
28 26 24 18 13 18 22 19 18 16 16 18 28 36 33 29 31 23 25 28
28 24 27 31 31 28 22 13 20 27 23 27 23 16 17 23 25 25 28 29

19 19 17 23 23 19 16 14 13 12 13 14 16 15 15 17 16 18 15 15
14 12 15 18 18 16 18 15 16 16 15 15 14 12 14 15 15 14 15 15
19 21 15 13 16 16 13 12 13 14 13 14 15 14 14 15 14 15 12 15
15 14 11 11 11 10 11 11 12 13 15 13 15 12 12 14 14 15 15 15
16 15 15 16 14 14 14 18 18 19 24 26 28 28 24 19 18 21 21 19
23 23 29 29 25 29 36 31 27 27 21 16 14 14 16 27 27 27 22 17
20 28 18 15 16 23 23 16 19 16 16 18 20 19 19 23 21 23 21 19
19 21 21 25 25 17 17 19 16 22 17 20 20 16 21 27 27 25 23 30

18 16 15 21 28 24 17 15 14 14 12 12 13 17 16 13 12 14 11 11
14 16 16 15 19 18 18 18 16 14 12 15 13 13 15 15 15 13 14 18
23 18 22 16 14 15 12 13 15 13 12 15 14 14 13 13 14 14 14 15
13 11 10 10 11 11 11 11 14 15 15 15 13 13 13 15 15 15 16 16
```

```
15 16 15 14 14 14 13 13 13 15 15 15 15 16 14 13 13 13 13 13
14 14 14 15 14 14 14 14 14 15 16 23 28 31 25 24 25 22 22 24
26 25 24 24 25 28 36 36 28 23 17 23 19 16 16 16 13 14 14 14
16 21 22 22 21 21 20 16 15 16 16 15 15 16 15 18 19 24 24 20
19 18 18 19 16 16 16 19 18 19 23 24 19 17 25 24 28 20 22 28

18 18 18 19 18 15 15 14 11 14 12 15 16 16 14 14 12 13 13 14
16 17 14 16 17 20 16 14 11 11 14 14 12 12 14 16 16 13 16 24
28 26 25 20 14 12 14 14 14 14 14 12 13 13 15 12 13 16 13 12
11 12 12 12 11 11 13 14 13 14 15 15 15 15 15 14 14 15 16 16
15 16 13 13 13 12 12 14 14 16 16 15 15 13 13 15 13 13 13 13
13 12 14 14 14 14 13 14 15 16 17 27 33 31 29 23 25 25 24 23
28 26 25 25 25 28 34 24 24 24 25 25 24 23 16 17 17 16 17
16 17 20 21 17 14 16 14 15 15 17 16 15 16 16 18 19 23 23 23
22 18 19 16 16 19 22 22 19 24 21 18 21 24 21 17 19 28 31 33

18 16 16 15 15 15 15 12 11 12 11 15 16 15 12 12 16 15 14 14
17 16 14 14 19 17 16 14 11 12 13 13 15 16 17 20 22 15 19 23
19 21 25 23 15 13 13 12 12 13 13 14 12 12 14 12 12 12 12 12
12 13 13 12 12 12 12 12 12 13 13 14 13 14 13 13 14 13 15 15
15 15 14 14 12 12 13 13 16 18 18 16 14 14 15 14 14 14 12 13
13 13 13 13 13 13 14 13 13 14 13 16 23 30 33 23 21 27 22 17
19 22 25 28 25 25 31 25 24 24 25 25 28 31 19 21 24 21 16 21
23 21 21 24 25 21 21 16 17 17 20 21 14 16 17 17 21 20 21 27
33 24 25 24 16 18 25 18 19 19 22 25 28 26 22 16 29 33 31 29

17 17 16 15 15 15 14 12 12 12 13 16 18 15 12 13 11 11 14 16
18 14 15 17 19 16 16 14 11 12 12 13 16 17 21 19 22 21 20
21 22 22 21 18 15 12 14 12 14 14 12 12 12 13 11 11 11 11
12 12 12 11 12 12 12 13 13 13 13 12 14 13 14 14 13 13 14 14
13 14 13 11 11 12 14 15 15 15 15 16 15 13 15 15 13 15 13 12
14 12 12 12 14 12 13 15 15 16 17 17 21 27 22 26 31 35 31 27
20 23 23 23 26 27 30 29 28 28 24 23 24 28 28 26 28 26 26 31
28 25 28 29 28 25 21 17 16 21 24 24 17 16 26 26 20 17 16 16
23 30 23 21 20 26 23 20 18 19 22 24 24 24 31 24 24 24 28 28

13 16 16 13 13 13 14 15 15 16 16 16 15 12 14 13 12 15 18 16
13 15 16 18 16 15 12 14 14 14 15 15 15 16 16 16 21 16 13 14
17 21 19 22 15 15 14 15 14 14 14 12 12 14 12 12 12 12 11 11
11 11 11 11 11 11 11 12 14 12 13 13 15 13 15 13 13 16 16 12
12 12 11 12 13 14 15 15 15 15 16 15 15 15 16 15 15 15 12 15
15 13 13 13 13 12 14 14 14 16 22 18 25 25 28 29 29 33 28 26
27 26 27 26 26 28 30 27 26 27 27 28 26 25 25 29 29 29 29 28
32 34 34 31 28 18 15 15 21 25 25 21 23 23 24 17 17 16 21 28
23 22 22 28 34 31 27 20 17 20 26 26 26 33 29 22 19 25 28 19

12 14 13 12 13 11  9 12 15 14 13 13 14 14 15 15 14 16 18 16
15 16 18 20 15 12 12 13 12 14 12 14 14 15 15 14 15 12 16 19
17 21 25 19 17 14 16 14 13 12 14 13 11 11 14 11 12 12 11 11
11 11 12 12 12 11 12 12 12 12 14 12 12 12 14 12 12 13 13 12
11 11 12 11 11 12 14 13 14 14 13 17 16 16 16 16 14 14 12 14
14 16 15 15 14 15 15 13 13 16 16 16 16 15 16 16 22 28 31 23
25 25 25 25 25 28 31 34 26 24 26 27 27 27 27 27 28 28 28 29
31 33 33 36 29 22 15 15 18 26 28 26 26 19 18 19 19 18 21 23
24 21 24 29 29 28 21 22 26 28 28 31 35 35 33 28 27 30 33 34

12 15 15 14 14 11 11 12 12 12 12 11 13 12 12 15 17 17 20 16
14 14 16 16 14 11 12 12 14 12 13 15 15 16 13 12 12 12 15 15
14 15 16 24 21 17 19 19 15 14 12 12 13 12 12 13 11 11 10 11
10 10 11 12 11 14 16 14 12 12 13 12 13 13 12 12 11 11 11 11
11 11 11 11 12 13 15 13 15 15 16 16 16 16 14 14 13 13 14 14
13 15 16 14 15 15 16 15 14 14 15 15 16 18 18 19 18 24 25 31
25 24 24 24 24 18 29 30 25 24 25 29 29 26 28 28 26 27 31 30
```

```
26 21 21 17 20 20 22 18 18 24 29 28 29 28 28 26 25 25 28 31
24 24 25 29 29 25 25 24 31 31 31 29 31 31 31 31 34 31 27 27

12 14 15 14 12 11 11 14 14 14 12 12 14 14 13 13 13 13 13 12
13 14 14 13 11 13 14 13 14 12 12 12 12 12 12 12 12 12 11 11
11 15 18 22 22 24 25 19 14 12 11 12 11 12 14 11  9  9  9 11
 9  9 11 10 13 15 14 11 13 12 12 14 14 14 12 11 10 10 11 11
11 12 12 11 14 16 16 14 14 12 13 13 15 15 15 13 16 13 14 14
13 13 13 13 16 15 13 13 14 15 15 16 19 19 18 16 18 16 25 28
31 28 24 24 28 23 29 18 22 28 28 29 28 28 28 25 28 24 28 28
26 20 17 17 17 27 27 20 26 28 28 30 31 36 33 31 36 36 36 28
22 19 25 31 28 31 31 29 29 31 28 28 31 31 33 31 33 37 33 28

12 16 16 13 12 11 10 12 14 11 11 12 14 12 14 12 12 15 14 12
12 13 13 13 13 13 12 12 13 14 14 13 13 13 14 12 12 12 12 11
16 25 24 32 25 21 24 20 16 11  9 11 11 12 12 10 10 11 10 10
11 11  9 11 11 11 12 12 12 13 14 13 13 14 10 10 11 11 12 11
11 12 12 13 15 13 13 11 11 12 12 14 14 14 12 14 16 15 13
12 15 16 13 13 14 13 14 14 14 14 15 16 15 16 17 20 20 15 18
24 31 36 33 29 25 28 31 26 25 26 28 28 28 29 28 28 25 17 16
16 16 17 17 19 17 17 17 20 23 27 28 33 30 33 34 28 26 28 33
28 23 31 36 37 37 31 25 26 24 22 25 25 31 31 31 31 36 33 29

14 14 14 15 14 14 12 12 15 13 12 12 12 12 12 12 12 15 14 12
12 12 14 12 11 14 14 14 13 13 14 13 12 12 13 13 14 13 13 14
20 25 28 24 21 21 24 24 24 18 18 15 11 10 11  9  7  9  9  9
 9 10 12 15 15 14 14 14 14 13 13 13 13 12 12  9 11 10 11 13
13 13 14 14 14 12 12 12 12 12 12 13 13 13 12 12 15 14 14 14
14 16 16 14 16 15 15 15 15 13 15 13 16 16 20 21 21 16 16 16
17 22 21 20 22 28 31 40 36 29 28 28 26 31 26 25 28 28 22 23
16 16 16 22 24 19 21 23 21 23 29 30 34 35 27 28 28 26 26 33
33 30 33 34 34 30 28 29 28 25 25 28 29 25 31 34 34 37 37 32

14 14 14 14 13 11 12 12 14 12 12 11 11 12 13 15 15 15 13 12
14 14 11 12 12 12 14 11 14 14 14 14 14 15 13 13 12 12 16 22
22 22 15 17 22 22 16 23 28 28 29 29 25 16 11 10 10 10 11 11
11 10 10 16 18 14 11 12 14 14 14 14 12 11 11 12 12 12 11 12
19 15 13 15 15 14 12 14 14 15 15 16 15 16 17 19 17 20 16 16
16 16 16 20 27 30 30 33 33 34 30 25 28 24 24 28 29 26 28 18
16 18 19 24 23 29 25 25 25 28 29 30 36 29 29 29 29 29 23 21
26 31 34 34 31 28 28 33 27 27 23 26 28 28 31 33 31 31 29 28

13 13 14 16 13 13 11 13 13 15 14 10 11 14 14 16 16 16 15 12
12 12 13 13 13 15 15 11 11 12 14 16 18 19 17 16 16 19 17 19
17 21 14 16 22 21 17 19 27 20 21 23 27 28 11 11 10 10 11 11
11 11 12 12 12 11 12 12 14 14 12 10 10 11 11 11 11 12 12 14
15 15 14 14 13 13 13 13 13 14 13 11 13 14 15 15 15 15 18 16
16 17 17 17 20 20 23 28 31 34 31 28 30 28 26 27 22 25 19 15
15 19 29 29 26 26 28 32 31 25 24 31 29 33 36 29 31 33 24 23
23 25 30 28 28 34 28 31 28 31 28 31 28 34 30 27 28 33 30 29

14 14 14 14 14 15 13 12 12 13 14 12 11 11 14 16 16 16 15 14
12 12 14 14 14 12 14 12 11 11 11 15 19 19 19 18 15 15 16 15
15 16 16 15 16 17 24 25 23 28 31 28 26 28 23 14 10 13 13 13
11 12 12 12 12 12 12 14 13 12 10 10 11 10 11 10 10 10 11 12
14 14 15 14 14 14 14 14 12 12 12 12 13 14 13 14 15 17 20 15
15 15 15 14 15 15 15 15 12 12 15 16 19 18 13 15 13 16 15 15
16 16 16 16 21 23 19 21 29 33 30 31 31 34 35 31 20 15 13 16
20 26 26 20 25 25 28 31 23 21 24 28 28 25 22 20 25 25 31 29
36 36 24 24 33 30 36 36 30 29 28 25 23 21 23 27 26 34 34 30
```

```
14 14 14 14 13 12 12 15 15 14 14 12 11 11 13 14 17 16 16 15
13 11 13 13 13 11 12 11 12 16 18 21 21 16 15 13 16 16 16 15
15 12 12 12 16 24 19 24 25 24 24 21 16 28 22 12 11 12 12 12
13 11 11 13 13 13 13 11 11 11 10 11 10 10 11 11 11 11 12 12
12 14 14 14 12 12 12 11 12 12 14 15 15 15 16 16 19 17 13 14
16 16 15 14 14 15 15 14 13 14 14 18 18 15 14 14 15 16 16 16
16 16 16 15 15 15 16 18 23 31 33 30 28 30 25 23 14 16 17 20
17 14 16 20 23 26 27 26 27 34 25 24 24 23 29 36 28 24 26 32
31 25 24 25 31 31 33 33 28 25 25 24 24 29 29 28 30 30 31 34

13 12 13 13 12 13 13 15 14 13 13 11 14 14 14 14 15 16 16 14
14 13 12 12 12 12 11 14 20 21 20 17 20 19 16 16 18 18 14 11
12 11 12 18 21 21 18 15 14 15 22 25 22 14 14 11 12 12 11 14
12 12 12 13 12 13 13 11 10 10 11  9 10 11 11 12 11 12 11 12
14 12 13 12 13 13 13 15 12 14 14 14 16 18 25 19 17 15 15 16
15 15 14 16 21 26 23 19 17 14 14 14 15 14 15 15 16 16 16 16
16 16 15 12 13 16 18 16 18 19 33 33 29 33 33 24 19 19 19 15
15 15 16 16 17 21 26 22 22 31 33 30 30 30 36 33 23 21 23 29
28 25 25 28 26 32 31 34 41 41 34 28 23 28 33 28 28 31 29 28

14 12 15 14 14 13 13 12 12 11 12 13 14 13 13 15 18 18 16 15
15 15 14 12 12 12 12 14 17 19 17 17 17 16 16 16 16 14 10 11
11 12 16 21 16 14 12 15 13 15 24 25 16 11 10 11 11 11 11 12
14 12 14 14 12 12 12 12 11 11 11  9 11 11 13 11 11 13 13 13
14 12 12 15 19 28 28 25 21 28 31 28 31 24 18 25 19 18 19 22
22 25 15 23 23 15 17 19 25 13 14 16 16 16 17 16 16 20 17 16
17 15 15 14 15 19 15 16 19 19 24 28 32 34 31 22 15 15 14 15
19 23 24 23 25 25 25 23 21 25 28 27 34 31 35 34 27 26 28 26
21 21 22 26 28 31 25 29 40 37 29 28 25 26 28 32 26 31 28 31

14 15 15 17 14 14 14 14 14 15 23 14 14 18 16 14 12 14 14 14
13 14 15 14 13 15 23 19 16 16 16 17 17 17 17 16 14 11 13 15
20 21 17 14 13 13 10 11 16 16 15 16 11 11 12 11 11 12 12 12
11 11 11 11 11 11 12 12 11 11 12 11 11 12 11 12 12 13 13 17
14 16 21 22 17 17 14 15 15 15 14 14 15 15 19 19 19 18 13 12
12 24 18 12 15 16 16 22 23 17 21 17 17 17 21 16 14 17 17 14
14 16 20 20 16 16 16 17 20 19 21 24 29 33 37 20 19 19 19 24
28 28 25 24 23 29 25 22 18 24 25 33 30 25 29 25 28 28 35 28
26 28 26 30 33 28 28 30 33 28 28 25 31 28 24 29 28 31 28 28

13 16 19 19 16 16 16 20 17 16 15 22 16 16 16 14 14 15 16 16
19 19 14 12 13 15 16 16 15 15 14 15 19 19 15 12 12 14 15 16
14 17 16 13 12 13 15 15 14 13 14 22 11 12 14 12 12 12 11 13
13 12 11 11 11 11 11 11 10 10 11 10 12 12 15 14 14 19 29 23
22 21 14 12 12 12 14 16 15 15 15 15 15 15 14 14 12 12 12 12
14 13 13 12 13 12 13 13 31 25 24 22 19 18 16 17 16 15 17 16
15 16 14 14 14 14 16 14 16 21 21 21 21 21 22 33 25 24 24 31
31 25 24 24 26 25 24 18 18 22 29 33 28 28 28 29 24 29 33 36
33 29 24 27 37 31 26 27 27 28 30 34 34 26 20 27 28 29 33 29

17 21 23 24 19 21 22 21 19 16 14 20 17 16 14 13 14 14 23 25
22 16 14 14 14 14 18 14 16 16 16 14 16 15 11 11 14 13 13 15
15 15 14 12 14 16 16 16 12 11 11 16 16 13 13 13 11 11 11 11
11 10 14 14 12 11 10 11 10 10 11 12 13 14 15 14 23 23 12 12
14 15 15 15 15 15 15 14 14 14 16 16 16 16 15 14 13 11 13 14
14 14 14 14 14 14 15 12 19 20 15 16 20 24 19 19 18 15 15 16
18 15 16 15 15 14 15 17 24 16 19 21 19 21 21 28 34 34 30 30
33 27 25 25 23 24 36 23 21 28 32 28 20 22 24 20 31 28 29 31
28 24 23 23 28 37 30 25 29 30 28 34 34 28 26 26 34 33 34 34

18 19 23 23 24 24 21 17 16 16 16 17 17 13 12 12 13 16 22 21
15 13 14 16 18 19 18 16 15 16 14 12 11 12 12 12 12 12 13 11
11 11 13 14 11 15 12 11  8 10 10 21 16 12 12 12 13 12 10 10
```

```
13 13 11 11 10 10 10 11 10 11 12 14 13 15 21 24 15 13 13 14
14 15 14 14 14 14 15 14 14 15 15 15 15 14 13 13 12 11 13 13
14 14 14 14 15 15 14 12 12 12 12 14 14 15 20 25 21 21 24 25
26 14 15 15 14 15 22 22 21 19 21 19 21 23 21 22 23 27 34 35
34 26 22 22 26 28 27 28 36 33 28 28 24 19 19 26 28 25 32 37
26 25 24 25 25 24 23 28 33 33 30 25 28 30 30 33 37 34 34 28

14 16 16 24 23 16 18 18 15 15 15 15 17 16 14 14 14 13 17 17
17 13 13 16 16 16 16 17 16 15 13 11 11 12 14 14 12 12 12 13
13 13 12 10 11 11 11 9  9  9  11 12 14 14 12 11 12 14 12 11
12 13 13 13 11 11 11 11 11 11 13 16 18 25 18 15 14 14 14 13
15 13 13 13 13 15 14 14 15 15 14 14 14 14 14 12 11 12 14 15
13 14 13 13 13 14 13 13 13 14 14 14 13 13 15 14 14 16 16 16
15 25 21 15 15 16 18 19 16 16 19 19 18 23 21 23 24 24 25 29
31 28 26 28 35 35 31 28 36 33 27 33 27 22 23 25 28 29 33 31
21 16 22 22 20 20 22 24 32 36 31 25 25 31 29 28 28 17 17 17

16 16 16 16 19 19 16 16 14 13 15 11 15 15 14 14 15 16 17 16
16 15 15 15 14 17 16 14 16 16 13 11 11 13 13 13 11 14 14 12
14 11 12 9  9  11 13 13 11 9  10 14 15 16 14 10 10 11 12 12
14 14 12 12 11 9  11 12 12 11 11 13 23 16 15 15 14 15 15 12
14 15 15 15 15 15 15 15 15 15 13 13 12 12 11 10 11 12 12 14
14 14 15 15 15 14 12 12 13 14 13 13 13 14 14 13 13 13 15 16
16 16 16 19 21 16 16 16 19 18 19 19 19 20 22 22 23 23 23 25
28 36 33 36 33 31 28 25 28 31 26 26 28 28 27 22 23 16 21 28
23 20 18 18 18 21 21 19 24 34 31 31 32 34 34 32 24 22 22 19

13 16 16 22 24 24 23 19 19 13 13 13 14 14 13 12 18 18 16 15
14 15 15 16 16 16 16 16 14 12 11 11 12 11 12 13 13 13 11 11
14 15 11 10 12 12 11 12 11 9  12 20 24 28 16 10 11 13 11 13
11 11 11 10 11 11 11 12 11 12 16 21 17 13 14 14 14 16 14 14
15 15 16 16 17 15 16 16 15 12 14 11 11 10 11 11 12 12 13 12
14 14 14 12 11 11 12 14 14 14 14 14 14 14 13 14 13 13 14
15 14 15 20 23 23 25 21 21 19 23 25 28 25 25 20 20 25 25 24
29 29 31 33 31 29 23 23 23 21 23 28 31 23 22 21 16 21 27 28
23 20 26 26 26 28 27 22 33 33 36 31 28 31 31 31 32 31 32 28

16 12 25 27 28 27 21 16 15 18 16 15 12 11 10 11 12 11 11 11
11 12 12 15 16 15 13 13 11 11 14 12 14 14 12 14 15 13 12 12
12 14 14 11 11 13 13 11 10 9  10 15 25 28 24 14 11 11 12 11
12 13 11 11 9  10 10 11 11 14 14 17 15 15 14 14 14 12 14 14
14 17 19 19 16 16 15 15 14 14 11 11 10 11 12 12 12 14 14 13
13 13 12 11 11 13 12 12 12 14 14 14 12 15 15 16 14 14 15 14
14 16 17 17 17 21 19 22 23 25 21 20 20 25 19 21 24 25 23 23
24 26 26 31 37 44 34 22 22 23 23 25 19 19 16 16 23 29 31 31
31 23 28 37 40 35 27 21 23 33 38 33 30 33 28 28 24 28 29 31

19 14 17 23 23 27 23 19 17 15 15 13 9  11 11 13 12 11 12 12
11 12 12 13 12 12 12 12 11 11 13 11 11 11 11 11 13 14 14 19
16 11 11 10 9  12 12 13 17 13 11 13 11 13 11 14 11 12 11 12
12 12 11 12 11 10 11 10 11 12 14 14 13 13 13 13 11 11 12 14
14 14 12 12 11 16 13 12 12 13 12 12 11 13 13 14 14 14 14 14
14 11 12 12 12 12 13 13 13 13 13 13 13 14 14 16 15 14 14 12
14 15 16 16 21 19 17 22 26 22 17 16 16 16 20 23 25 23 22 27
23 24 28 33 31 33 42 28 26 21 21 25 28 26 24 25 28 25 28 31
36 33 31 36 40 40 33 29 26 34 31 37 34 27 27 28 27 28 28 30

19 15 17 23 24 25 23 15 17 16 12 12 13 13 13 13 13 13 13 13
13 13 11 10 12 12 11 11 12 11 11 11 12 14 13 14 14 15 15 13
14 13 11 11 11 12 16 17 12 12 11 11 11 11 11 12 11 11 13 11
11 13 11 10 11 11 11 10 11 11 12 12 12 12 11 11 12 13 15 15
11 13 13 13 11 12 12 12 11 11 12 12 12 12 13 12 13 13 13 11
10 13 13 13 13 14 14 15 14 12 12 14 13 12 12 12 12 12 11 12
```

```
11 11 14 15 16 22 21 25 17 16 16 17 17 17 17 19 17 21 21 23
25 33 33 28 33 38 47 42 36 25 29 36 28 26 26 28 28 26 32 31
29 33 36 33 42 45 42 33 31 36 36 33 23 19 27 28 34 31 28 34

18 18 16 18 16 15 16 18 15 14 12 12 12 12 12 12 12 12 13 14
12 12 14 13 11 11 13 15 15 14 14 15 16 15 14 12 18 18 14 16
16 16 14 11 13 14 15 14 13 14 12 11 10 10 11 11 16 14 14 12
12 13 12 10 11 11 10 11 10 11 11 12 14 12 14 12 12 13 13 11
11 12 12 11 11 13 13 11 11 11 13 13 12 14 14 14 14 11 12 11
11 12 13 12 12 13 14 14 13 11 11 13 12 14 12 12 14 12 12 13
12 12 12 12 16 20 23 28 31 33 28 24 19 19 17 23 23 23 23 22
27 35 35 31 31 37 38 43 45 43 34 28 34 24 28 29 29 28 36 31
20 20 25 26 34 39 31 37 37 33 31 33 24 25 28 29 29 31 36 40

15 16 15 16 16 16 15 12 13 11 11 13 11 10 13 12 12 12 14 15
16 16 13 14 14 13 16 16 14 15 15 15 15 16 14 14 16 16 15 14
15 16 14 14 14 16 16 13 11 15 15 11 9 9 9 11 14 14 14 14
12 12 12 12 12 11 12 11 11 11 11 14 14 14 13 11 11 14 11 11
12 12 12 11 11 11 12 12 12 12 12 13 13 11 13 13 11 11 12 12
12 12 12 14 14 13 13 13 12 12 12 12 13 13 14 14 14 15 15 14
14 14 14 15 12 15 21 21 20 24 25 28 22 16 15 18 24 29 33 25
29 29 29 31 36 40 34 34 40 43 40 45 45 34 28 22 27 30 33 33
19 19 29 36 36 40 37 26 32 32 34 39 32 26 25 33 36 33 31 31

15 15 14 14 14 12 14 11 11 13 12 12 12 12 13 13 15 15 14 13
15 16 15 15 17 17 16 16 14 14 14 14 14 13 13 15 14 14 13 13
15 14 15 14 15 16 15 12 13 11 10 11 11 10 13 11 10 11 11 10
10 10 11 12 14 15 12 12 12 12 12 13 13 13 12 11 11 11 13 13
13 10 11 11 12 12 11 12 12 14 12 12 12 12 11 10 11 11 10 13
13 13 14 14 14 14 11 11 12 14 12 13 14 13 14 21 27 23 15 14
14 13 14 14 13 15 15 16 18 23 24 25 20 21 24 25 24 31 37 29
29 36 42 36 37 42 33 33 33 31 31 36 35 31 23 26 27 23 21 25
17 17 28 41 38 27 29 23 28 33 36 25 25 28 32 39 34 31 31 31

13 13 13 11 11 11 13 12 12 14 12 11 12 14 13 13 13 15 16 15
15 18 18 15 16 16 15 15 15 14 14 15 12 12 11 14 13 12 11 12
14 16 14 14 15 16 13 9 9 9 10 11 12 15 15 13 11 11 12 12
12 10 10 10 11 10 11 11 13 14 14 14 14 12 12 11 13 13 13 13
11 11 11 11 11 11 13 13 13 14 14 12 14 12 12 11 11 11 11 13
12 12 13 12 11 11 11 11 13 13 13 14 15 15 15 16 17 16 23 17
13 14 16 14 14 14 16 16 16 20 23 29 29 24 19 21 23 21 37 31
19 20 24 31 34 37 29 31 36 31 25 33 45 42 37 31 28 25 25 26
27 22 26 37 40 37 21 22 33 36 33 27 28 29 31 31 31 31 29 31

12 13 12 12 11 12 11 11 13 13 13 13 14 15 14 18 19 18 16 15
15 16 16 18 18 18 15 15 15 14 11 9 10 10 9 12 12 11 11 12
12 12 12 9 11 11 11 9 10 11 10 14 13 15 13 12 11 11 12 12
11 11 12 12 12 11 11 11 12 14 13 13 13 10 11 12 14 14 12
11 11 12 12 12 13 13 13 13 12 14 13 13 14 11 10 11 11 12 12
12 12 12 12 11 10 12 12 12 12 13 13 14 14 14 16 16 16 16 15
15 15 15 14 15 13 13 14 16 16 22 23 28 34 28 22 23 27 23 23
19 28 29 36 31 28 32 37 39 28 26 28 37 36 37 42 40 37 31 24
23 25 29 31 31 29 31 31 35 37 37 34 34 33 28 30 33 33 34 30

15 14 11 12 12 14 12 13 13 13 13 13 16 16 17 17 17 16 15 15
15 15 16 14 12 12 11 12 10 10 11 10 12 11 10 10 10 9 10 10
11 11 10 8 11 11 11 9 12 12 12 17 16 13 13 13 13 11 11 10
10 11 10 11 11 10 10 11 11 12 13 12 11 10 13 13 11 11 11
11 11 12 12 12 14 14 14 14 13 13 12 12 12 12 11 11 13 13 13
13 13 13 11 10 12 14 12 14 14 13 14 14 14 14 14 16 18 16 14
13 13 13 13 15 14 12 15 15 17 23 21 27 35 37 34 35 30 23 22
28 34 28 22 26 36 37 33 29 33 33 25 31 34 28 28 34 37 24 29
29 28 29 33 36 29 33 36 36 29 33 37 34 37 37 37 40 43 37 33
```

```
11 13 13 11 13 13 15 14 11 12 14 15 16 19 19 17 17 19 16 13
16 15 15 13 11 16 11 10 10 11 12 11 12 14 10 11 10 11 10 11
11 13 11 11 10 11 11 11 12 14 16 15 16 14 16 13 12 11 11 11
11 11 11 11 10 10 10 11 12 15 12 11 12 10 11 11 12 11 11 12
12 13 13 13 13 13 11 13 13 12 14 14 14 14 12 11  9 12 12 13
13 12 12 11 10 11 14 14 13 14 14 14 16 14 14 14 12 13 12 13
12 12 12 12 13 13 14 14 15 16 19 23 28 29 28 33 36 35 28 23
23 27 23 26 36 41 33 23 28 33 34 33 29 36 33 29 31 37 41 44
44 44 44 41 39 33 36 40 42 40 37 29 40 42 42 40 36 33 36 35

13 13 12 11 12 15 13 11 13 14 16 16 16 21 21 17 16 17 15 14
13 13 12 12  9 11 11 10 11 13 13 13 14 14 12 12 11 10 11 14
12 10 10 12 12 13 16 13 14 15 14 14 11 13 15 15 12 11 11 11
10 11 11  9 11 11 11 11 11 11 13 13 13 14 13 11 11 11 11 11
11 12 13 13 13 12 12 12 12 11 11 13 13 13 10 10 11 12 14 14
12 12 11 11 11 12 12 13 13 13 13 14 14 13 11 13 14 15 15 14
12 12 15 16 13 14 13 13 12 13 16 18 22 24 19 18 18 31 25 29
28 24 29 29 35 35 23 21 27 28 27 34 34 33 30 33 28 27 33 31
28 28 31 33 31 39 34 31 31 26 25 19 25 31 31 33 28 28 28 36

15 16 16 18 19 18 19 15 13 12 15 15 19 18 15 13 13 13  9 12
11 10 11 10 11 11 11 12 13 14 14 14 13 14 15 14 11 13 11 11
10 11 12 12 16 16 15 16 18 15 15 13 13 13 13 11  7  9  9  9
10 11 11 11 12 12 15 11 12 13 11 12 11 12 13 11 11 11 11 11
11 14 14 12 12 14 12 12 12 12 12 12 12 11 10 10 11 13 13 13
11 10 11 12 12 14 14 14 15 13 13 13 13 12 12 13 15 15 14 13
13 17 23 18 16 15 14 14 15 13 14 18 18 19 16 16 19 29 33 37
29 19 23 25 19 17 17 24 31 33 27 35 37 37 28 22 28 30 27 34
33 25 22 25 29 25 24 25 29 24 21 21 31 31 28 31 31 32 28 29

15 20 17 17 16 19 19 15 14 14 14 15 16 13 11 11 11 11 10  9
 9 11 11 11 15 15 11 12 16 16 14 14 15 16 16 17 14 13 14 13
11 13 16 20 17 15 15 16 15 15 14 15 14 12 11 10 10 10 10 10
 9 11 10 10 10 11 18 16 17 15 15 15 11 12 13 11 11 12 12 12
12 13 13 14 14 13 14 14 14 14 12 11 11 11 11 11 11 13 12 11
10 11 11 11 11 11 13 13 14 14 14 14 14 12 14 14 13 13 13 13
11 13 16 25 36 38 22 13 15 15 15 14 18 19 19 16 19 18 24 31
26 28 25 23 23 18 28 26 28 24 33 29 37 40 36 33 29 27 22 31
35 34 28 21 22 28 30 28 27 25 20 25 29 29 31 28 29 28 32 28

12 17 17 14 16 15 15 15 14 15 16 15 16 11  9 11 12 11 11 11
12 13 16 19 18 15 14 14 15 15 14 15 16 16 15 14 12 15 14 14
13 14 17 16 14 16 14 14 14 14 11 13 14 12 11 10 11 11 11 10
10 11 11  9  9 13 12 22 16 15 16 22 14 11 12 12 12 11 14 14
12 13 13 13 13 13 13 13 13 13 11 13 11 11 11 11 12 12 14 10
11 11 11 11 12 12 13 12 12 11 13 11 13 13 14 14 14 14 14 12
11 12 16 19 31 31 19 14 14 16 14 14 14 16 17 22 23 23 24 22
19 22 23 20 18 20 31 28 19 19 16 18 19 26 31 36 36 25 28 29
29 29 28 31 22 34 37 35 31 28 22 25 25 27 25 25 27 21 28 29

25 19 16 16 16 16 16 14 14 16 17 11 10 10 13 14 14 16 14 16
18 16 15 14 14 13 13 15 15 15 13 18 15 13 13 15 15 14 14 15
14 14 15 14 15 14 13 13 12 12 12 12 10  9 10 10 10 10 10  9
 9  9 10 11 12 11 12 13 11 11 15 20 25 11 11 11 13 14 14 13
14 14 14 14 14 12 11 12 12 12 12 12 11 12 13 13 13 11 10 11
11 11 12 12 12 12 12 11 12 12 12 11 12 13 14 14 14 14 11 11
13 15 17 16 23 24 17 16 14 12 13 13 13 14 16 25 25 21 17 20
17 21 18 16 18 19 22 22 23 21 21 19 21 32 32 25 36 33 36 36
36 31 29 33 33 36 33 29 24 23 19 22 21 26 22 22 21 22 30 28

19 15 16 16 15 15 18 18 16 19 18 13  9 11 12 14 17 16 20 16
16 15 14 15 15 15 15 14 15 15 14 16 18 13 13 16 15 12 12 13
```

```
14  14  15  14  14  14  14  14  14  14  12  14  12   9   9  11  11   9   9   9
10  11  11  10  13  11  12  14  12  11  14  19  24  15  11  11  12  13  13  13
13  13  13  13  11  11  10  11  14  11  11  11  12  12  12  13  13  12  11  11
11  11  13  14  13  14  13  13  12  12  12  12  14  14  14  14  14  13  11  10
12  13  15  15  19  18  18  29  14  16  17  15  14  16  19  19  22  26  23  23
22  17  21  20  21  22  22  22  21  23  18  22  24  28  23  22  25  26  26  28
34  32  25  31  31  31  31  31  31  31  19  19  23  23  24  23  24  23  22  26

20  16  16  15  15  14  16  16  15  15  11  14  10  11  16  16  15  15  14  14
14  14  14  12  13  14  14  14  14  15  14  14  14  12  12  11  11  11  12  12
11  12  13  12  13  15  15  13  11  10  11  11  12  10  10  10  11  10  10  11
11  11  12  11  11  12  11  13  10  11  15  13  14  18  11  12  11  12  12  14
12  12  12  13  11  12  12  10  11  10  10  10  13  13  14  14  15  12  11  12
12  11  12  12  13  13  13  13  13  14  14  14  13  14  14  12  12  10  12
15  19  18  18  14  16  14  16  37  28  19  15  14  15  15  19  19  21  25  28
24  21  21  21  19  17  17  21  19  17  16  20  20  25  23  21  28  29  28  29
28  28  31  21  25  28  28  31  34  28  23  22  24  19  19  19  19  23  21  21

15  16  16  16  15  16  18  15  15  18  11  13  15  16  14  13  14  15  14  14
14  14  12  12  13  14  14  13  13  12  12  11  10  13  13  14  14  15  12  14
14  15  14  14  15  13  12  13  13  11   8   9   9   9   9   9  10  10  10  11
12  12  12  12  12  11  11  12   9  12  13  12  11  11  11  11  11  11  13  12
12  12  12  11  12  12  12  12  10  10  11  13  13  14  13  13  11  10  10  10
12  14  15  15  14  14  12  12  13  13  13  13  13  12  14  14  13  11  11  14
15  23  16  14  10  16  16  20  19  19  18  16  16  16  16  18  23  23  23  22
19  21  24  23  19  19  17  17  16  16  16  19  22  28  28  21  21  20  22  25
27  28  31  25  23  23  24  28  25  25  28  25  20  16  19  19  24  18  16  16

17  15  15  18  18  18  19  14  16  15  13  13  13  13  12  12  11  13  13  13
11  13  11  11  12  12  12  14  11  11  11  12  12  14  14  16  16  16  15  14
15  14  14  13  15  16  15  14  11   9   9   9   9  10  10  10  10  10   9   9
10  13  11  11  11  12  12  11  11  12  12  12  11   9  11  11  11  11  12  16
14  14  13  13  13  14  12  13  10  11  12  12  14  13  12  11  11  11  10  11
10  11  14  11  11  13  13  14  14  12  14  14  14  12  13  13  14  13  11  12
13  16  16  14  15  20  17  18  18  18  18  18  24  31  18  20  24  19  16  20
20  22  24  26  28  23  26  28  17  17  21  24  25  28  28  23  26  23  21  21
23  27  28  27  34  34  23  21  25  25  29  29  25  19  22  22  20  19  21  20

21  17  16  20  21  20  15  13  10  10  11  11  10  10  10  11  11  11  10  11
10  10  10  11  11  11  10  10   9  10  13  11  12  14  14  14  14  14  14  14
14  14  14  14  15  14  13   9   9   9   9   9  11  10  10  11  11  11  10  11
12  12  12  12  12  13  10  10  11  11  10  10  11   8  11  11  11  11  11  12
13  12  12  12  13  12  11  11  11  11  13  14  13  14  11  11  11  11  11  12
11  12  13  13  12  13  13  13  13  14  14  13  11  13  14  14  15  12  12  11
12  13  14  14  10  17  19  17  17  16  22  25  22  20  23  16  14  14  15  10
16  18  21  24  25  28  31  31  23  26  22  18  24  29  28  25  24  21  21  29
37  31  27  27  34  37  28  21  19  25  25  22  22  25  23  23  16  18  19  19

19  21  17  19  14  12  11  11  11  11  11  10  11  11  10  11  10  11  10  11
11  11  11  10  10  10  11  10  11  11  14  14  13  11  11  10  11  15  15  15
15  15  14  16  14  13  13  11   9  11  10  10  11  11  13  13  14  12  14  12
12  12  12  11  12  11  11  11  11  11  10  10   9  10  11  10  10  10   9  12  12
12  16  16  12  12  12  11  11  11  12  13  13  11  11  11  11  13  15  14
14  14  14  14  14  15  13  13  12  13  13  12  13  14  13  11  10  11  11  14
14  15  16  15  15  19  16  16  16  21  35  35  26  16  15  15  15  14  15  15
18  22  23  22  18  18  19  28  28  26  24  19  20  19  28  26  22  26  36  36
28  23  23  23  25  21  19  17  22  31  31  31  27  22  22  22  21  20  21  17

24  23  19  15  12  12  12  12  10  11  12  12  12  12  12  12  11  11  11  11
13  13  14  11  11  12  12  10  10  11  11  11  12  12  11  12  13  13  14  14
13  13  14  16  15  14  11  10  12  12  12  11  12  14  14  14  14  13  11  13
13  13  11  11  12  11  10  11  10  11  11  11  10  10  10  10  10  11  10  11
13  13  14  14  14  12  11  11  12  12  12  12  13  12  11  11  12  12  12  15
```

```
15 14 13 14 14 14 15 14 14 14 14 12 14 12 13 11 11 11 11 12
14 13 14 15 15 16 17 15 16 17 23 28 28 16 16 13 13 14 14 14
16 16 21 23 20 20 20 23 28 36 28 19 22 18 28 34 21 24 25 25
25 26 24 19 23 22 19 19 16 24 29 33 29 29 24 22 22 22 23 19

13 12 15 10 10 14 13 11 11 11 12 11 12 10 11 12 12 12 12 11
11 11 13 11 10 13 13 11 10 10 10 11 10 11 12 11 10 12 11 12
11 12 12 14 14 13 13  9 11 11 12 14 14 14 14 14 12 13 13 12
12 13 13 14 13 11  9 10 10 10 10 11  9 11 11 10 10 11 12 13
12 12 12 13 12 14 13 11 10 11 11 11 12 12 11 12 12 12 14 13
14 13 13 14 10 13 14 14 14 11 13 14 13 12 11 12 12 12 12 12
13 13 12 13 15 15 16 15 16 18 26 28 18 19 14 17 16 15 16 16
17 16 17 19 21 19 19 20 23 25 25 21 20 16 22 25 28 36 31 19
22 24 19 25 28 24 25 24 22 26 28 29 33 37 37 23 24 19 24 24

12 12 11 12 12 11 10 11 11 13 13 13 10 11 11 12 12 11 11 12
11 12 13 12 11 12 12 11 11 11 13 13 13 13 10 11 11 10 11 12
11 11 11 12 12 12 11 12 12 13 11 13 11 11 10 10 12 14 12 14
14 14 12 12 11 11 11 11 11 11 10 10 10 10 11 11 13 12 12 14
12 12 15 14 12 12 15 12 11 11 12 13 11 13 13 11 10 13 12 12
14 14 12 14 14 13 12 12 13 12 12 12 13  9 10 11 13 13 13 12
14 15 14 15 15 20 18 15 18 19 15 12 14 14 14 18 16 16 19 16
16 17 17 21 16 16 17 16 17 19 23 17 17 16 16 14 16 16 20 25
24 24 23 24 20 31 25 20 25 25 20 26 34 37 29 31 33 31 26 26

13 12 12 12 12 11 10 12 11 15 15 13 11 11 11 13 11 10 10 11
11 14 15 12 12 12 11 11 12 11 11 13 13 11 11 11 11 11 13 11
10 10 11 10 10  8 10 11 11 11 11 11 10 11 10 10 11 11 11 11
11 11 11 11 10 10 11 10 10 11 11  9 11  9  9  9 11 14 15 15
13 13 13 13 11 11 11 12 14 14 12 12 11 11 11 12 11 11 11 11
10 10 10 10 11 14 12 12 12 12 12 11  9 11 12 12 12 12 12 14
14 15 15 14 15 16 16 15 16 14 12 15 13 15 16 19 18 18 18 16
18 16 16 14 13 15 16 21 23 23 21 19 16 17 17 17 19 19 19 19
22 23 22 22 22 22 23 19 24 22 19 25 28 28 32 37 31 28 25 24

13 13 13 13 11 11 13 13 12 15 12 11 12 12 12 11 11 12 11 11
12 13 14 13 11 10 11 11 11 11 11 12 14 12 14 11  9 11 14 13
11 11 10 10 10 10 10 11 11 11 12 12 12 11 11 11 11 12 11 12
10 10  9  7  9  9 10 10 10 10 11 11 11 11 11 11 12 12 11 11
12 13 12 12 13 10 10 11 13 14 15 11 11 11 12 11 10 11 12 11
11 12 12 13 11 10 11 13 13 10 11 11 10 11 12 12 12 12 14 13
13 13 14 13 13 15 14 14 14 11 13 16 16 16 15 19 18 14 16 16
18 18 20 19 25 19 16 19 18 16 22 23 29 17 24 28 25 19 19 24
17 16 19 19 21 28 28 27 22 16 17 20 23 20 36 36 36 36 28 24

14 14 13 13 13 13 13 15 15 13 11 13 13 11 11  8 11 12 12 12
15 12 11 11 11 10 11 12 11  9 11 14 14 13 10 10 12 12 12 12
11 11 11 11  9 11 11 12 10 11 11 11 11 10 11 12 11 11 10 10
11 10  9  9  9 11 10 11 11 11 11 11 11 11 11 13 12 12 12 12
14 12 11 11 11 12 14 12 11 11 13 13 13 11 11 13 11 11 11 11
12 12 11 11 12 12 13 12 10 11 12 13 14 16 15 13 13 13 14 14
14 15 12 14 13 14 14 13 13 16 16 17 17 15 21 17 16 20 18 19
19 19 22 22 16 20 24 22 20 16 16 22 23 19 24 24 24 19 22 28
23 23 17 17 24 23 23 23 23 27 21 17 21 20 22 25 33 28 23 25

15 14 12 12 14 14 15 14 17 19 14 12 14 13 11 11 10 11 11 13
13 14  8 11 11 11 11 11 11 10 12 13 12 11 10 10 13 13 14 11
11 11 11 10 10 12 12 14 13 11  9 12 12 12 12 13 13 11  9  9
 9 10  8 10 11 12 11 12 12 11 11 12 12 12 12 11 14 14 13 13
13 10  9 10 11 12 12 12 11 12 13 14 13 12 11 11 11 11 13 11
13 14 13 14 12 14 12 12 14 12 11 12 12 15 13 13 13 13 14 13
15 16 13 13 14 15 16 14 14 19 17 14 14 14 14 13 14 16 20 16
15 16 16 20 22 15 18 19 22 16 16 22 22 22 25 24 26 25 20 24
```

```
22 23 24 18 16 18 16 17 25 25 17 17 21 26 17 16 21 26 26 26

16 11 11 15 18 18 18 16 18 16 15 14 15 15 12 11 11 12 11 11
11 13 11 11 11 13 11 11 11 12 12 14 11 11 11 12 12 12 10 10
11  9  9 11 11 11 11 11 13 12 10 11 11 12 12 11 13 11 11 11  9
11 11 11 11 11 11 11 11 13 11 12 12 14 12 12 12 12 12 12 11 11
11 10 11 10 11 11 11 13 15 13 12 14 15 14 12 12 12 11 12 12
13 12 12 13 15 14 13 14 15 15 14 12 15 24 23 14 14 14 13 12
13 15 13 13 15 16 16 14 13 16 16 16 23 17 16 16 15 16 19 19
16 14 14 14 10 20 16 15 20 17 17 20 25 29 29 24 25 23 22 19
20 25 22 18 24 26 16 15 16 16 18 18 16 21 16 16 23 23 23 24

11 11 15 16 19 19 15 16 15 18 15 14 13 12 12 14 12 10 11 12
12 12 12 13 13 16 13 13 13 11 13 13 13 13 12 12 10 10 12 14
10 10 11 12 11 11 12 11  9 10 11 11  9 10 11 11 11 11 11 12
12 11 12 13 11 11 12 12 11 13 11 13 13 13 13 12 14 12 11 11
11 10 12 15 13 11 15 20 18 11 11 10 11 13 11 10 12 11 11 14
15 15 15 14 16 16 27 23 14 13 14 14 21 27 15 14 14 12 14 14
14 14 14 15 13 13 13 12 12 12 15 19 22 24 22 19 16 17 19 19
17 21 21 16 14 14 17 17 21 21 22 15 20 20 17 17 23 28 28 33
31 28 25 23 28 25 20 18 18 18 15 15 16 18 19 24 23 18 16 25

16 14 16 19 19 19 16 15 15 15 15 18 15 13 13 13 10  9  9 10
12 12 12 15 15 14 16 12 13 13 13 13 13 12 14 11 10 10 13 13
11 11 10 11 12 11 11 10  9 10 11 11 11 12 10 10 11 11 10 10
11 12 14 14 15 14 12 12 12 12 12 13 13 12 11 11 13 13 10 10 10
11 11 11 12 12 11 12 12 10 12 11 11 11 11 13 18 13 15 19 15
11 15 19 15 15 21 33 23 15 15 14 14 14 16 16 16 13 14 14 14
14 11 13 14 14 14 12 14 11 12 12 12 13 16 19 22 16 19 18 18
22 22 18 15 16 21 29 28 24 24 17 22 16 17 17 17 19 22 28 34
34 36 33 28 25 25 25 29 23 18 18 19 16 18 19 22 24 22 15 23

15 16 17 16 14 16 16 15 14 12 12 15 13 13 13 12 12 11 11 11
11 11 11 11 11 13 16 15 14 15 15 15 14 13 13 11  9  9 11 11
10 10  9 10 10 10 11 10 11 11 11 11 12 11 12 13 11 11 11 11 10
10 10 10 11 11 10 10 12 10 11 11 12 11 11 10 11 11  9  9 11  9
11 10 11 10 10 11 10 10 11 11 11 10 10 12 19 24 31 26 25 18
22 26 16 14 14 16 22 16 15 15 15 19 18 15 16 15 13 13 14 14
12 12 11 11 13 13 11 13 13 14 14 14 15 15 18 23 15 15 16 25
25 15 16 19 23 23 28 26 24 24 15 16 15 14 15 15 15 23 23 34
37 34 27 26 23 23 27 25 22 20 21 23 19 19 25 28 23 22 18 19

13 13 16 14 13 13 13 14 13 13 14 15 15 15 15 15 16 15 15 12
13 12 11 12 11 11 13 14 15 16 15 16 14 14 11 10 11 11 11 12
12 11 11 11 11 11 11 10 13 10 11 11 14 14 14 11 10 11 11  8
10  9 10 10 10 10 10  9  7 10 10  9  9 10 11 11 11 11 10 11
11 12 12 11 12 11  9 11 10 11 10  9  9 11 17 25 29 25 25 28
16 19 25 20 16 13 13 13 13 15 15 15 16 15 15 13 15 15 15 14
12 12 12 12 12 12 12 12 11 13 13 15 16 20 17 17 18 16 16 15
15 15 15 24 31 26 24 20 19 24 24 19 15 14 14 13 14 16 21 24
23 24 23 23 23 27 27 23 22 16 19 28 27 34 36 27 20 20 22 16

19 18 16 15 12 13 13 13 13 13 12 14 16 17 20 20 17 17 16 15
16 15 12 12 14 14 12 12 11 13 16 15 11 11 10 10 10 10 11 12
12 11 11 11 11 11  9 11 11 12 11 11 12 14 13 10 13 13 13 11
10 11 10 10 10 10 12 10 10 10 10 10 10 10 11 11 11 10 10 10
11 11 12 12 11 12 12 10 11 11 11 11 11  9  9 23 30 30 30 34
27 15 25 16 14 12 12 14 12 12 13 13 12 13 13 12 13 13 13 11
11 11 11 12 14 14 14 14 14 13 13 14 14 16 14 14 15 14 13 15
15 17 16 15 16 18 19 18 22 18 20 15 15 19 15 13 13 15 16 16
19 24 24 22 25 24 19 19 17 17 19 19 23 27 23 17 22 21 16 16

14 14 14 14 12 14 14 12 12 14 16 16 14 16 17 12 12 12 17 22
```

```
23 21 20 22 23 14 14 14 12 15 14 11 10 10 10 10 10 11 13 10
 9 10 10 10 11 11 10 10 11 12 12 12 13 12 12 11 12 13 12  9
10 10 10 10 10 11 11 10  8 10  8  8 10 11 12 12 10  9 10 10
11 11 11 10 10 13 10 11 11 11 11 11 11 12 16 19 26 27 22 21
26 14 13 11 13 14 14 14 14 12 12 12 12 14 15 13 13 12 11 11
12 11 13 13 14 13 14 15 15 15 15 15 15 14 12 13 12 12 14 22
22 22 15 15 16 20 20 15 15 15 15 15 16 16 15 18 13 18 24 31
32 26 22 22 23 22 16 14 15 15 16 15 14 15 19 15 12 13 12 14

12 15 13 12 11 13 11 10 15 19 19 16 15 14 11 10 12 12 12 12
13 13 19 21 17 21 16 13  9 13 14 13 10 11 11 11 12 11 11 10
11 11 10 10 11 10  9  9  9 11 11 10 12 12 11 11 12 12 14 13
11  9 11  9 11 11 11 11 10 10 10 10 11 11 14 14 10  8 10  8
10  9  9 10 11 12 11 10  9  9 11 14 13 11 14 12 12 12 14 12
14 17 11 13 17 16 12 13 14 14 14 13 13 13 11 14 12 11 12 14
12 12 12 12 13 13 13 13 15 15 14 14 13 11 11 12 12 14 15 23
16 18 19 16 16 16 19 19 14 15 15 16 22 17 15 16 19 15 16 19
18 15 14 22 20 16 15 16 22 31 14 14 18 14 13 15 15 16 19 21

14 14 15 14 11 11 12 16 20 19 16 15 11 11 11 11 11 14 12 11
12 10 11 11 10 19 17 11  9 11 12 13 11 11 13 13 13 10  8 11
11 11 10 10 11 11 10 10 11 10 11 10 11 10 11 11 13 13 11 14
10 11 10 10 10 10 11 12 12 11 11 11 11 11 13 14 11 10 10 10
 8 10  8  8 10 11 11 11 11 10 11 13 11 10 11 10 10 10 13 13
13 14 12 12 10 18 16 14 13 12 12 12 13 13 13 11 11 11 13 13
13 12 14 12 12 14 12 15 12 12 12 12 12 11 11 11 13 13 14 22
18 16 23 21 21 19 15 15 15 16 16 16 19 17 17 22 21 25 20 20
16 20 33 15 14 12 14 16 23 24 24 16 19 25 22 20 20 14 18 18

15 14 14 10 12 16 19 18 16 12 12 12 13 12 11 11 13 15 14 13
11 10 11 11 11 11 10  8  8 10 12 13 13 12 12 12 11 10 11 11
11 13 11 11 10 11 10 11 11 11 10 11 11 11 12 12 12 12 10  9
10 10 10 10  9 10 11 12 12 11 11 10 11 11 12 12 12 11  9 10
11 11  9 10 11 11 14 14 14 14 11 11 10 10 11 12 12 16 15 13
10 10 11 14 13 11 11 14 14 14 12 14 14 13 12 11 12 13 12
11 13 13 13 13 13 13 12 12 14 12 11 11 12 12 13 13 15 18 16
16 16 18 22 24 15 15 14 14 19 23 19 19 31 40 22 22 22 19 19
22 22 30 22 22 23 25 25 30 33 36 31 31 28 22 16 16 16 19 16

14 16 13 13 14 20 20 20 15 13 12 12 12 14 12 12 15 15 15 13
11 12 13 11 11 10  9  9  9 14 18 16 15 14 11 11 12 11 12 12
12 12 13 13 13 11 10 10 13 11 11 10 10 11 14 12 11 12 11 10
10 10 11 10 10  9 10 10 11 11 11 11 11 10 11 12 14 12 12 11
11  9  9 11 11 13 13 15 15 13 10  9 10 11 12 14 15 18 15 24
12 13 13 15 15 11 12 14 14 14 13 13 15 14 12 12 12 12 12 12
12 13 13 12 12 13 12 11 13 14 13 11 13 13 14 14 15 15 16 18
16 19 18 19 19 16 13 13 16 18 18 19 23 31 37 37 28 23 23 23
31 28 21 31 35 31 28 28 31 36 36 28 27 27 27 20 18 18 19 19

15 15 14 12 13 17 17 17 17 13 11 11 13 14 14 14 16 18 21 19
16 15 16 14 12 11 13 13 16 18 12 13 10 10 11 14 16 14 14 14
14 12 11 12 12 11  9 12 12 12 12 11 11 11 13 13 11 11 11 10
 8 11 11 10 10 10  8  9 10 11 11 12 10  9  9 11 11 11 11 11
11 10 10 10 12 18 13 12 12 12 11 11 11 11 11 11 11 11 16 20
21 16 16 16 14 12 12 15 15 13 13 12 12 12 11 10 11 13 13 13
14 14 12 14 12 14 14 12 13 14 13 11 12 12 13 14 16 16 15 15
15 16 16 15 15 12 12 14 13 16 18 18 16 18 19 36 36 29 24 23
22 24 19 18 20 31 24 18 31 40 37 28 31 28 31 27 17 17 17 20

15 16 15 18 21 19 19 16 14 12 11 12 14 16 17 16 16 17 21 23
20 20 17 15 15 12 12 11 12 11 11 11 11 12 13 15 15 13 13 11
10 11 13 11 10 11 12 14 14 12 12 12 12  9 11 12 12 11 10  9
11 11 10 10 10  8 11 10 10 10 11 12 10  9 10 11 12 12 11  9
```

```
 9  9  7  9 10 16 15 12 12 12 12 12  9 11 12 12 12 12 14 15
15 14 13 11 11 11 12 14 14 14 14 14 14 11 10 12 12 12 12 14
13 14 13 13 13 11 12 14 14 12 12 12 12 13 13 14 16 13 14 14
15 15 14 13 14 14 14 14 16 14 15 16 16 19 24 26 24 25 25 18
20 19 16 18 18 13 13 15 37 31 29 29 31 37 33 19 19 19 14 19

18 19 18 18 18 15 16 15 13 12 12 15 16 18 16 15 13 13 16 16
16 14 13 16 17 16 14 13 10 13 10 14 14 12 14 14 16 16 12 11
12 12 12 10 10 10 10 13 13 11 12 12 11 11 10 11 11 11 12 12
11  9 11 11 10 10 10 10  9  9 10 12 11  8 10 12 14 12 10 11
10 10  9  9 11 13 10 10 13 11 13 13 14 11 11 11 12 12 10 11
12 14 13 13 12 13 11 11 11 14 14 14 13 12 10 11 12 14 14 13
12 12 12 13 12 12 13 13 13 11 11 13 13 12 14 15 15 15 14 12
14 13 13 12 12 12 13 14 14 14 14 15 16 17 21 18 28 28 24 28
24 26 22 13 15 15 15 13 19 26 31 33 29 29 40 24 23 21 19 24

18 18 16 19 18 13 13 13 11 14 13 14 15 16 15 11 12 14 15 15
16 15 11 12 16 22 19 12 12 12 11 13 14 13 13 13 12 12 11 14
12 12 12 12 10 10 11 11 12 10 10 10  9 10 10  9 11 10 11 12
11 12 12 11 13 12 11  9  9  9 11 10 10 10 13 21 28 21 11 12
11 10 10  8 10 13 10 11 13 12 12 11 13 10 11 11 10 11 11 11
12 14 14 12 12 12 12 13 13 12 13 13 12 11 11 13 11 11 13 12
12 14 12 12 12 12 12 12 12 12 12 12 13 13 13 15 16 13 11 11
14 14 14 12 14 14 15 14 14 14 14 16 17 16 16 17 21 23 28 30
34 33 37 29 19 16 15 15 16 19 25 26 26 22 24 37 31 22 15 23

17 15 16 16 15 14 12 11 12 15 13 13 12 10 11 11 15 15 11 11
13 11 10 11 19 18 12 12 14 12 12 13 13 12 12 13 13 11 13 13
13 13 12 12 10 11 11  8 10 10  9 10 11 10 10 10 13 11 11 11
11 11 10 11 11 12 11 10 10 11 12 19 19 19 26 17 14 11 13 11
10 10  9 13 12 11 11 12 12 11 12 11 10 11 12 10 10 10 14 13
14 13 13 14 14 12 14 14 12 11 11 12 11 11 12 13 11 11 14 14
14 13 13 15 13 14 14 14 12 14 12 12 13 13 13 12 11 12 11 11
13 11 11 11 13 13 16 15 15 16 16 15 12 16 16 14 16 23 31 27
30 30 33 33 27 21 14 16 14 16 24 29 29 24 24 22 19 18 22 20

21 14 14 13 13 14 15 11 11 12 12 11 11 12 11 15 13 12 11 11
11 12 13 14 22 23 14 11 13 12 12 12 12 12 11 12 12 12 12 13
12 12 13 10  9 10 10 10 10 10 11 10 10 11 10 10 10 10 10 10
 9 10 11 10 11 13 10 10 10 11 19 16 10 11 12 10 11 12 11 12
11  9 12 14 11 11 13 13 11 11 13 12 11 12 11 11 10 10 11 11
11 11 11 11 13 15 13 13 11 11 13 15 12 12 14 12 11 11 12 12
13 13 14 13 13 13 13 13 13 14 13 14 12 14 14 12 11 12 12 12
13 12 12 12 12 13 14 14 14 13 13 13 13 14 18 21 18 16 19 28
27 26 26 27 28 23 19 15 15 14 14 16 25 25 24 24 29 19 15 15

21 19 14 12 14 13 14 12 10 10 10 10 11 10 10 11 11 11 11 11
14 19 13 15 16 16 15 12 12 13 13 11 10 11 13 12 14 11 12 12
12 11 13 11  9 11 11 11 12 10  9  9  9 10 11 11 11 12 10 10
10 10 11 12 11 11 12 13 13 11 11 10 11 10 10 10 10 12 14 11
11 11 12 15 16 12 12 11 11 12 11 11 11 10  9 10 11 11 11 11
11 11 11 11 12 13 12 12 12 12 12 11 14 13 10 13 14  9 12 12
14 14 14 14 14 12 12 12 12 14 16 14 13 13 13 11 11 13 13 12
12 12 14 14 14 15 15 12 12 12 12 12 15 13 16 16 16 19 22 19
25 19 19 19 21 23 28 17 16 13 14 13 13 17 23 23 21 17 15 21

21 16 16 16 16 16 14 11 11 12 13 11 11  9 10 11 11 13 15 20
20 16 14 12 12 12 12 12 12 11 11 12 11 12 11 11 11 11 10  9
 9 10 10 11 10 11 10 11 12  9  9 11 12 12 12 11 11 11 11 11
10 11 11 12 23 19 15 15 12 11 11 12 11 10 11 12 10  9  9  9
10 10  9 15 11 11 12 11 11 11  9 11 11  9 11 12 12 10 11 11
11 13 11 13 12 12 14 14 12 11 11 11 11 11  9 10 11 11 13 13
13 13 13 13 14 12 12 14 14 14 14 13 13 13 12 12 12 13 13 13
```

```
13 14 14 15 15 15 14 14 14 12 14 15 16 20 19 16 19 25 18 15
15 14 22 23 19 22 25 28 25 19 18 18 19 23 26 23 22 17 16 27

14 16 16 15 14 13 11 11 11 11 11 12 12 14 11 11 12 14 13 14
16 15 11 13 15 15 14 13 12 11 12 11 11 11 12 12 12 11 10 10
10 10  9  9 10 10 10  9 11 11 11 12 14 12 12 11 11 12 12 11
11 11 11 14 17 14 13 13 13 11 12 11 10 11 12 14 10 10 10 10
11 10 10 10 11 13 11 10  9 10 11 11 10  8 10 11 11 12 12 11
11 11 11 11 13 13 14 13 11 10 10 11 12 11 11 10 11 12 13 12
12 12 12 13 13 13 13 14 14 14 14 14 12 12 14 12 11 11 13 13
13 13 13 14 16 14 14 14 14 14 14 14 15 16 18 19 18 15 14 13
16 18 22 26 31 26 29 33 29 26 22 18 15 21 19 15 15 15 18 24

15 16 18 16 15 13 10 10 10  9 10 10 10 13 14 12 12 14 14 12
12 14 13 13 12 12 14 13 13 13 11 10 13 13 11 12 14 11 10 11
11 10  9 10 10 11 10  9 10 10 11 11 13 13 11 12 11 10 10 10
11 10  9 11 14 13 12 12 12 13 13 10 10 11 13 11 10 11 11 10
11 12 11 10 10 11  9  9 11 11 11 10  9  7  9 11 11 12 11 10
10 11 11 11 11 12 11 12 12  9  9 11 11 13 11 11 13 14 14 12
12 15 15 14 15 13 13 13 16 15 13 14 13 13 11 14 14 13 12 14
14 15 14 14 15 16 13 12 14 14 14 16 15 16 16 16 16 14 13 15
19 16 21 25 29 25 22 25 31 26 24 22 18 19 22 19 15 15 16 14

14 16 16 14 12 11 10 11 10 11 12 12 13 14 14 11 11 11 11 11
14 11 11 12 14 16 21 12 12 12 11 12 12 13 13 13 11 10 11 10
11 12 11 11 11 10 10 10 12 12 12 12 11 10 10 11 10 10 10 10
10 11 14 15 12 12 12 12 13 13 12 11 12 11 11 11 13 13 11 13
11 10 11 11 11 10 10  8 11 13 11 10 10 10 10 11 11 10 10 10
10 10 11 11 12 12 12 11 11 11  9 11 11 12 12 11 11 13 13 13
13 11 10 12 12 16 16 15 16 15 12 13 13 15 18 15 13 13 13 13
14 14 14 14 15 15 15 14 15 15 18 14 16 16 17 16 14 12 14 17
16 17 17 17 20 19 23 33 31 29 31 28 19 18 18 18 15 13 12 13

23 17 15 14 13 11 10 11 11 14 14 12 14 12 12 13 10 10 11 11
11 13 11 13 13 15 19 12 11 14 12 14 14 12 12 14 13 11  9  9
11 11 11 10 10 10  9 12 12 12 12 12 12 11  9 10 11 10 11 10
11 11 13 13 13 13 13 13 12 11 11 12 11 11 11 11 12 12 11 12
11  9 10 10 10  9 10 10 11 15 12 11 12 15 11 11 11 10 10 11
12 11 10 13 13 11 11 11 11 11 11 11 12 15 16 13 13 13 12 13
13 13 12 11 11 13 13 14 13 13 15 14 12 14 14 12 13 13 12 12
13 15 13 13 15 14 14 15 15 16 15 18 16 18 18 14 14 14 12 14
16 21 17 13 16 22 21 27 30 28 33 30 16 16 16 18 18 16 19 12

23 22 16 16 14 13 13 14 13 13 14 13 11 12 11 11  8 10 11 11
10 11 11 11 13 13 20 16 13 13 13 13 14 14 14 14 11 11 11 10
 9  9 11 11  9  9  9 11 13 13 13 13 10 11 11 12 12 12 10 11
11 10 10 11 12 12 12 11 10  9 10 10 10 11 10 10 10 11 11 11
11 11 11 11 11 11 11 11  9 11 13 10 11 14 15 14 11 10 10 10
11 11 14 12 12 10 10 11 12 12 11 13 13 13 14 14 14 14 12 12
11 12 11 12 12 13 13 14 13 12 13 13 14 15 14 14 12 14 14 14
15 15 15 15 15 15 15 15 16 16 18 16 16 15 14 13 13 15 16 16
15 16 15 15 16 16 17 23 27 26 26 27 25 17 16 17 15 15 15 16

24 19 18 16 14 12 13 12 12 13 12  9  9 10 10 10 11 11 13 11
11 11 10 12 18 21 16 11 12 11 11 11 11 11 11 11 10 10 11 10
11 10 10 10 11 11 11 12 12 12 12 11 11 10 10 11 10 10 10 10
10 10 14 12 12 12 12 13 10 13 11 11 11 11 10 10 11 11 11 10
10 11 12 11 12 11 10 10  8 11 11  9 11 11 12 11 11 11 13 14
13 11 12 12 10 11 11 11 12 12 11 11 12 11 11 14 13 13 10 10
11 13 12 12 14 14 14 14 12 12 12 13 13 13 14 13 11 11 14 14
15 16 15 15 16 16 18 16 18 18 18 18 15 12 11 13 13 15 14 14
14 15 19 19 15 15 19 25 28 29 21 23 21 19 16 21 19 16 22 17
```

```
29 23 15 11 12 14 14 14 12 11 11 12  9 11 12  9  9 11 11 11
10 10 13 15 16 21 16 10 11 12 11 12 11 11 11 11 10 10 10 10
10 10 10 10  9 11 14 14 12 14 14 12 10  9 11 11 11  9  8 11
11 11 13 13 11 11 11 11 10 12 10 10 11 11 11 11 11 11 11 12
11 10 11 11 10 10 10 10 10 11 14 11 11 11 11 12 11  9  9 11
 9 11 11 10 11 14 14 13 13 12 12 12 12 12 11 12 11 10 11 12
12 12 11 11 14 15 14 13 13 14 15 14 14 14 14 15 13 13 13 12
14 14 16 14 10 17 16 16 16 16 16 16 12 11 11 11 12 13 12 13
13 15 16 20 18 16 19 18 19 18 19 19 21 18 19 16 16 19 22 27

25 15 15 15 12 12 10 11 11 10 11 11 11 12 12 12 11 12 12 13
13 12 13 12 12 17 22 21 13 11 11 11 11 12 12 12 11 10 11 10
10 10 10 10 10 11 11 13 13 13 11 10  9 10 10 11 10 11 10 10
 9  9 11 14 13 13 12 11 10 11 11 11 10 11 13 11 11 12 11 11
11 11 11 10 10 11 10  9 11 10  9 10 11  9 10 11 11 10 10  8
10 11 11 12 12 12 12 13 12 11 13 11 11 14 11 11 10 10 11 11
14 12 12 12 13 15 13 12 13 13 14 14 14 14 15 14 15 15 15 14
14 15 16 16 14 17 17 17 17 14 14 11 11 11 11 13 13 15 14 11
14 15 14 15 22 19 16 15 13 16 16 22 19 22 26 23 15 15 25 24

14 15 15 15 15 13 13 13 12 13 12 10 11 11 12 11 11 11 14 11
11 11 11 14 16 19 22 21 14 11 13 13 13 13 14 13 11 10 10 10
11 12 14 12 12 12 12 12 13 12 11 10 10 10 11 11 12 11 10 11
10 14 18 14 14 12 11 11 12 11 11 11 13 11 10 11 11 11 11 13
11 12 11 10 11 11 10 10 10 10  9 11 10 11 11 10  9  7 10 13
13 12 14 14 12 12 12 12 11 12 13 13 13 13 10 11 11 10 10 11
11 12 12 11 12 12 12 12 13 12 12 12 12 12 13 14 14 13 13 13
13 14 15 16 18 18 16 15 12 13 12 11 12 12 11 12 14 11 11 13
14 15 14 14 15 18 19 19 15 15 19 25 24 24 25 22 16 18 19 19

22 21 16 20 15 15 14 16 18 22 15 10 11 11 10 11 11 10 12 10
10  9 11 16 14 19 19 14 12 14 12 12 12 13 12 12 12 12 11 11
13 13 14 13 13 12 14 14 11 11 10 10 11 11 11 11 11 11 11 11
11 13 15 14 14 13 12 11 11 12 11 11 11 12 12 12  9 11 11 11
11 11 10 11 11 11 11 11 11 11 11 10 10 10 11 10 10 10 10 12
12 13 23 24 11 11 13 11 12 12 14 14 12 12 13 12 12 12 11 11
12 13 13 11 11 13 13 11 12 12 12 12 12 12 14 16 15 16 15 15
16 15 15 16 16 16 15 13 11 12 11 12 11 11 12 12 12 12 11 11
13 13 14 14 14 15 17 22 23 17 18 25 23 24 29 33 29 13 15 19

13 16 17 23 21 23 27 23 22 20 20 14 15 12 12 11 11 11  9 10
11 13 16 15 11 12 16 12 10 11 11 12 12 12 14 14 14 11 11 12
12 12 12 12 11 13 13 13 10  9 10 11 10 10 11 11 10 11 11 10
12 13 12 12 11 11 11 11 11 11 10 10 10 11 11 10 11 11 10 14
16 15 18 13 12 12 13 12 11 13 13 11 11 11 12 12 12 14 12 12
12 12 12 12 11 12 12 11 13 11 10 11 13 13 15 15 16 16 15 14
15 14 15 16 15 15 13 11 11 13 11 11 11 13 13 11 12 12 11 11
14 14 15 16 14 16 16 14 13 16 22 22 22 27 25 22 23 16 12 18

19 25 28 26 26 26 23 19 17 21 22 22 20 17 15 14 17 11 11 15
16 12 12 12 12 11 11 12 12 11 11 11 12 12 11 12 11 12 14 14
12 12 11 12 13 12 11 11  9  9  9 11 10  9 11 10 10 10 10 10
15 23 15 15 12 12 10 10 10 10 11 11 12 10 10 11 11 11 13 12
12 11 11 11 11 12 12 11 11 11 11 12 11 11 11 10 11 11 11 11
12 14 12 11 11 11 11 12 11 11 11 12 11 11 11 13 13 13 14 14
12 14 12 11 11 12 12 12 12 11 12 12 12 14 15 15 16 15 16 15
14 15 14 14 14 11 11 12 11 11 12 15 24 24 16 10 11 11 13 13
14 13 15 14 14 14 14 14 14 17 17 19 16 14 14 16 25 20 16 17

14 15 16 24 18 18 16 18 16 15 16 16 17 16 12 14 14 20 22 25
17 13 14 13 11 11 12 12 12 12 12 12 12 13 13 13 11 11 10 11
11 11 10 11 11 12 11 11 11 11 11 11 11 11 11 11 11 12 11 10
```

```
11 13 15 15 13 12 14 11  8 11 12 11 12 11 11 11 11 11 11 11
11 10 10 11 11 11 11 11 11 11 11 10 10 11  9 11 11 12 11  9
11 10 10 11 11 11  9  8 10 12 15 12  8 10 10 11 12 12 13 13
11 13 13 13 11 13 13 12 12 12 12 14 14 14 16 16 14 13 14 14
13 13 14 11 13 11 11 13 11 11 12 12 18 19 15 11 11 12 12 13
13 12 13 13 13 13 14 15 15 16 19 21 16 16 15 21 21 27 26 21

16 18 18 14 12 12 11 11 12 11 14 15 16 18 16 18 18 19 21 22
21 13 12 13 11 13 11 11 13 14 12 11 14 14 14 11 10 10 10 10
11 10 10 10 10  9 11 10 11 11 12 12 12 14 12 12 12 12 11 12
17 17 12 12 12 13 13 13 11 10 10 11 11 11 11 11 11 11 11 11
11 10 10 10 10 11 10 11 10 10 10 10 10 10 11 12 12 12 11 11
 9 11 11 11 11 12  9 10 11 13 13 11 10 10 11 12 12 14 15 15
12 11 11 11 12 12 12 13 13 13 13 13 13 14 14 15 15 14 14 14
12 13 13 11 11 12 12 12 11 11 10 10 11 11 11 11 12 12 12 12
14 12 12 12 12 13 15 15 13 16 19 14 15 18 14 13 33 28 23 18

16 16 16 12 13 12 13 13 13 12 12 14 16 15 16 18 22 23 23 16
12 14 14 13 12 11 12 12 13 13 13 13 11 10 10 10 10 11 11 11
10 11 11 10 11 10 10 11 10 10 10 10 12 11  9 10 11 12 12 12
11 11 11 12 12 12 13 12 11  9  9 11 11 11 11 11 11 13 13 11 10
10 10 10 12 11 10 11 11 11 11 10 10 10 13 13 11 11 10 11 11
15 16 12 11 11 11  9 11 12 12 12 11 11 13 14 14 14 14 11 15
14 11 12 11 11 12 12 12 12 13 12 13 15 14 14 13 11 10 10
12 11 11 12 11 12 12 11 11 11 12 12 11  9 13 13 11 11 13 13
11 11 12 14 15 16 15 12 15 12 12 13 13 18 18 18 19 22 26 23

15 16 15 14 14 14 16 19 19 13 15 11 12 13 15 18 22 24 24 18
12 12 12 14 12 11 12 12 14 14 14 16 14 12 11 11 10  9 11 10
10 11 11 11 11 12 11 11  9 10 10 11 10 10 11 10 10 10  9  9
 9 11 11 12 14 12 12 11 10 11 11  9 11 11 12 12 12 13 11 10
11 14 14 11 13 11 11 11 15 14 15 11 10 10 11 11 11 10 10 11
10 11 14 14 10 16 11 12 12 11 11 11 11 12 13 13 13 13 12 13
11 11 13 14 11 11 10 12 10 11 11 11 10 13 12 10 10 11 12 12
11 11 11 11 10 11 11 11 11 12 12 11 11 10 11 12 12 13 13 13
12 14 16 14 14 14 13 13 12 12 14 15 16 19 16 13 18 25 24 18

16 14 15 15 15 15 14 15 18 18 16 15 14 15 18 22 25 25 24
18 13 12 12 12 11 11 12 12 15 15 12 11 10 10 11 11 11 11  9
 9 11 11 11 11 13 13 11 10 10 10 11 10 11 10 10 10  9 10 10
10 13 13 12 11 11 12 11 10 10  9 11 12 12 12 14 14 14 13 12
12 13 14 13 16 11  9 11 11 13 15 11 10 11 11 12 11 11 11 10
10 10 10 10 11 12 11 11 11 11 11 10 11 12 14 12 14 11 11 12
11 11 11 12 12 11 11 13 13 13 13 11 10 11 11 11 10 12 14 11
12 12 12 11 11 11 11 11 13 10 10 11 11 11 10 11 12 12 14 14
14 12 13 13 13 13 12 12 13 13 14 17 17 16 15 15 15 18 25 24

22 16 16 16 17 17 16 14 20 22 22 22 21 21 23 22 28 25 22 15
14 14 12 12 11 11 11 10 12 11 10 10 10 11  9 11 11 11 11 11
10 11 11 11 12 12 13 12 11 10 11 10 11 10 10 10  9  9 10 11
16 16 14 12 11 11 12 10 11 12 10 11 11 11 12 12 11 12 14 14
15 22 15 12 17 12  9 11 13 12 11 10 10 10 11 11 13 11 12 10
11 10 11  8  8 11 11 13 12 10 11 10 11 13 11 10 11 11 14 12
12 11 12 12 11 11 12 12 12 12 11 12 13 11 10 11 14 13 10 12
11 11 11 12 12 12 11 11 11 11 11 12 12 13 11 13 11 13 13 11 10
12 14 15 14 12 12 14 14 12 13 13 14 16 16 16 17 16 22 30 27

23 18 16 22 23 25 23 18 16 17 23 23 26 26 26 25 25 16 11 13
13 13 14 12 12 11 11 11 10 10 11 11 11 11 11 10 10  9 10 10
10 10 11 12 12 12 12 11 11  9  9 11 11 12 11 11 11 10  9 15
22 22 14 11 12 11 11 12 11 10 12 10 10 11 11 11 10 13 11 12
14 14 12 12 15 12 11 12 14 12 11  9  9  9 11 11 11 12 11 11
10 10 11 10  9 10 12 12 11 14 14 11 12 11 11 12 12 11 11 11
```

```
11 10 11 11 11 12 12 12 12 11 10 10 11 11 11 11 12 12 11 11
11 11 10 11 11 11 12 11 11 11 15 18 16 11 12 12 13 13 13 13
13 13 14 13 14 16 16 15 14 12 14 16 18 18 21 22 22 19 14 13

26 23 26 26 26 26 24 23 16 22 24 25 24 24 29 27 26 26 21 13
12 12 14 13 11 10 10 11 11 12 11 11 11 11 11 11 11 11 11 10
10 11 11 10 11 12 12 12 10 10 10 12 11 10 11 11 11  9  9  9
12 12 12 13 13 13 13 14 11 10 10 10 10 10 10 10 11 11 11 13
16 12 12 12 11 10 11 14 11 10 12 10 10 12 12 12 14 12 13
11 11 12 12 11 10 11 14 13 10 10 14 12 11 11 11 11 11 12 11
12 12 11 11 12 11 11 11 11 11 11 10 10 12 11 11 12 11 11 11
 9 11 11 11 12 12 11 13 14 15 22 22 20 12 14 15 15 15 15 15
12 12 13 13 15 13 15 16 15 14 15 16 18 18 19 21 19 15 14 14

26 26 26 25 25 24 25 24 28 28 26 23 22 23 28 25 24 18 12 12
12 14 13 12 11 11 11 11 11 10 11 10 11 11 10 11 11 10  9  9
10 11 11 11 11 12 11 11 10 11 10 10 10 10 10 11 11 12 11 12
12 12 12 12 13 13 14 13 12 10 10 10 11 11 14 15 14 14 14 16
12 12 14 15 13 11 11 12 11 11 12 14 14 12 11 12 12 14 12 11
10 14 12 11  9  9 12 13 17 11 13 14 11 11 11 10 10 11 11 12
12 11 10 12 12 11 11 10 11 10 12 10 10  9 10 11 11 11 12 12
11 12 12 14 14 13 14 13 14 22 27 23 21 20 20 17 20 15 15 15
14 15 15 16 14 12 15 16 15 15 15 15 18 16 14 13 13 14 15 18

31 29 24 22 24 24 25 26 20 16 22 24 22 22 16 16 16 14 12 12
12 14 12 11 12 12 11 12 11 11 12 12 11 11 10 11 10 10 10 10
11 10 11 12 12 12 11 12 11 11 10 10 10 11 10 11 11 11 11 11
11 11 11 11 12 14 12 14 12  9 11 12 12 11 12 13 13 14 15 15
15 13 13 14 12  9 10 12 14 14 13 15 13 12 12 12 12 11 11 11
12 12 10 10 10 12 12 14 15 14 11 11 12 13 12  9 11 11 11 11
13 11 11 13 12 10 10 10 12 11 10 12 11 10 12 12 12 12 10 10
10 10 11 13 15 18 19 16 21 28 28 23 22 21 22 23 23 23 23 16
16 16 20 17 20 17 14 15 18 16 16 18 18 16 11 11 13 15 18 18

23 20 21 25 27 25 27 28 22 19 24 23 22 23 18 12 12 12 12 12
13 12 12 11 11 11 10 11 11 10 11 11 12 12 11 12 11 11 11 11 12
11 10 11 11 11 13 11 13 11 11 11 10 10  9  9 10 10 11 12 12
12 11 11 11 11 11 11 12 11 11 11 11 12 12 14 11 11 12 13 16
16 13 13 13 11 10 13 14 13 14 15 15 12 12 11 11 12 13 11 11
13 13 12 11 12 12 12 14 11 12 11 16 24 18 14 11 11 11 11 11
11 11 12 11 11 11 10 11 11 11 13 12 11 10 11 14 14 14 12 12
13 12 12 12 13 18 18 23 28 24 19 19 18 15 15 15 15 16 19 17
14 17 17 17 17 17 16 20 20 16 14 15 11 14 11 14 16 18 18 15

20 23 26 27 27 27 28 23 16 17 20 21 27 25 23 14 12 11 12 12
11 11 11 11 11 11 10 11 11 11 11 11 11 11 12 12 12 12 12 11
12 11 11 12 12 12 12 11 11 11 10 10 10 10 11 10 11 11 12 12
12 12 12 12 11 11 11 10 10 10 11 12 12 11  9 11 12 16 16 14
14 14 12 13 11 11 12 12  9 14 13 13 11 11 11 11 12 11 12 14
18 15 14 12 11 12 12 13 19 12 14 14 18 24 23 15 11 11 10 12
12 11 11 11  9  9 11 11 11 12 13 10 11 13 13 13 11 14 12 14
15 12 11 11 10 22 22 19 15 15 16 19 18 14 15 14 15 15 16 15
16 16 18 18 18 17 17 16 14 13 13 12 11 14 16 21 16 14 16 12

19 24 28 25 24 25 27 23 23 20 17 20 23 27 17 13 14 13 13 11
11 11 11 12 11 11 11 11 12 11 12 11 10 11 10 11 11 10 10  9
12 12 11 11 11 11 12 12 13 12 11 11 12 11 11 14 14 13 14 14
14 12 12 14 12 12 11 11 10 11 12 12 11 10 10 10 12 14 11 11
12 14 12 15 12 10 10 11 11 11 11 12 13 13 12 12 13 13 13 14
15 13 11 11 12 15 11 14 14 13 11 11 12 12 13 24 14 11 11 11
12 11 11 11 11 10 11 11 12 12 12 11 11 12 11 12 13 13 11 11
11 11 11 13 12 14 14 14 14 16 15 13 15 15 13 12 13 15 15 15
15 16 16 16 15 18 18 18 15 14 12 12 12 12 13 14 17 17 16 17
```

```
23 23 24 24 26 26 24 24 23 19 18 22 25 23 23 16 13 14 13 12
11 11 11 11 11 11 11 11 11 11 10 11 10 10 10 10 10 10 11 10
10 10 10  9 10 10 10 12 11 11 12 12 11 11 12 11 11 12 13 13
12 14 13 11 11 11 13 13 10 10 11 12 11 11 11 13 13 12 11 11
13 15 12 11 11 10  9 10 12 12 14 15 16 14 14 14 12 13 12 16
14 13 12 15 17 15 11 15 11  9  9 11 12 11 18 24 31 20 18 20
18 15 11 11 10  9 10  9 10 11 11 14 12 14 12 11 12 11 11 12
11 11 12 12 13 11 11 13 13 13 13 15 15 15 16 16 15 14 15 15
16 18 16 18 18 16 18 16 14 13 11 11 14 12 16 18 18 19 16 16

24 24 25 26 25 23 26 26 24 15 23 26 25 22 16 14 15 12 11 12
12 11 11 11 12 11 10 11 10 10 11 11 11 11 10 10 10  9 10 10
10 10 10 10 10 11 11 10 10 12 12 11 10 11 11 11 12 14 14 14
12 12 11 11 12 12 13 11 11 13 14 13 14 14 12 12 12 11 12 14
12 11 10  9 10 11 12 15 15 14 14 12 11 11 11 12 12 12 12 14
15 16 12 13 13 11 17 16 10  9 11 14 14 13 13 23 15 15 19 10
22 12 10 11 11 11  9 10 10 11 11 10 11 10 12 14 10 11 12 12
11 12 14 12 12 12 13 12 12 12 14 14 14 15 14 13 14 15 15 16
18 16 16 18 18 18 15 12 12 11 11 11 14 16 18 15 16 15 16 16

25 28 28 24 22 22 25 25 19 15 19 22 26 23 15 15 14 11 11 12
11 12 12 12 12 12 11 11 11 11 11  9 10 10 10 10 10 10 10 11
10 10 10 11 14 19 12 12 12 12 12 10 11 11 10 11 10 12 12 12
14 12 10 11 12 14 12 13 12 11 11 11 12 12 14 13 10 11 15 15
11 10  9 10 12 12 12 12 13 11 11 11 10 11 11 11 12 11 11 12
12 11 16 14 11 12 12 12  9 12 12 11 13 12 13 11 10 11 10 11
10  9 10 11 11 10  9 10 11 11 11 10 11 10 10 12 11 11 12 12
10 11 14 12 12 12 15 14 11 12 12 13 13 13 13 13 13 14 15 15
14 15 15 14 15 15 12 12 12 10 11 12 15 16 19 16 16 18 19 19

25 27 27 21 20 18 23 22 19 16 16 16 28 25 15 12 11 10 10 11
11 11 10 10 11 11 14 15 11 10 11 11  9 11 11 11 11 14 17 13
13 13 14 20 23 16 15 14 14 12 12 11 10 11 11 11 10 11 11 18
11 11 10 12 11 12 10 11 12 12 12 11 11 11 11 11  9 11 12 14
14 10 10 13 13 11 11 11 12 11 11 10 11 11 12 12 12 12 12 13
12 12 15 18 12 10 11 12 10 12 12 12 12 12 12 13 11  9 11  9
11  9  9 11 14 10 10 11 11 11 10 11 11 11 12 12 11 11 11 11
11 13 13 12 15 12 10 11 12 11 11 12 14 12 12 12 12 13 12 13
13 13 12 12 14 14 11 11 10 11 11 14 14 16 18 16 18 18 19 16

26 23 23 26 20 17 22 22 20 20 20 22 19 14 14 11 12 12 11 11
11 12 16 18 11 14 11 12 14 11 11 11 11 10 11 12 15 19 14 16
21 22 22 27 20 14 14 15 14 11 11 11 11 11 10 11 11 11 15 16
11 11 12 11 10 11 10 11 11 12 12 11  9 12 12 15 15 14 15 14
12  9 11 12 12 12 12 13 11 10 10 13 13 13 12 11 14 14 12 14
12 15 20 15 10 11 11 11 10 11  9 10 10 11 10 11 11 11 11 11
14 11 16 22 23 22 19 19 11 11 11 11 10 10 11 11 11 12 11 10
11 11 11 13 12 12 11 12 12 11 11 10 12 12 11 11 12 12 14 12
14 14 14 13 12 12 11 11 12 12 11 14 22 22 20 16 16 18 19 23

24 22 24 24 21 17 17 17 14 14 14 21 20 16 14 11 11 11 12 14
14 16 22 23 16 16 18 18 20 20 20 10  9 10  9 12 18 23 23 24
24 28 25 24 19 17 13 14 13 13 12 11 10 10 10 10 11 11 15 19
18 15 11 11 10  9 11 12 12 12 13 13 12 10 11 14 11 11 12 11
14 12 12 12 11 10 11 11 11 11 11 12 13 13 13 14 15 14 14 14
11 11 11 12 11 14 11  9 10 10 10 10  9 10 11 10  9 10 11 10
11 19 25 28 22 28 29 28 22 22 21 16 13 14 21 11 11 11 13 11
10 11 11 12 14 12 12 11 12 11 11 12 13 12 12 13 14 12 12 14
12 11 11 12 12 11 12 14 15 15 12 12 14 22 22 19 16 22 22 20

24 24 24 23 22 19 22 16 15 16 18 13 14 14 13 11 11 12 11 13
13 20 22 21 22 22 22 24 25 24 22 19 12 15 13 18 22 19 22 24
```

```
28 26 19 14 12 18 15 14 16 14 12 11 11  9 11 11  9  9 11 23
27 25 17 14 14 13 11 11 12 12 12 14 12 11 10 10 10  9  9 10
10 10 11 10  7  9 10 10 10 11 12 12 12 12 12 12 13 13 14 13
12 10 11 10 11 15 13 11 10 10 10 11 11 11 11 11 10 10 10 10
10 14 18 23 23 24 28 28 29 29 25 22 11 11 12 22 22 22 16 12
13 11 11 11 11 11 13 13 11 14 12 12 14 11 12 14 12 12 13 13
12 12 12 12 11 11 11 12 14 12 12 12 14 18 18 16 16 17 16 14

31 28 26 22 22 18 16 16 18 16 18 15 15 14 14 15 16 13 14 13
16 22 20 20 17 25 23 21 22 21 24 24 24 23 24 23 22 22 24 25
20 15 12 12 12 14 14 12 11 11 11 10 11 11 11 11 11 12 12 16
22 28 27 21 13 11 13 13 13 13 13 13 10  9  9  9 10 10 10 11
10 10 10 10 16 10 10 10 11 11 12 12 12 14 14 14 12 14 12 10
 9 11 11  9  9 11 11 13 11 11 10 11 11 10 11 11 11 11 11 11
12 12 15 20 24 20 15 26 26 23 14 10 10  9 18 23 18 12 12 12
12 12 13 13 13 11 11 12 13 13 13 13 10 11 13 14 14 12 14 15
14 14 12 13 18 15 12 12 15 14 14 16 16 15 14 15 18 18 16 16

28 31 28 25 22 19 19 20 22 25 16 15 13 15 18 18 22 23 25 24
22 16 16 14 13 14 16 16 14 14 22 25 27 27 28 27 17 22 18 18
16 15 15 14 13 16 13 11 11 11 12 11 10 10 10 11 11 10 11 11
14 24 29 19 12 12 12 13 12 13 12 12 11 10 10 11 11 11 13 11
10 11 11 11 11 12 11 12 13 12 11 12 12 12 15 15 14 11  9  9
11 12 11 10 11 11 11 11 11 12 12 11 11 13 11 10 11 11 11 11
12 16 18 16 24 24 14 13 13 13 12 11 11 13 12 15 19 18 14 12
11 15 15 15 14 12 14 12 12 13 13 13 13 12 12 16 15 16 16 13
13 14 12 12 12 14 15 16 18 20 24 25 24 22 22 19 15 15 18 19

30 30 27 28 31 29 25 23 25 25 25 19 13 16 18 19 20 26 28 23
19 16 16 15 15 16 16 15 14 14 16 20 17 22 27 27 27 13 13 11
13 14 14 13 16 15 11 12 15 11 11 12 10 10 11 10 10 11 10 10
10 11 26 29 12 12 12 15 15 15 14 14 13 11 11 12 11 11 12 11
11 13 13 11 13 13 11 12 14 12 12 12 12 12 12 13 11 10 10 10
10 10 10  9 10 10 11 11 12 12 11 11 11 12 11 11 11 12 11 12
15 17 15 15 13 14 14 12 14 15 15 16 14 12 16 19 16 15 22 24
13 11 12 14 12 12 11 12 14 14 14 14 14 12 12 14 14 13 13 14
14 13 13 14 15 16 17 23 23 22 24 24 24 22 22 22 16 18 22 20

22 27 32 34 30 28 23 20 27 25 23 15 15 16 23 31 26 24 18 16
13 13 13 14 15 14 14 13 13 15 14 12 18 22 25 28 20 13 12 12
12 13 13 13 11 11 11 11 11 13 11 12 11 10 11 11 14 11 10 11
11 26 34 22 11 14 15 12 12 12 12 15 14 12 11 11 12 14 12 12
12 12 12 12 12 11 11 11 11 13 13 13 14 12 11 10 10 10 11 11
10 10 10 10 10 13 11 11 11 11 12 11 11 12 10 10 11 10 11 12
14 13 13 12 12 14 13 13 13 13 11 14 17 15 14 11 12 23 33 31
12 12 13 12 12 13 15 12 12 14 12 14 11 11 14 15 16 16 16 15
15 12 13 17 19 16 13 13 17 22 21 20 21 21 21 19 19 28 25 16

23 24 26 34 31 29 29 29 26 30 22 16 21 20 20 20 29 19 14 15
18 19 12 11 13 12 12 13 12 13 14 16 24 24 28 31 24 14 14 14
14 12 12 13 12 11 11  9 11 11 11 10 10 13 14 14 18 11  9  9
14 33 42 28 12 12 12 15 15 12 15 11 12 12 11 11 11 11 11 11
12 12 11 11 14 13 12 12 12 12 12 12 14 14 13  9 10 11 13 11
10 10 11 11 10 13 10 11 10 11 11 12 12 10 10 11 10 10 10 11
12 11 12 14 14 12 14 13 12 13 13 12 11 15 13 11 11 13 16 16
16 12 12 14 15 14 12 13 12 13 13 12 13 19 12 15 19 18 15 14
14 16 16 18 10 16 14 11 13 14 16 14 14 14 16 21 23 25 21 16

26 23 24 28 25 25 29 23 22 27 27 17 12 11 12 11 17 15 16 21
23 25 16 15 14 15 15 14 15 15 16 18 20 26 25 20 29 16 11 11
11 11 11 14 14 11 11 10 11 13 11 11 11 11 12 12 11 10 10 13
25 43 34 21 15 12 12 14 14 15 16 15 12 13 13 11 11 11 10 10
11 10  9 11 12 14 12 12 12 11 14 15 13 13 14 11  9  9  9 11
```

```
11 11 11 11 11 10 11 10 10  9 11 12 11 10 11 11 10 10 10 11
11 12 12 14 14 12 12 14 12 14 15 12 12 12 12 14 20 13 11 11
14 13 13 13 14 13 13 14 14 18 24 23 15 16 22 19 18 18 16 13
15 15 15 15 14 12 11 14 15 15 15 14 15 16 15 16 19 22 19 14

18 24 28 28 26 23 29 26 18 12 12 14 12 14 12 12 13 16 17 21
22 22 17 21 20 20 23 23 23 24 28 24 25 23 18 22 22 16 12 11
11 11 10 10 10 10  9 10 11 10 11 11 11 10 10 10  9  9 11 20
23 27 27 13 13 13 13 11 13 20 18 16 14 15 14 11 10 11 11 11
 9  9 11 13 12 12 11 10 11 12 14 15 14 14 11 10 11 11 12 12
11 11 11 11 11 11 10 10 10 10 11 11 10 10 10 10 10 11 11 12
12 12 12 11 11 11 11 10 11 11 11 11 14 12 12 16 18 16 14 12
13 12 11 12 12 13 13 13 14 15 14 14 16 22 22 22 22 15 16 16
15 15 18 18 19 15 19 15 14 12 14 12 11 12 14 15 16 16 16 15

24 16 20 26 31 26 20 15 14 13 14 15 15 14 16 12 12 12 15 16
19 23 22 22 20 17 20 21 28 32 32 28 27 28 23 18 22 15 11 12
12 10 10 10 10 10 10 11 11 10 10 10 10  9  9 10 10 11 12 19
25 29 24 16 13 13 14 14 16 20 14 14 15 14 13 11 11 11 12 10
 9 11 12 12 12 11 11 10 11 13 13 12 11 12 10  9 10 10 11 12
12 12 12 11 11 11 11 11  9 12 12 10  9 10 11 13 10 11 12 12
12 12 12 12 12 12 13 12 11 11 11 10 11 11 10 12 11 11 12 12
14 11 10 12 14 15 13 13 13 12 13 12 16 21 21 21 20 17 14 15
16 24 29 14 12 14 12 18 16 15 12 12 11 13 12 14 14 16 16 12

27 24 22 23 24 23 16 16 15 15 19 24 24 20 22 16 14 13 13 18
24 24 24 23 25 23 22 19 22 27 27 21 22 22 22 23 28 25 15 12
10 10 10  9  9  9 11 11 11 10 11 10 10 10 10 12  9  9 14 23
24 23 18 14 14 14 16 16 14 14 17 14 13 14 16 16 13 11  9 10
10 13 13 14 14 14 12 11 12 14 14 12 12 11 11  9 10 10 10 12
14 11 10 11 11 11 11 11 12 14 12 12 11 11  9  8  9  9 10 10
11 13 13 13 14 14 12 11 11 11 14 12 11 11 11 11 11 10 12 10
11 14 16 16 15 14 14 14 14 14 12 11 12 20 21 22 17 14 12 12
15 14 14 14 13 13 14 14 12 14 12 11 12 12 15 11 12 12 12 12

23 23 23 25 22 15 21 18 22 23 24 23 22 16 19 13 13 13 13 13
26 23 22 19 22 22 23 22 24 19 14 16 24 18 21 22 27 21 12  9
11 10 10 10 11 12 12 11 11 12 14 19 22 22 28 15 16 22 25 24
19 20 14 14 14 14 15 16 14 15 15 16 16 14 15 15 12 11  9 11
11 11 11 11 11 11 11 14 12 11 11 11 11 10 10 10 10  9 11 12
11 12 13 13 14 11 11 14 14 13 11 11 12 12 14 12 12 12 15 13
12 13 13 13 11 15 15 12 12 12 11 11 12 15 15 14 15 14 12 14
14 12 13 13 12 12 13 13 11 11 11 14 14 14 12 14 15 15 12 12

28 28 34 31 23 16 16 25 25 23 21 17 15 15 15 18 19 16 14 12
15 13 13 16 18 16 16 22 22 18 13 11 22 23 14 14 14 12 12 12
14 11 11 12 13 11 11 12 12 15 16 22 25 25 21 29 24 22 25 24
16 19 15 12 13 19 18 13 12 16 13 14 15 18 14 11 11 10 11 12
12 11 12 13 13 12 12 13 12 11 13 10 11 13 13 14 13 11 14 12
10 10  9 10 10 11 12 12 11 10 11 11 12 11 10  9  9 11 12 11
12 12 12 12 12 12 11 11 14 13 11 11 13 13 11 13 11 10 11 11
12 12 11 10 12 11 12 11 10 10 11 12 11 11 11 11 11 12 12 15
23 23 18 16 16 15 12 12 13 12 12 13 13 11 13 16 14 14 14 13

24 23 23 24 24 18 19 17 17 14 13 14 13 13 15 20 22 22 22 17
11 12 14 15 19 19 22 20 22 16 11 12 12 16 14 13 13 13 13 11
13 15 15 15 14 12 12 14 13 16 22 27 26 26 20 16 14 16 27 28
22 16 28 18 14 15 12 12 12 13 16 18 19 19 18 15 11 10 11 14
13 11 11 14 12 11 10 10 11 11 11  9 11 11 11 11 11 11 13 10
 9 10 10  9 11 11 12 12 11  9 11 12 12 10 10 10 12 10 12 15
12 11 11 12 14 12 12 12 14 12 11 11 13 13 13 13 13 11 11 11
11 11 10 11 11 11 10 11 11 10 11 12 11 10 11 11 11 11 12 15
```

```
16 18 18 22 26 24 16 18 12 15 16 14 12 16 20 21 21 13 12 12

15 16 18 18 18 22 24 22 18 19 19 16 18 13 14 17 22 23 26 16
16 15 16 17 20 23 23 22 16 12 14 11 11 19 13 11 12 11 11 12
14 14 14 11 14 19 18 18 24 31 31 22 22 16 17 16 13 13 22 23
17 23 22 21 20 12 12 14 14 15 16 16 16 15 12 12 11 11 11 10
10 10 10 10 11 11 13 11 10 10 11 11 12 12 11 11 11 12 11 11
11 11  9 11 11 11 13 10 10  9 12 14 12 12 12 12 11 11 11 11
13 12 12 11 11 12 12 11 11 10 14 15 15 14 12 11 10 11 13 13
11 12 11 12 11 13 11 10 10 11 11 11 11 11 12 12 11 11 12 12
11 11 12 15 16 16 12 14 14 12 14 12 14 15 15 15 14 16 14 13

19 18 18 19 24 26 18 16 18 22 26 31 31 16 18 15 18 23 22 23
23 22 23 23 22 23 27 28 28 20 11 11 11 12 12 12 11 11 12 14
15 19 16 13 18 19 19 18 18 22 19 15 15 15 15 15 14 15 19 18
15 14 14 16 17 17 14 13 15 16 15 14 14 14 14 12 12 10  9 10
12 11 11 12 15 12 12 11 10  9 13 11 10 10 11 11 11 11 12 11
11 11 11 11 12 12 12 12 11 10 13 14 15 15 14 10 11 10 10 11
11 11 11 13 12 11 11 12 11 11 12 12 12 12 11  9 11 12 14 14
11 11 12 12 12 12 12 12 11 11 12 11 10 11 14 14 13 13 11 12
11 12 15 15 11 18 16 13 13 13 13 13 15 14 15 14 12 14 12 12

19 18 19 23 21 24 22 18 16 16 16 22 28 31 33 26 22 22 23 22
22 24 24 24 25 28 27 27 23 21 14 13 12 12 12 12 12 11 12 15
16 19 22 25 26 28 29 18 16 18 16 15 15 16 15 15 13 11 15 14
16 14 12 14 15 15 14 14 13 14 13 14 13 12 11 11  9  9  9 10
11 11 11 11 11 11 12 12 12 11 10 11 12 12 10 12 10 13 15 14
10 10 11 12 12 11 10 11 11 11 11 11 11 11  9  9  9 10 10 11
11 13 11 11 11 11 11 11 12 14 12 12 12 12 10 10 12 14 14 11
 9 10 10 11 14 14 12 11 11 11 12 12 12 12 14 14 13 13 14 13
14 13 13 14 14 12 15 15 14 15 16 18 15 19 18 13 12 12 11 11

22 17 16 16 13 21 16 15 23 28 23 25 31 31 32 32 28 24 22 19
16 15 16 26 20 25 23 19 19 18 15 14 13 13 13 11 12 12 13 20
21 20 22 28 30 27 28 23 22 16 16 16 15 16 13 16 20 24 20 19
15 11 15 19 16 11 11 14 14 14 14 14 12 12 11 11  9  9  9 11
12 11 10 10 11 11 11 11 11 10 11 11 11 11 14 11 11 11 12 10
11 10 10 10 10 11 11 11 11 10 10 11 10 10 10  9  9  9 11 12
12 12 11 13 13 13 13 11 13 12 12 11 11 10 11 12 13 13 11 11
10 11 11 12 11 12 10 11 11 11 12 12 12 14 15 12 12 12 12 13
12 13 13 11 14 14 15 16 20 21 20 14 12 14 18 14 11 11 12 11

22 21 17 14 12 11 11 11 15 17 17 25 29 33 37 40 29 21 16 13
13 12 18 24 22 23 22 23 16 14 13 13 14 15 12 12 14 15 14 16
16 17 17 23 28 27 27 21 14 14 16 16 21 18 24 28 28 25 23 19
13 12 11 12 12 12 12 13 13 13 15 14 10 11 11 11 10 10 11 11
11 11 11  9  9 11 11 12 10 10 10 11 11 11 14 11 11 11 11 11
11 12 15 11 10 11 10 10 11 11 10 10 11 10 10 10 10 11 12 11
11 12  9 11 11 11 11 12 11 12 13 13 11 11 11 11 13 12 12 12 11
11 11 11 12 13 12 11 12 12 11 10 11 12 14 15 15 12 14 14 14
14 14 12 11 11 12 14 14 16 16 11 13 13 13 11 11 10 10 11 11

18 19 16 12 11 11 13 14 13 16 27 34 32 36 38 34 23 15 18 16
15 16 23 24 18 19 20 24 19 13 13 12 11 11 11 11 13 13 15 15
14 14 16 28 28 24 22 20 12 16 20 20 21 25 27 28 30 28 23 23
18 16 14 11 10 11 11 12 12 13 13 11 10 11 11 10 11 10 11 10
11 11 11 11 11 11 11 12 11 11  9 11 12 11 11 14 11 11 10 11
11 11 12 14 12 12 11 12 11 11 11 11 11 11 11 10 11 10 10 11
11 11 11 10 11 12 12 11 10 13 13 12 11 12 11 11 13 13 11 10
11 11 13 14 12 11 10 11 11 11  9 10 12 13 15 15 12 12 12 11
11 14 14 12 12 12 12 12 14 11 12 13 13 13 12 12 11 11 10 13

15 15 10 13 11 12 15 16 23 31 31 34 35 35 31 31 27 16 16 15
```

```
20 22 17 13 15 15 16 15 14 14 14 11 12 11 11 11 11 12 14 15
14 16 22 23 24 22 15 15 18 15 15 16 21 14 22 31 27 26 26 22
20 15 15 12 11 11 11 11 14 12  9 11 10 11 10 10 10 10 12 20
16 16 18 19 19 22 13 11 10 11 11 11 12 15 12 11 11 12 11 11
11 11 11 11 10  9 10 11 10 11 11 10 11 10 11 11 10 10 11 11
11  9 10 11 10 13 11 10 11 15 15 14 14 12 12 13 13 12 12 11
11 11 13 14 11 11 14 16 13 12 12 12 12 11 11 12 11 15 11 10 12
13 13 13 11 10 11 12 12 14 16 14 15 12 12 11 12 11 11 13 14

13 13 15 15 14 14 11 11 16 19 22 28 29 29 28 24 25 23 14 13
16 23 21 13 13 13 13 15 14 13 11 11 12 12 11 10 11 14 12 12
12 13 13 15 13 22 14 14 15 14 14 15 22 22 25 28 19 23 25 26
23 26 23 21 20 16 11 11 12 12 10 10 10 10  9 10 11 18 24 25
26 25 25 26 25 25 24  9 11 10 10 14 22 19 12 10 11 11 12 12
12 12 12 12 11 11 11 10 10 11 10 10 10  9 11  9  9 10 12 14
12 11 12 11 13 11 11 11 11 10 11 13 13 13 14 14 12 12 12 11
11 12 13 13 13 12 11 12 14 14 14 13 13 13 13 11 11 12 12 14
14 14 14 10 10 10  9 11 11 12 12 12 12 11 11 10 10 12 12 14

15 14 14 11 13 12 11 15 22 24 18 22 24 23 18 15 23 29 23 18
15 16 18 15 13 14 13 13 13 12 11 11 11 13 11 11 13 13 14 12
14 14 14 14 12 13 13 12 13 13 13 16 13 13 13 15 16 24 24 25
22 22 24 25 28 23 11 12 12 12 11 11  9 10 10 11 12 20 23 23
25 25 25 24 25 29 28 20 18 15 15 20 22 20 18 15 13 19 26 19
12 12 12 12 12 11 11 11 11 11 11 11 11 11 10 10 10 14 15 12
12 12 12 12 14 12 12 12 11 11 11 12 12 19 31 10 11 13 10 11
10 13 12 12 14 14 11 12 15 14 13 14 12 12 13 13 13 14 15 15
15 14 11 10 11 11 11 10 11 12 12 12 13 11 10 11 13 11 14 14

13 11 14 11 14 18 24 24 24 21 13 13 12 11 11 18 24 19 19 16
13 13 13 14 14 12 15 14 14 14 11 12 11 12 12 12 12 13 13 14
15 14 11 13 12 12 14 14 12 12 14 15 13 16 19 19 19 22 24 23
22 26 16 16 14 14 12 12 11 10 11 11 11 11 11 12 12 14 20 23
25 27 25 27 28 27 24 24 25 25 25 25 23 24 24 24 26 24 19 13
11 14 13 10 10 10 12 11 11 11 11 11 11 11 11 12 12 14 23 21 12
12 11 12 11 12 13 11 11 11 10 14 24 37 36 25 13 10 11 11 11
11 15 11 13 11 10 11 11 18 23 22 15 12 12 11 13 13 13 12 12
13 13 11 11 11 11 11 11 13 12 12 12 11 12 12 12 18 13 11 11

16 13 13 22 28 25 20 20 20 15 16 14 12 12 12 11 19 15 15 13
12 13 12 16 19 23 22 14 13 13 14 15 14 12 12 12 14 13 13 17
17 14 13 13 14 15 15 14 14 14 14 14 15 15 18 21 24 24 25 25
22 16 12 12 11 13 11 10  9 11 11  9 11 11 11 12 12 14 15 23
26 26 27 26 23 23 25 27 25 25 25 23 23 24 24 25 22 15 14 16
12 12 11 11 10 10 11 10 10 10 11 11 11 12 12 14 14 18 15 11
11 12 11 11 11 12 12 11 11 11 15 20 21 25 36 25 16 14 18 23
19 15 13 11 12 12 12 13 14 22 18 11 11 11 11 12 11 15 14 14
14 12 13 11 11 12 11 11 11 13 13 11 11 11 10 11 15 12 11 14

22 18 19 21 27 27 26 17 13 14 13 14 13 13 14 14 12 12 14 14
14 12 11 13 16 20 20 18 13 11 13 11 11 11 11 11 11 13 14 16 23
19 15 15 15 12 13 14 13 13 13 13 15 15 14 14 16 22 23 23 19
19 18 12 12 11 11 11 11 11 10 11 11 10 11 10 11 11 11 13 19
23 19 22 23 24 24 26 27 23 23 23 23 26 23 27 27 17 14 12 14
14 12 11 11 11  9 11 10 10 11 11 12 10 11 11 11 14 15 11
11 10 11 11 11 11 11 11 11 11 11 11 14 21 28 30 30 32 34 30
34 14 14 14 14 15 15 12 18 16 13 11 10 11 10 13 13 13 11 11
13 13 14 11 10 11 12 14 14 11 12 14 12 13 14 12 13 11 10 10

16 19 24 25 24 22 15 14 12 13 12 13 13 13 13 14 13 13 13 13
11 11 12 12 12 12 12 10 11 11 11 10 11 13 13 12 13 13 18 23 19
22 18 15 14 14 12 12 14 14 15 13 13 13 13 13 21 23 27 28 20
14 13 11 11 12 12 11 11 10 10 10 10 10 10 11 11 12 12 13 11
```

```
13 19 23 23 23 25 25 24 25 25 25 25 22 22 23 26 21 13 11 11
11 10 11 10 10 11 11 10 10 11 12 15 13 12 13 16 19 19 18 10
10 11 10 11 10 10 11 10 11 11 12 12 14 23 21 17 14 16 17 22
34 36 32 27 26 14 12 15 14 12 11 12 12 11 11 13 12 13 12 13
12 10  9 10  9  9 11 10 12 12 15 18 14 12 12 12 12 11 12 11

18 16 23 24 19 15 13 13 13 14 14 12 12 11 11 13 12 13 12 12
11 11 13 11 13 11 11 13 13 11 11 11 11 12 12 11 15 16 16 18
19 18 19 14 14 13 13 13 13 13 15 15 16 18 16 18 19 22 20 12
12 12 12 12 11 11 10 10 10 10 10  9 10 11 11 11 11 12 10 10
11 13 19 26 28 23 24 28 28 26 26 26 23 24 24 25 24 22 11 11
12 11 11 11 11 11 11 10 10 10 12 22 12 14 18 23 28 28 23 12
11 11 11 12 11 10 10  9  9 10 13 14 16 15 14 12 12 14 15 19
22 26 31 26 17 12 12 12 14 12 11 12 14 12 12 12 12 12 14 14
14 12 11 10 10 10 11 12 11 11 13 15 15 11 13 12 14 14 14 14

21 15 16 15 12 12 12 12 13 13 11  9 10 11 11 11 11 12 12 11
11 11 11 12 12 12 12 11 11 11 11 11 11 13 11 12 14 14 15 16
16 15 20 18 13 12 12 12 13 15 16 16 23 24 26 28 22 19 23 22
23 22 15 11 12 11 11 11 11 11 11 11 13 11 10 11 13 11 11 14
16 19 21 23 26 26 26 28 31 31 25 23 23 26 26 24 23 14 12 14
11 11 12 12 11 13 16 14 16 16 20 25 27 27 28 32 28 25 14 15
16 12 11 10 10  9 10 11 11 12 12 11 14 13 11 13 16 24 36 23
31 36 36 31 15 11 13 12 12 12 12 13 13 12 14 11 11 14 15 15
11 10 10 11 11 11 12 13 12 15 13 12 15 19 16 18 19 22 15 13

14 24 23 14 11 12 14 15 13 15 18 18 20 24 11 14 13 15 14 11
14 12 12 12 15 14 12 11 12 12 12 11 12 12 12 11 13 13 13 11
13 13 18 18 19 18 15 15 14 15 18 24 25 26 25 24 26 23 18 16
18 15 14 11 12 12 12 11 11 11 11  9 13 17 14 20 21 13 14 16
20 20 15 15 26 29 31 31 28 25 23 22 24 25 26 25 25 18 11 10
11 10 13 16 14 18 28 29 31 29 29 26 28 28 28 28 28 31 15 12
15 12 11 11 11 10 10 11 11 11 12 11 12 11 11 13 22 32 31 37
33 33 31 23 14 13 12 14 14 14 12 12 14 12 13 13 12 13 13 14
14  9 10 10 12 12 15 14 16 15 14 11 11 15 22 18 16 15 15 13
```

Index